The Quantum of Explanation

The Quantum of Explanation advances a bold new theory of how explanation ought to be understood in philosophical and cosmological inquiries. Using a complete interpretation of Alfred North Whitehead's philosophical and mathematical writings and an interpretive structure that is essentially new, Auxier and Herstein argue that Whitehead has never been properly understood, nor has the depth and breadth of his contribution to the human search for knowledge been assimilated by his successors. This important book effectively applies Whitehead's philosophy to problems in the interpretation of science, empirical knowledge, and nature. It develops a new account of philosophical naturalism that will contribute to the current naturalism debate in both Analytic and Continental philosophy. Auxier and Herstein also draw attention to some of the most important differences between the process theology tradition and Whitehead's thought, arguing in favor of a Whiteheadian naturalism that is more or less independent of theological concerns. This book offers a clear and comprehensive introduction to Whitehead's philosophy and is an essential resource for students and scholars interested in American philosophy, the philosophy of mathematics and physics, and issues associated with naturalism, explanation and radical empiricism.

Randall E. Auxier is Professor of Philosophy at Southern Illinois University, Carbondale. He is the author of *Time, Will and Purpose: Living Ideas from the Philosophy of Josiah Royce*.

Gary L. Herstein is an independent scholar with research interests in the philosophy of science, logic, and American philosophy. He is the author of *Whitehead and the Measurement Problem of Cosmology*.

Routledge Studies in American Philosophy

Edited by Willem deVries, University of New Hampshire, USA and Henry Jackman, York University, Canada

1. Intentionality and the Myths of the Given
   Between Pragmatism and Phenomenology
   Carl B. Sachs

2. Richard Rorty, Liberalism and Cosmopolitanism
   David E. McClean

3. Pragmatic Encounters
   Richard J. Bernstein

4. Toward a Metaphysics of Culture
   Joseph Margolis

5. Gewirthian Perspectives on Human Rights
   Edited by Per Bauhn

6. Toward a Pragmatist Metaethics
   Diana B. Heney

7. Sellars and Contemporary Philosophy
   Edited by David Pereplyotchik and Deborah R. Barnbaum

8. Pragmatism and Objectivity
   Essays Sparked by the Work of Nicolas Rescher
   Edited by Sami Pihlström

9. The Quantum of Explanation
   Whitehead's Radical Empiricism
   Randall E. Auxier and Gary L. Herstein

The Quantum of Explanation
Whitehead's Radical Empiricism

Randall E. Auxier and
Gary L. Herstein

Taylor & Francis Group
LONDON AND NEW YORK

First published 2017 by Routledge

2 Park Square, Milton Park, Abingdon, Oxfordshire OX14 4RN
52 Vanderbilt Avenue, New York, NY 10017

Routledge is an imprint of the Taylor & Francis Group, an informa business

First issued in paperback 2019

Copyright © 2017 Taylor & Francis

The right of Randall E. Auxier and Gary L. Herstein to be identified as authors of this work has been asserted by them in accordance with sections 77 and 78 of the Copyright, Designs and Patents Act 1988.

All rights reserved. No part of this book may be reprinted or reproduced or utilised in any form or by any electronic, mechanical, or other means, now known or hereafter invented, including photocopying and recording, or in any information storage or retrieval system, without permission in writing from the publishers.

Notice:
Product or corporate names may be trademarks or registered trademarks, and are used only for identification and explanation without intent to infringe.

Library of Congress Cataloging-in-Publication Data
Names: Auxier, Randall E., 1961– author.
Title: The quantum of explanation : Whitehead's radical empiricism / by Randall E. Auxier and Gary L. Herstein.
Description: 1 [edition]. | New York : Routledge, 2017. | Series: Routledge studies in American philosophy ; 9 | Includes bibliographical references and index.
Identifiers: LCCN 2016048943 | ISBN 9781138700161 (hardback : alk. paper)
Subjects: LCSH: Whitehead, Alfred North, 1861–1947. | Empiricism.
Classification: LCC B1674.W354 A99 2017 | DDC 192—dc23
LC record available at https://lccn.loc.gov/2016048943

ISBN: 978-1-138-70016-1 (hbk)
ISBN: 978-0-367-25849-8 (pbk)

Typeset in Sabon
by Apex CoVantage, LLC

Contents

Preface — vii
Acknowledgements — xiii

   Introduction — 1
1. Reading Whitehead — 23
2. Whitehead's Radical Empiricism — 39
3. The Logic in Metaphysics — 50
4. The Quantum of Explanation — 66
5. Extensive Connectedness (The Metaphysics in Logic) — 82
6. The Principle of Relativity — 98
7. Genetic and Coordinate Division and Divisibility — 112
8. The Problem of Possibility — 143
9. The Algebra of Negative Prehension — 175
10. The Nature of Naturalism — 193
11. Synoptic Pluralism and the Problem of Whiteheadian "Theology" — 221
12. Possibility and God — 240
13. God's Mortal Soul — 270

Notes — 297
Index — 353

Preface

This book has a long history, more than a quarter century. There is no way to bring something quite so ambitious to fruition without incurring countless debts. We will do our best to discharge some of these debts in the narrative that follows, and others will be mentioned among the notes. This book has the peculiar added feature of having two primary authors. Yet, large parts of this book existed in draft material and even highly developed studies *before* it dawned on the two authors that they were working on the same book. The story of how it came to look as it does, and which of us bears primary blame for its inevitable flaws, may be of at least some transient interest. We both endorse everything the book says, and without reserve. But that is not the story. In fact, we both *wrote* this book, together, and there are many parts regarding which we can no longer recall who produced the initial draft.

We have very different backgrounds and life experience, and we know different things (with large overlaps). Still the book is genuinely *ours*, not just a collaboration but a single view of all the subjects treated here, shared by both of us, without need for exceptions. We had no disagreements to negotiate, although we had to teach one another many things. One could say that Herstein is the "teacher" in the history and philosophy of mathematics and science, while Auxier is the "teacher" in the history of religious and theological thought. Both of us are historians of philosophy and serious in our studies of metaphysics, epistemology, and logic. We both formalize our thought systematically and situate it historically. Our complete lack of disagreement seems improbable, but, as we like to say, everything that is actual is thereby also possible.

The thinking herein and the Whitehead interpretation we offer reach back to our first (independent) readings of Whitehead in the early 1980s, Herstein at Occidental College and Auxier at (what is now called) the University of Memphis. We each had to teach ourselves Whitehead, receiving encouragement, sometimes, from our philosophy instructors, but neither of us had the advantage, or the burden, of a specialist in Whitehead or process philosophy among our instructors. We both had the benefit of outstanding teachers in systematic and historical philosophy, teachers who emphasized intellectual honesty and scholarly rigor, and who enabled us to recognize our strengths and limitations. Both of us were convinced from our independent reading that if anyone had come close to answering the questions that we couldn't leave alone—time, possibility, order—it was Whitehead.

We both knew that it would take a long time to understand and digest his views, and we went immediately to work on the project of truly understanding Whitehead. We consulted secondary literature, for help, and we found some. This was especially true of Donald Sherburne's *Key* and, for Auxier, also the first edition Elizabeth Kraus's guide called *The Metaphysics of Experience*. Herstein stayed mainly with primary texts during these years as he pursued a career in computer technologies, even taking only one book, *Process and Reality*, on a vacation to Tahiti. Perhaps this is the best beach reading for some tribe of truth-seekers, but the tribe would be small. Although we were too inexperienced to put our fingers on the causes, our intuition was the same: much of the commentary failed to reflect the richness and full intellectual depth and connectedness of Whitehead's ideas. For many years we were thus obliged, independently, to wrestle with the primary texts, and that is what we did.

Although Herstein is the senior in years on earth by about five, Auxier's circumstances enabled him to pursue the standard academic path and hence to arrive somewhat earlier at the threshold of a writing project on these ideas. The earliest portions of what became this book were drafted in 1989 while Auxier was still in graduate school. An essay called "Concentric Circles: An Exploration of Three Concepts in Process Metaphysics" was drafted as part of the "take away" from a seminar on the philosophy of time with Charles Sherover, and it was published in 1991 in the *Southwest Philosophy Review*. It contains the earliest version of the argument about possibility, potentiality, and actuality that constitutes an important part of this book. An examination of that (now ancient) essay will reveal that it is the work of someone unseasoned and too ambitious, but the core view, now both chastened and (one hopes) deepened by experience, reflection, and study, is given here, in chapters 8 and 9.

The argument has grown in prospect since we began working together on these issues in 2000. But we know we are not finished with these ideas. We have begun writing on a project we are calling *The Continuum of Possibility: Models and Modality in Process Metaphysics*. Perhaps we are too ambitious still, but if the reaper spares us, we will offer at some time (when it is ripe) a fuller formalization and working out of the formal sketch in chapter 9 of this book. We intend to make good on the suggestion that algebraic thinking can handle time in relation to possibility just as Whitehead so boldly asserted, and we are audacious enough to believe we can make some progress in developing these ideas in new ways, and to do so by re-reading the history of algebraic thinking.

At Emory, Auxier found professors who allowed him to pursue these topics in a dissertation on process metaphysics. Jude Jones, a student of

Elizabeth Kraus, was slightly ahead of Auxier at Emory, providing a context for exchange and creating conditions for later collaboration on Whitehead. But the topics that interested Jones and Auxier, although both were motivated to investigate metaphysics and value, turned out to be almost without overlap in emphasis at that time. Jones's book *Intensity: An Essay in Whiteheadian Ontology*, based on her dissertation, remains (in our view, although we do not assume Jones would automatically agree) mainly a treatment of the higher phases of experience (and how they come into being) that we can endorse with practically no modification. It is, in our terms, a book on concrescence. The higher phases of experience, exemplified in the human kind of consciousness and self-reflection, is exceedingly rare in the universe and, from the standpoint of cosmology, is of almost no importance, apart from providing evidence for how actuality can achieve astonishing intensity. Yet, intensity and concrescence are everything *to us*, as human beings.

We do not pursue these questions in this book, partly because we think Jones has done an admirable job, and partly because we think philosophers are bad about exaggerating the importance of human experience in the grand scheme of things. One thing we both admire about Whitehead is his unflinching vision of the cosmos as a place that *allows* for experience of our kind, but doesn't necessitate it or do very much to support or perpetuate it. Jones has said what needs saying about this kind of experience and we don't have anything to add, at this point. Still, the project we are working on to follow this volume is a treatment of concrescence that supplements our focus on transition and the "fact of the world," in the study before you now.

Auxier wrote the first draft of chapter 12 in 1992. Both Don Emler and Leo Werneke read versions of chapter 12 and affirmed the direction of the research. However, early exploration into the process community, as it existed then, alerted Auxier that his views went against the grain of the prevailing lines of interpretation. After some correspondence with several opinion leaders in the area, it became clear that it would not be possible to work out these ideas in the existing journals, controlled as they were by people who would demand greater conformity to the current conventions. There was solace in conversation with other young scholars, especially Bill Myers, of Birmingham Southern, and the occasional supportive senior colleague, especially Bill Garland from the University of the South and Pete Gunter from the University of North Texas. It would be necessary to approach the Whitehead community indirectly and by slow degrees. The community of process philosophers was, at that time, small enough to enforce an orthodoxy.

Auxier decided to focus on Hartshorne's thought as an alternative to disputing with powerful people over Whitehead. This turned out to be a fortunate turn. Hartshorne was still alive and vigorous and happily accepted Auxier's invitation to travel to Oklahoma and to speak and teach. Following that adventure in the spring of 1993, Auxier was able to travel to Austin numerous times and carry out his arguments with Hartshorne himself about how to interpret Whitehead. Hartshorne not only did not demand agreement from younger scholars, he insisted upon vigorous dissent wherever it could be supported by serious argumentation. One result of these exchanges was the essay that became chapter 13 of this book, first written in 1995.

Philosophers who spent serious time on Hartshorne's thought were not numerous, but were welcoming and open-minded. In particular, Auxier found valuable feedback from Don Viney, Barry Whitney, and Dan Dombrowski. Mark Y.A. Davies, whom Auxier had known at Emory, joined Auxier in a collaboration on Hartshorne's thought in 1994, and came to Oklahoma City University in 1996, providing real discussion. These exchanges resulted in a number of articles and the publication of Hartshorne's correspondence with Edgar Sheffield Brightman. In addition, through Davies's connection to Robert C. Neville, Auxier came under the influence and eventually the tutelage of one of the finest critics of process thought and original thinkers of that older generation. Neville generously read and responded to Auxier's correspondence during those years and continues to provide valuable resistance and corrective critique to these ideas.

Herstein's academic career has been what is euphemistically called "nontraditional." Starting at the University of Southern California, Herstein put himself through school by working full time at the Jet Propulsion Laboratory's Space Flight Operations Facility. Herstein's interest in science and mathematics had always been exceptional, and this job was in many ways the perfect opportunity for him. He was privileged to watch the very first images slowly scroll across the monitors in the Mission Control Center during the Voyager 2 encounter at Jupiter, and the Voyager 1 and 2 encounters at Saturn. Because his technical skills liberated him from the common vocational justification for higher education and, in many ways as importantly, confronted him with fundamental and patently concrete issues in the problem of knowledge, Herstein changed his original major from computer science to philosophy. During this time, Herstein had the good fortune to take Bas Van Fraassen's graduate seminar in modal logic, which was Herstein's first encounter with formal reasoning beyond the binary systems he'd worked with in computers. For various personal reasons, Herstein later switched schools from USC to Occidental College.

Herstein's first encounter with Whitehead occurred in the early 1980's, when he was an undergraduate at Occidental College. Herstein's philosophy professor at the time, William Neblett (who taught at Occidental College from 1965 until his retirement in 2004), wrote his Master's thesis on Whitehead. He knew Whitehead's process thought. By luck, a copy of the Corrected Edition of *Process and Reality* appeared in the Occidental bookstore. Herstein purchased it on the spot. Over the next two decades, Herstein carefully studied that volume many times. But because of his life outside of formal academics, he never encountered the secondary literature, and was thus able to form his own thoughts and ideas about Whitehead's metaphysics.

Herstein formally entered the scene in 2000, to work toward his Ph.D. in philosophy. That was the same year that Auxier moved to Southern Illinois University at Carbondale, where Herstein was now enrolled. They met several months earlier at the 2000 meeting of the Society for the Advancement of American Philosophy in Indianapolis. Auxier gave a plenary talk in which he did two things that determined Herstein to meet him: Auxier spoke of knowing what his cats were thinking, and then highlighted Herstein's three favorite philosophers in a single sentence: Dewey, Cassirer, and Whitehead.

Herstein's numerous classes and conversations with Auxier created a focus on Whitehead and his natural philosophy, especially as articulated in the sorely neglected triptych of works on the subject that Whitehead published between 1919 and 1922. Herstein defended his dissertation, with Auxier as the chair of his dissertation committee, in May of 2005. The dissertation, *Whitehead and the Measurement Problem of Cosmology*, was published a year later by Ontos Verlag (later acquired by De Gruyter) as a book by the same title. Herstein left Illinois to pursue an itinerant life as a visiting professor around the country.

While Herstein was at SIU Carbondale, he and Auxier spoke many times about the possibility of a collaborative book addressing some very significant shortcomings in the secondary literature on Whitehead. Ideas were bounced back and forth, which led to papers written and presented at conferences by each of them (providing draft material for a book only vaguely conceived). Finally, with a number of such studies accomplished, structures were proposed and modified by e-mail. But the project never achieved the status of a real outline, and remained little more than a shadow. Difficulties arose due to the size and complexity of the project, not to mention its ambitiousness. Distance was not the issue it would have been in previous decades, but hindsight informs us that we never could have finished this project while living in different places.

Herstein returned to Southern Illinois in 2011, where he could reinvent himself as an initiate in the new mendicant order known as the independent "modern" scholar (that is, homeless except for the patronage of others.) But with the problems of time and space no longer obstructing them, Herstein and Auxier directed their efforts upon the current project. After some struggle, a real plan for the book began to emerge. That initial outline, hammering the completed drafts into an order that made narrative sense, and recognizing what parts remained to be written, survives in large measure into the final book. From the fall of 2011 until the early winter of 2013, Auxier and Herstein met at least once weekly for a full afternoon, with specific assignments in between—they took turns carrying the weight of writing between meetings depending upon whether the subject matter was closer to Herstein's or Auxier's expertise. The first full draft was finished in time for a sizable weekly reading group consisting mainly of SIU graduate students, but also a few who had graduated and other interested parties, who read and reviewed the manuscript along with a reading *Process and Reality*. This was very helpful in testing the narrative.

Following this experience we were able to work our way back through the completed manuscript during the fall of 2013. We made every effort to digest every article that had a bearing on what we are saying, and we made many adjustments as we read new secondary works. There were several important discoveries for us during this period. Neither of us had read F. Bradford Wallack's book on Whitehead prior to this time, and it turned out to be a tremendous find in the literature. This work had, we believe, been effectively buried by the narrowness of the approaches dominant in the last 35 years, a perception we found confirmed when we contacted Wallack herself. Also added to our reading during this time was Catherine Keller's fine work in process theology, which reminded us of the vitality and relevance of these ideas when taken in a theological light. If we seem dubious about process theology in this book, we want to assure everyone that this impression is due to our own misgivings and not a reflection on the fine work done by so many people in the train of John Cobb, Jr.

Some readers will be curious as to which of us wrote which parts of this book, and indeed, one kind and anonymous reviewer insisted upon such information. We realize that co-authored books in our discipline raise eyebrows: "Surely one or the other is responsible for *this*," (Russell but not Whitehead, Dewey and not Tufts or Bentley, and so forth). It is a question that simply makes us smile, since there are so many parts regarding which we couldn't tell you ourselves who ought to be thought of as the author. This is particularly true of chapter 9, which is perhaps alternating sentences as one or the other of us worked our way through the difficult formalisms. If an accounting is required, Auxier drafted the Introduction, and chapters 1, 2, 7, 8, 11, 12, and 13; Herstein drafted chapters 3, 4, 5, 6, and 10. We truly wrote 9 together, alternating weeks with one or the other being responsible for a section, and it wasn't written in the order it now appears. We did all the editing and revising together, across a table, and no changes were made without the consent of both. We admit that it is peculiar that two such different people should find themselves so much and so easily in agreement on such wide-ranging issues. We have no explanation and we find it remarkable ourselves. Our writing styles (and senses of humor, and penchant for polemical expression) are also similar, almost indistinguishable, which is no less curious to us, since our backgrounds really are quite different. But it is worth repeating that whatever is actual is possible, and this is actual. The mystery lies in over-thinking it, not in the occurrence of it.

Acknowledgements

Herstein: Anyone who has even contemplated a work of this scope and size knows what an almost impossible task it is to acknowledge all of those persons who contributed to it. High on this list are all the members of my dissertation committee: Randy Auxier, Janice Staab, Larry Hickman, Pat Manfredi, Jurek Kocik. They all shared my conviction that the line of inquiry I was pursuing was worthy of the effort. Among the general groups that need mentioning, the Whitehead reading group that helped us out with an early draft of our manuscript in 2013 at SIU Carbondale, and the Society for the Study of Process Philosophy, for their interest and encouragement for papers that later became sections of chapters. Individuals include Michel Weber, Johanna Seibt, Jude Jones, Tim Eastman, Michael Epperson, Ronny Desmet, George Shields, John Cobb, Hank Keeton, among many others.

Finally, very special thanks must be given to Patrick and Antoinette Mulholland, without whose constant support I would never have been able to participate in this project. I dedicate my portion of this work to Patrick, who did not quite get to see it finished.

Auxier: In addition to those listed by Herstein, most of whose names would be repeated on my list, a number of people have read full or partial drafts of this manuscript and have contributed helpful comments that led to its improvement. First and foremost there are the students in the seminars I have run over the last 16 years at Southern Illinois University Carbondale. A number of these persons have gone on to carry out substantial research in process philosophy. Some of the work was completed during a sabbatical supported by SIUC in the spring of 2014. A number of colleagues also read the manuscript, at one stage or another, in whole or in part, and these include Robert C. Neville, David Connor, Bogdan Ogrodnik, Lukas Lamza, Dennis Soelch, Aljosche Berve, Jan Olof Bengtsson, Helmut Maassen, various members of the Institute of American Religious and Philosophical Thought, including Donald Crosby, Wesley Wildman, Charley Hardwick, Nancy Frankenberry and the late Creighton Peden; various members of the Society for the Study of Process Philosophy, including (again) Jude Jones. Also of help were Ted Calhoun, Keith Robinson, Duston Moore, James McLachlan, Gary Dorrien, and Gregory Budzban. The community of scholars associated with the American Institute of Philosophical and Cultural Thought (especially John Shook and Larry Hickman) and the International Forum on Persons (especially Richard C. Prust, Tom Buford, Richard Beauchamp, and James Beauregard) have been an ongoing source of conversation and encouragement in this work. At different times the Philosophy Departments of Indian-Purdue University Fort Wayne, Luther College, and also the Society for the Study of Process Philosophies have invited our work for presentation and have responded to versions of this work. The Society for the Philosophy of Creativity heard two early chapters and I must thank Pete A. Y. Gunter particularly for responses there and at meetings of the Southwestern Philosophical Society. Many ideas were worked through in conversations with Ralph D. Ellis and Bill Myers (and his former students) over the course of three decades. The Center for Process Studies at Claremont has never failed to provide support and response, especially John Sweeney; Roger Mark Dibben of the Whitehead Research Project in particular has been a valuable conversation partner. And of course, thanks are due to Routledge and Andrew Weckenmann and to the series editors, Henry Jackman and Willem de Vries.

Finally, since in this book I have personally come full circle in the studies that genuinely began when, as child, I listened to my mother, Eileen Gunter Auxier, explaining "creative motion" in music, I dedicate my portion of this work to her memory, and to my father Charles David Auxier, and the beautiful music they made together

Introduction

The central aim of this book is to make an argument, as persuasively as possible, for the concept of explanation—in all domains that human beings study—defended by Alfred North Whitehead. The authors believe, and passionately so, that the central insight of Whitehead's philosophy has been passed over, misinterpreted, and forgotten even by his most sympathetic followers. The idea is both very simple and hard to keep in mind because it runs contrary both to our native habits of thinking and to our long-established ideas about what constitutes a satisfying explanation. Because its implications are far-reaching, subtle, disturbing, and complex, this approach to explanation has not taken hold, even among process philosophers. That must change.

In this Introduction and the first two chapters of this book, we will be offering an overview of Whitehead's thought and our path through it. The detailed arguments await later chapters, so we encourage readers to be patient in forming conclusions about the assertions we make early on. Our aim is to bring readers (including advanced ones) into the context and language of process philosophy and only then to offer the rigorous case that serious readers expect. Included is a thorough discussion of issues associated with the theological reading of Whitehead that has played so prominent a role in process philosophy in the last half century. We are neither anti-theological nor anti-religious, and we do not share the views of those who would remove either "God" or "eternal objects" from Whitehead's ontology. Yet, we are confident that most readers of Whitehead who have applied his ontology to theological problems will find what we have to say here unwelcome. This discussion must be deferred until the more straightforwardly philosophical business has been accomplished, but we do not think of this part of the book as an add-on. Getting Whitehead's ideas about God right is just as important as getting his ideas right about actual entities and possibilities. Part of "getting God right" involves getting nature, as Whitehead understands it, right, and thus, the discussion of God is deferred until late in the book, since nature is our main focus.

Among the techniques we employ throughout the book is to present various illustrative examples of a qualitative type. Our intentions with these often somewhat "homey" stories are both to cultivate our readers' intuitions around subject-matters that are often extraordinarily abstract and to do so by situating those ideas in more tangibly concrete examples. In this regard, we note at least three different modalities of "making sense" that humans deploy against any puzzle, only two of which are typically called upon in contemporary philosophy: There is what might be called "empirical adequacy"—does observation and experience support what is being claimed in the relevant situation? There is "logical coherence"—do the various claims being asserted hang together logically? But there is also what we will refer to here as "narrative intelligibility"—does the story, *qua story*, make sense? Narrative intelligibility is not what one would normally acknowledge as an "argument," in the logical or philosophical sense. Rather, it is an illustration that brings the argument to life, concretizes it.

There seems to be an aversion, bordering on anaphylaxis, amongst philosophers (even and especially Whiteheadian philosophers) to turn to this third form of making sense.[1] This strikes us as ironic, since narrative intelligibility was one of the primary modes of philosophical discourse used by Plato. We are here recalling Whitehead's famous epigram about footnotes to Plato. Yet even Whitehead was generally reluctant to use stories and illustrative examples in his philosophy even though, as his beautiful biographical essays illustrate, he was a lovely storyteller.

We do not pretend for a moment that we are comparable to such storytellers as Plato and Whitehead. We do, however, insist that stories—even *our* stories—*are* an essential part of making sense of the world, and that includes Whitehead's philosophy. We appeal here to the authority of Iris Murdoch, and her carefully presented argument that some ideas (especially in metaphysics and morals) are so complicated that the only way to approach them is indirectly, via metaphor and narrative.[2] One might note here that Murdoch was herself one of the better storytellers of at least the last 100 years.[3]

Start Me Up

Our central idea is that *concrete existence explains the abstract aspects of experience* and not vice-versa. The unexpected characteristic of our experience is that it abstracts from the flux, not that it flows concretely, which we expect. This sense of the term "abstract" means something like, "creates a stable space," but spaces are created by the variability in the flux itself. No space is wholly stable, as far we know. Where the pulses of the unfolding creative advance approach light speed, we trade one *kind* of stability for another, the stability of relatively constant acceleration for the stability of repeated vibrations at various rates. But wherever there *is* variability, something like a space is created in the communication of what pulses with greater zest and frequent repetition as compared with what "moves" more slowly.[4]

To illustrate: we can observe the difference between light and sound, taking as a case study the way these two varieties of energistic order are differently propagated through typical earth conditions. Obviously, for all their differences as *organizations* of energy, sound and light are *experienced* together by us, through a synesthesia that our bodies accomplish. Yet, as a general fact about the universe, sonic waves vibrate at a level that is more difficult to propagate, compared with light. Sound is "heavy," entropic, requires a dense medium for its propagation, and dissipates into the background radiation after a short endurance. Light, by contrast, is economical, lively, relatively easy to propagate, does not require a dense medium for its propagation, and can maintain its physical structure for very long epochs of duration.[5]

When either sound or light interacts with the complex electromagnetic gravitational macrosystem we call "earth," under the right circumstances, *locations* arise that reveal structures even more fleeting, even less stable than sound itself. The likelihood, in the grand scheme of the universe, of the occurrence of a laser light, multimedia show *with* the music of the Rolling Stones, interacting on some beach in South America, well, let's just say that arrangement of light and sound energy is cosmically improbable.[6] No, it is not merely improbable. An infinity of infinities of infinities, raised to the power of the infinite, in terms of possibilities, is needed to bring together just *this* confluence of variable temporalities and their complementary spaces.[7]

A Bigger Bang

We choose this example because some people claim that the February 18, 2006 concert by the Rolling Stones in Rio de Janeiro, was the largest rock concert in history, attended by some 1.5 million people. All shows in "The Bigger Bang" tour began with a huge multimedia screenshow, and lights and pyrotechnics, depicting the Big Bang and tracing all of time and the creation of every actual space from the beginning down to the improbable moment when the first chord of the song "Start Me Up" is struck on Keith Richards' Telecaster, in the dark, and what follows is a silence, a pulse, a repetition, a count, a drum roll, and, well, a Bigger Bang, as the lights explode in a blinding flash and Mick Jagger is illuminated in the midst of a leap as the masses erupt into an ecstatic presence that really defies description. As they say, "you had to be there" (not that we were).

This is an "event" in the best Whiteheadian sense. It is complex, but it is also a unity, an enjoyment, and a satisfaction of all sorts of variable pulses of light, sound, gravitational influence, electromagnetic pattern, inertial eddies in the flux that gather in the actual world to which they belong, evaluate that world, and add to it a unique synthesis. In this case, the "event" involves over a million human beings, some very loud sounds and very bright lights, a beach, the edge of a continent, a city, and a durational epoch, ragged at the edges but definitely describing a span.[8] But the point is that one does not, and indeed, *never could* explain all that was absorbed and released, in all of its temporal, spatial, physical, and conceptual modes, at that time and place by appealing to some collection of abstract ideas.

The Big Bang, as a concept, is *explained by* the Bigger Bang, the event on the beach, and not the other way around. Notwithstanding our habitual appeals to the abstractions of science for the meaning of the word "explanation," the *true* explanation of *any* physical reality *is* the concrete event that it surrounds. The reason to adopt this path of explanation is very simple: *whatever is actual is possible*. That is the reason that explanation really *has to* proceed from the actual to the possible, at least to the extent we can expect to explain things beyond our ken, such as the character of possibility *as such*.

One does not need an "event" so complex as the Rolling Stones' Bigger Bang to make the point about explanation. Something as simple as dropping the pencil from one's hand to the ground is equally (and uniquely) an "event." While we are in the habit of appealing to, for example, the "laws of gravity" to "explain" what happened, the real story goes the other way around. The unique event *is the reason* for our appeals to abstract and general ideas, such as "laws," insofar as those general ideas *have* an explanation—and it is they, *not* the falling of the pencil, that *need* an explanation. It is hard to get used to this "appeal to the concrete" as *the* explanation, and one might complain that subsuming concrete events under general categories just *is* what "explanation" *means*. But this view is empirically indefensible. There is no denying the importance of the uniqueness and unrepeatability of any concrete event. Simply to set aside, in advance, everything that makes an event unique, singular, and hence universal, is to treat as concrete that which is abstract and vice-versa. It is not explanation.

The Poverty of Philosophy

It is true that we don't ever understand concrete actuality in its fullness, but we can get at what is actual with greater confidence than we can expect to gain access to some sort of universal necessities that *make* a unique event come to pass *just as it has*. That latter quest is a fool's game. Universality, so far as we have immediate access to it, has to do with the logical meaning of what is singular, unrepeatable, unique. The problem of finding the universal and necessary is tricky, as every philosopher in the West since Plato has known (and those in the East have known it still longer). Given the limits of understanding with which we begin, it is wiser to generalize with the greatest care than to leap on a floating universal and proclaim it the ground of all existence. That kind of God, or principle, or law, or force is no ground at all. It is a fiction, a puff of human desire adrift in a cosmos indifferent to human depth, need, and yearning. It is, as Socrates reports in the speech of Diotima, a vagrant at the doorstep of Being. It is *not* an explanation.

No one has ever been as careful to avoid bad generalization and floating universals as Whitehead was. Existence, in its uniqueness and full concreteness, *is* what it is, and no matter how simple the event is (and the Stones' Bigger Bang is pretty complex, for an event, with nests and nests of temporal eddies and ephemeral tensions), the simplicity or complexity wouldn't change the main issue. The "event" could be the shedding of an electron by a radioactive element, or something simpler still. "Events," in Whitehead's sense of the word, are not of any particular size, complexity, simplicity, or duration. In all cases of the event, its uniqueness places it beyond the reach of the abstractions to which it gives a ground, just as such uniqueness places it beyond the full comprehension of any other event. As far as we can tell, empirically, the cosmos seems to be made up of events. They can be explained in as many ways as we can devise to generalize from them and across them, but those generalizations are more likely to lead us astray than to lead us toward the best kinds of conclusions, unless explanation moves from the concrete to the abstract. The best explanations are those that invigorate and add intensity to our experience, and that sort moves thus.

We think that our main point bears repeating. *What is actual is possible.* This is the first law of metaphysics. And the actual is our best guide to the possible, but it is still a poor and unreliable guide, because generalizing from the "event" in February of 2006 to explain the Big Bang is *easy* compared to answering the question about *what else* was genuinely possible that *did not* happen on that day or any other day (or eon, or cosmic epoch). Our access to the structure of possibility is usually mediated by our powers of abstracting from the concrete in ways that respect the limitations imposed upon us, and our desire for explanations, *by* the actual. We face a trilemma: either (1) deny the reality or existence of possibility, or (2) reduce possibility to the likely or the inevitable, and then try to get our most general concepts to serve as explanations, or (3) treat the mode of existence of the possible as essentially abstract. All of these ways of handling the problem of possibility have been tried in the history of metaphysics. None has succeeded.

A Fourth Conception of Being

Whitehead suggested a fourth way, a way that leaves the relation of the possible and the actual fruitful, creative, dynamic, interesting, and open: the actual, in its full concrete uniqueness, *is* the explanation of anything and everything that can be explained. The actual offers us a glimpse of the possible, its structure and its meaning, if not all of its determinations. The Bigger Bang, or any other "event," rightly understood, rightly situated in its own actual world, *is* the reason for that world, insofar as that world *has* a reason. In explanatory power, the event we have called "the Bigger Bang" beggars the concept of the Big Bang. The Big Bang is just an idea to guide how we may *think about* what happened in the unimaginably remote past. The Bigger Bang actually happened in February of 2006.

But how do we get at The Bigger Bang, or any other "event," and *use it* to explain its "world"? The limitations are, unhappily, more numerous than the opportunities, but the most important limitation is self-limitation. Favoring just one type of generalization, such as counting or measuring, or formalizing in some sort of natural or devised language, and insisting that this or that method is right for answering *every* kind of question, is a great failure of self-limitation. If such a perfect method of generalization exists as will answer to every variety of our *desire* for explanation, it has not been discovered, and it is not likely to be discovered by claiming we already possess it when clearly we do not. Such exaggeration is intemperate and unwise. But Whitehead found a way to secure our norm of self-limitation, while yet asking the profoundest question: given that there is no single method of generalizing from the unique event to its world, is there not, accessible to our imaginations and reasonings, a basic *unit* of explanation? Is there not something irreducibly common to every desire for every sort of knowledge? Whitehead answers in the affirmative. *That unit is the actual entity*. This is what we mean by calling it "the quantum of explanation."

Explanation and Its Discontents

This has been a hard pill for philosophers after Whitehead to swallow. They insist upon thinking of measurement, and "laws of nature," and other numerical devices as "explanations," when in fact it is simple common sense to recognize that *what is unique*, as every fully concrete actuality is, *cannot be explained*, and that means that if it is explanation we seek, then the *explanans* must be the unique event, and the *explanandum* the world to which the event belongs. We do not have to give up scientific and traditional "explanations," but these explanations are not fully concrete—they are general, abstract. They are *not* universal. The universal, insofar as it plays a role in genuine explanation, *is* the singular event that concretizes its entire world from one perspective, exhibiting in a definite way one genuinely possible example of order in our cosmic epoch.

Thus, what stands in need of explanation is not the way existences and the flux pass, it is why *anything* in the flux is sufficiently stable for a difference to arise in its "passage" between its "transition" from one condition to another, on one side, and its "concrescence," its internal "valuation" of the "factors" it takes in from its "actual world," and then "grades for relevance," judges, in many modalities, and finally "expresses" in its "achievement." The quantum of explanation, i.e., the actual entity, does the main part of the work in helping us analyze and evaluate all this.

We draw on a vocabulary here that Whitehead developed in the course of numerous writings over 40 years. This vocabulary has confused, daunted, and deterred readers, and it has otherwise prevented his ideas from settling into the culture at large. These important ideas haven't even entered the more rarified discussions of physicists and mathematicians and philosophers, except in pockets and nooks. The words are more like the ornaments on a Christmas tree of prose than ideas we can pin down. What, after all, is this chap on about? There are good reasons for the specialized vocabulary Whitehead chose and refined. It is not easy to understand, still harder to learn and use, and once mastered, sadly effective at cloistering the exchange of ideas and keeping serious discussion limited to the initiates. We cannot, in this book, wholly reverse 80-plus years of scholastic specializing. But we aim to provide a guide that will enable reasonably devoted readers to follow our discussion as it moves increasingly into the world of Whitehead's curious language.

A Linguistic (Re)Turn

The language is not the only challenge. Whitehead's ideas are themselves both novel and difficult, conceptually speaking. A good grasp of the history of science, mathematics, and logic is needed for a thorough understanding of Whitehead. We cannot wholly remove this barrier, but we have supplemented our arguments with extensive examples and illustrations of the most difficult points. Yet, we really must stress that Whitehead's concept of the "actual entity" is not a bit of physical existence. It is a conceptual tool that helps the inquirer arrest temporal passage and the flux of the physical universe.

The concept of the actual entity is the most important in a collection of logical tools doing this work of holding the object of inquiry in stasis while it is analyzed. That "holding" cannot be achieved without some distortion entering the picture, but nothing will yield to analysis without some means of rendering it an object of study. How the inquirer achieves this condition of "arresting the object" has a great influence upon the success or failure of an inquiry. Whitehead's approach to the problem is original and worth understanding. The logical and analytical tools he develops have certain virtues not possessed by those of his predecessors in natural philosophy and cosmology. His tool kit is also adaptable to problems further removed from the areas with which he was best acquainted, and he brought them to bear on questions of religion, civilizational development, the history of science, education, and a few other areas in scattered essays. But his abiding concern was the study of the physical universe and the development of analytical tools for that kind of study.

Clouds of Mystery Pouring

Unfortunately, the majority of Whitehead's interpreters have been confused about what Whitehead was doing and how he was doing it. Most insist that the "actual entity" is a part of the physical universe, the universe of *events*. They usually do not grasp that the "actual entity" is a philosophical concept set out, along with others ("eternal objects," "prehensions," and "God" being other principal concepts, but one might add nexus, and the ontological principle) for the sake of *analyzing* the physical universe, and especially analyzing *events*. The actual entity is not a unit of change. The event *is*.[9] The event is impossible to study without some kind of abstraction because it is unique, and yet, in its uniqueness is a logical singularity, and singulars function as universals in logic. Every event has an explanatory power, and in analysis, the "actual entity" conceptualizes, irreducibly, that explanatory power. Every event is *more* (indeed, infinitely more) than the actual entity that is used to analyze it, but properly limited, the actual entity introduces a minimal and non-vicious kind of distortion into the event. With proper self-limitation, the event will be recoverable from the analysis, intact, recognizable, meaningful, and now seen in some of its modes of generality, implication, promise.

Some interpreters have gone so far as to claim that Whitehead's ontology of events was abandoned and replaced by an ontology of actual entities. This is unsupportable on the basis of the text. Many *ad hoc* hypotheses, speculations, and assertions must be maintained in order to argue that Whitehead has two different ontologies, while massive amounts of clear evidence must be ignored and distorted to read Whitehead this way. The simple truth is that those who assert such a view haven't understood *what* Whitehead was doing and *how*. That is a forgivable mistake. Whitehead is hard to understand. But the relation between an event and the actual entity isn't that difficult to understand, if taken alone; still the logical tools Whitehead develops for his analysis are organically related, and in fact there is far more to the event than *any* set of logical tools can hope to describe. And even with Whitehead's best tools, we get at only a limited collection of characteristics of an event. Fortunately, since some problems are more pressing than others, we don't need the event in its concrete fullness for most of our purposes. So long as we can come by results that answer to our most immediate needs for knowledge, the philosophy of organism has served its purpose.

Thus, with the inquiry into cosmology that is entitled *Process and Reality*, Whitehead takes on his biggest project. His analytical tools have been developed over nearly 40 years. They are brought together in a way that leaves intact the aspects of the event that are most salient for metaphysical description, and secondly for the study of physical nature as a system of order. The outcome is a description of the cosmos as a "creative advance." These analytical tools, most importantly the actual entity, the actual world, the eternal objects, prehensions, God, propositions, generic contrasts, and the principles by which they are related, are *chosen* by the inquirer, *not necessitated* by the physical universe.

Confusion on the Ground

Assessing the event suggests (rather than demands) a tool kit and a path of inquiry. The norms of excellent inquiry vary according to the kind of knowledge the inquiry is expected to produce and its purposes. All successful inquiries intensify our experience and are "adventures," but no inquiry, for Whitehead is measured by the achievement of some kind of certainty or closure. We want the event because the event *is* experienced. We are wise when we grasp, in advance, that the event explains the world, not the other way around. Thus, in a metaphysical description of cosmological order, the actual entity explains its actual world, not vice-versa. The actual entity is not, however, the *whole* explanation of the event, considered as a part of the cosmic order. Indeed, we strive, as a norm in inquiry, for parsimony, and we find that our desire for explanation requires not only the actual entity, but those other entities listed above, along with a number of principles and limitations. If we should develop and employ a single unneeded tool for the inquiry, the inquiry would fail in parsimony and elegance, and might also introduce undesirable complexities into our descriptions. So we do our best to develop and employ the optimal tools, which means the fewest and the best adapted to the demands of the event, considered as a part of, and an explanation of the physical universe *as a cosmos*, a system of order.

None of the tools of analysis Whitehead uses in *Process and Reality* is entirely new for him. Most of his analytical tools were developed in the course of his career. Some he adapted from other inquirers. Many interpreters have not paid close enough attention to the genesis of these tools, while others have not understood how the tools change when adapted to new inquiries and other purposes. Similar names for these tools of analysis across inquiries lead philosophical interpreters to look for a development and deepening of Whitehead's understanding of his own philosophical concepts through the decades. It does not occur to most readers that Whitehead's level of "commitment" to his analytical tools is limited to their usefulness *in a given inquiry*.[10] He was a mathematician. For such people, concepts can be defined and redefined as context and purposes require. Philosophers simply don't understand: "First, let the actual entity be the event in its actual world; *now*, we adopt a rule that only one actual entity is fully concrete, and the rest are occasions; *now*, let there be a limit to the actual world in two modes, God as the limit of its actuality and eternal objects in the mode of possibility. . . ." Such is mathematical reasoning. We are now in a position to ask "what is the event?" by asking "what is the actual entity and how is it the reason for its actual world?" That is math.

It is useless to ask: "Are there *really* actual entities?" Clearly, in one sense yes and in another sense, no. They exist as analytical tools, and that is a *real* existence. But they aren't little temporal atoms or puffs of energy that *make up* the physical universe. That is a silly way to think about the universe and it confuses the order of analysis with the order of existence. Both are real, but one is the thing to be explained and the other is the tool of explanation—and the world explains the tool. Yet, the choice of tools is not arbitrary. The success of inquiry, the norms of inquiry, the choice of analytical tools, and the aptitude of the inquirers are all tied to a real universe that will not yield to just any old hypothesis or assertion. This is not child's play. The universe is patient of analysis and indeed the discovery of tools *for* that analysis thrives on errors and on the discovery of error, as indeed all development does, but our logical tools, especially, are refined only with great effort and care. We attend to the finest shades of relevance, applicability, adequacy, rigor; and with effort, insight, and good fortune, we do better rather than worse in achieving our purposes. Humility before the enormity of the task is *de rigueur*.

Over the course of many inquiries, we begin to learn something about the relations between our analytical tools and the universes of meaning that explains them. There is something that ties our processes of thinking, inquiring, and learning to the whole we are trying to grasp, in fits and starts. We don't achieve much, but we wouldn't get anywhere at all if the connection between our efforts to know and the objects of our knowledge were not linked in a sympathy not of our making. Knowledge, for all its fallibility and pluralities, is both *real* and *of* the real. This dawning awareness is not an occasion for beating our chests in triumph. But it is a nice reason to say that "every atom belonging to me, as good belongs to you," so let us sound our barbaric yawps and then "loafe" and invite our souls. *That* is success in inquiry. The connection of our thinking to what is real, including us, is our song of ourselves, and *it is enough*. We don't have to be gods. Let us see whether we can even succeed in being decent *animals*, shall we?

A Different Look

There are a number of features of our book that distinguish it from all predecessors. The first is the central idea that gives the book its title, and which we have just rehearsed in preliminary form: the actual entity *is* the quantum of explanation, and not just in *Process and Reality* or in Whitehead's other inquiries, but in *all* inquiries that are essentially logical or mathematical; wherever we seek order, there *will be* a quantum of explanation, and it *will be* some version of the actual entity uniquely situated in its actual world. This is a very great discovery Whitehead (with a few others) made, a permanent advance in science and philosophy, and a way to overcome 2500 years of confusion that has haunted natural philosophy and most other sorts of explanatory thought. We now know how to avoid taking our abstractions for concrete events, but even Whitehead's followers have mostly failed to understand this most important discovery. Because his interpreters have failed, the wider world has not learned of Whitehead's achievement. It is a matter of indifference to us whether Whitehead gets *credit* for this, and we think he wouldn't have cared either. What is important to us is that the world of explainers should *wake up* to the fundamental requirements of successful explanation, so far as the human race has yet discovered those requirements. The work has been started, both by Whitehead and a number of other inquirers, all of whom could easily carry the title of "radical empiricists."

The second distinguishing feature of our book, and hence the subtitle, is both an interpretation of Whitehead as a radical empiricist and a defense of radical empiricism itself. Along with William James, Dewey, Bergson, and to a great extent Cassirer, Peirce, and Royce, we situate Whitehead with a group of thinkers all of whom could easily accommodate Whitehead's results without introducing any great alteration of their own inquiries. In the cases of Bergson, Cassirer, Peirce and Royce, the (mathematico-logical) habits of inquiry are even similar, if less formalized. We offer no extended discussion of any of these radical empiricists and quasi-radical empiricists here, but we have discussed and defended them elsewhere and will continue to develop these historical theses as time and life permit. Within these pages we have attempted to document exhaustively both what radical empiricism is and how Whitehead is committed to it. We are far from the first to notice Whitehead's radical empiricism, or his kinship to the list of thinkers above in philosophical orientation. But this book is by far the most extensive treatment of the thesis and seeks to settle the question.

The third distinguishing trait of this analysis is its defense of what we are calling Whitehead's "radical realism," especially as connected with the method of extensive abstraction and his lifelong concern with the problem of space. Whitehead's objections to Kant's reduction of the problem of space to the transcendental conditions of presentation in cognition have not been well understood or widely discussed. Part of the reason is that one has to spend a good deal of time with Kant first, and the secondary literature is of little help, since Whitehead didn't know that literature and read Kant on his own. His readings of Locke and Descartes and Leibniz and Plato and Aristotle are similarly untutored and thus unencumbered with the categories and schools of interpretation that filter the educations of nearly every philosopher these days. But Whitehead's objection to Kant is especially important because Kant initiates a style of thinking about space that has placed it under artificial restrictions relative to the genuine reach of our analytical tools. A wide reading of Kant, beyond the first *Critique*, shows that the limitations of scientific knowledge are not as severe as the Kantians have generally argued, but that doesn't matter, since Whitehead really *is* concerned with the scientific and philosophical act of knowing, and that act of cognition *is*, according to Kant, limited by the *a priori* conditions of space as the empirical form of outer sense intuition and also by the formal constraints of space as a pure intuition in the *analysis* of sense. This view is what Whitehead criticizes.

Whitehead has the criticism right, in terms of understanding Kant, and he objects that Kant's constraints are too severe. Yet, to see where the problem lies with Kant and his dubious gift to his successors, one must make a clear distinction between extension and space, granting "space" more or less to Kant (although there is more to the form of space than Kant expected), and looking for something prior to the scientifically mediated processes of counting and measuring, but also for something still thoroughly intelligible to the finite mind. As with the quantum of explanation, the discovery of the irreducibility of "extension" in every inquiry contributes something permanent to our adventures in knowing. But there is a difference. Where the actual entity is an analytical tool, extension is the real and existing continuum that gets analyzed by any and every tool. If there is any old-style ontological claim in Whitehead's philosophy, it is that the extensive continuum *exists*, but even here he circumscribes the claim by "our cosmic epoch," which is the hypothetical whole, characterized as the "undivided divisible" (another name for extension), which is the presupposition of our cosmic epoch. Whitehead's concern to theorize extension as a strategy for handling "the problem of space" was a permanent feature of *all* of his theoretical work (not just his philosophical writings).

The final version of the theory of extension ends up as Part IV of *Process and Reality*, after having been drafted, using various tools of analysis, several times over Whitehead's career. It was first projected as Volume Two of his *Universal Algebra*, then as Volume Four of *Principia Mathematica*, but Whitehead was never satisfied with his analysis. He ended up expressing it as a fairly simple axiomatic system in Part IV of *Process and Reality*, but there were still problems with it (which we will discuss in some detail). But even with those problems solved, as they have been, the true importance of Whitehead's theory of extension hasn't been understood at all by philosophers (although some computer scientists and mathematicians have seen promise in it). But philosophers have treated this theory as a sort of incomprehensible "add-on" to *Process and Reality* and have failed to grasp that the first three parts of that book exist *for the purpose of* situating the theory of extension as a piece of natural knowledge. Combined with the natural philosophy in Whitehead's three books between 1919 and 1922, the theory of extension is asserted as a real piece of natural knowledge. Although Jorge Nobo has preceded us in understanding how the theory of extension belongs with and to Whitehead's cosmology, his account is limited to the condition that the world be a solidarity, which is one of the analytical constraints of *Process and Reality*, but with importance for all of Whitehead's inquiries into the natural world. Thus, we defend a version of Whiteheadian realism that stands alone both in the field of Whitehead interpretation and also in the philosophy of science itself. We make bold to suggest it may supplant the mainly verbal debate between those who call themselves realists and those who call themselves anti-realists in the present.

A fourth feature of our interpretation that is new is our way of distinguishing the levels of generality in Whitehead's major philosophical works. We believe that many readers have not noticed that the theory of prehension, the theory of perception, the theory of transition and concrescence, and the theory of cosmic epochs are different *layers* of theory, with different rules and results and constraints, and operating at different levels of abstraction. The organic character of Whitehead's philosophical inquiry can disguise the real differences in his complementary and mutually inter-dependent lines of reasoning. And his habit, as a writer, of reminding and digressing and hopping around to point out an implication in his current line of reasoning for something he has said earlier, will say later, or has discovered during a different inquiry, adds to the confusion. Yet, Whitehead's use of terms is quite precise. There are reasons, as we will explain, for his habits. They frustrate not just the casual reader but also the seasoned and devoted ones. We will provide the needed guide in our early chapters for recognizing the different roles played in the philosophy of organism, as in Whitehead's other inquiries, by this kind of multi-layered approach to generalization.

The explanation of these relatively simple differences in levels of generality yields a clearer understanding of Whitehead's genuinely difficult philosophical method of genetic and coordinate analysis. A fifth feature of our book not found in others is a thorough explanation of how this kind of algebra actually works in *Process and Reality*. It works similarly in Whitehead's other inquiries, but our presentation here assumes that if this method can be understood in the most challenging instance, which is *Process and Reality*, then seeing how it operates in Whitehead's other books will not be as difficult. We do not have space here to carry out a full exposition of the method of genetic and coordinate analysis in all of Whitehead's books, but we hope other researchers will find occasion and energy to do so, and indeed we may also do some of this work in our own future efforts.

Finally, there are numerous suggestions, assertions, surmises, conjectures, and opinions contained in this book that are ours rather than Whitehead's. We have not been shy about commenting on the philosophy and science of our own time, and indeed, the *reason* we read and write on Whitehead at all is for the sake of adding something to the adventure of knowing for *our* time and for the future. Hence, there is (for example) contained here a thorough, and we hope devastating, critique of contemporary model-centric thinking in the interpretation of science. We believe Whitehead would agree with all of our assertions, but of course they have to be considered on the weight of our lesser authority. Still, we think that we are making an addition of some merit to our own context. The same may be said for the original parts of the account of possibility both in cosmology and in theology contained here. We have tried to be explicit when we speak for ourselves and not as interpreters of Whitehead.

About Whitehead Scholarship

This brings us to the aspect of the book that some readers may find least appealing, which is the situating of our work within the context of Whitehead scholarship. In an earlier draft of this book, we had far more extensive (and critical) engagement with that literature. However, we think that the summary treatment that survives in our final edit should suffice for scholarly purposes. Following the references in our notes, a motivated reader can glean how we depart from the main lines of interpretation, and from some more than others. We have here contented ourselves with a summary of what we once had pounded out in great detail. Such are the realities of publishing in the twenty-first century.

We conclude this Introduction with a summary of how our efforts stand with regard to the major works of Whitehead scholarship. Readers who are not concerned with this part of the chore should skip to chapter 1.

The Contemporary Revival

Whitehead's thought has been enjoying a widespread revival. A sizable group of specialists in Continental philosophy has been engaged in studying and writing on Whitehead's connections with Bergson and Gilles Deleuze[11] in ontology, Bruno Latour in the philosophy of technology and critique of culture, and Isabelle Stengers, whose unusual combination of concerns and talents has made her the soul of the revival within these circles.

Also there have been independent rumblings by such writers as Steven Shaviro, who imaginatively reconstruct a world in which Whitehead, who *deserved* the role (in Shaviro's view) that was accorded to Heidegger instead.[12] Also from this domain comes the "object-oriented" philosophy and "speculative realism" of Graham Harman. Whitehead has successfully entered the blogosphere and the world of new media.[13] Shaviro and Harman also press the relation between Whitehead and contemporary Continental thought. There are numerous young scholars operating within and at the edges of the traditional academic institutions who are doing exciting and novel work.[14]

Finally, we mention Roland Faber's appointment as Executive Co-Director of the Center for Process Studies at Claremont and his various projects (the Whitehead Research Project, especially), and the emergence of Catherine Keller as a major voice in feminist theology and in social ethics.[15] They have influenced and reinforced the initial (Continental) direction of the Whitehead revival. Also, Whitehead Societies have appeared in Germany, France, and several Central and Eastern European nations, and interest has always been lively in Belgium, supplementing the strong interest that has existed for over 3 decades in China, India and Japan. In short, Whitehead is on many lips at many meetings and is being widely studied, referenced and re-introduced into teaching and discussion. We are encouraged by all this activity.

From a Logical Point of View

Isabelle Stengers rightly argues (but does not always remember) that Whitehead's general *approach to* philosophy is more closely akin to the methods of philosophers before Kant than after Kant, but he has the added advantage of access to more updated science and a greater number of mathematical and logical tools (some invented by Whitehead himself) than the Moderns possessed. The idea that philosophy ought to be approached by creative formalization of its problems is an old-fashioned idea, these days, and philosophers are rarely trained in dealing with questions of the foundations of mathematics and logic that animated not only Whitehead's time, but the two centuries between Descartes and Kant.

Unhappily, several generations of Whitehead's interpreters have been trained in a narrow logic and a version of the history of philosophy according to norms that would fall significantly below those common in Whitehead's day, tempting interpreters to concentrate on the parts of the text that interest them, while leaving the parts they do not understand to *different* specialists. There is no harm in this, in the retail sense, it being like the ancient Indian fable of the blind men and the elephant, except that if no one can see the whole elephant at all, this leaves no one to *tell* the story.[16] Similarly, then, Whitehead's *philosophical* methods are also poorly understood, especially the relation between the tools of genetic and coordinate analysis, on one side, extensive abstraction on the other, and the role played in both of these methods by creative formalization, in the quest for *philosophical* knowledge (which is quite distinct from other kinds of knowledge, for Whitehead).

Developmental Challenges

Whitehead's *development* as a philosopher is also a matter of confusion due to much semi-professional biographical research and associated speculation (sources are regrettably slender), coupled with a basic failure on the part of such historical researchers to understand Whitehead's *permanent* philosophical concerns with the problem of space, and with the various versions of a non-metrical theory of extension he worked toward his entire life.[17] Thanks to off-handed, inaccurate, and irresponsible remarks made along the way by Bertrand Russell, many interpreters have acquiesced in Russell's "authority" and distorted the way Whitehead thought about the relation between mathematical order and knowledge.[18] Very few interpreters grasp how Whitehead's investigations in the foundations of mathematics led him to favor algebraic over geometrical explanations (although Stengers has this right, in our view). But many interpreters, and nearly all *casual* readers of Whitehead, still confuse the narrow logic of *Principia Mathematica* with the broader logic of metaphysics and thus exaggerate Whitehead's commitment to the project of reducing mathematics to an axiomatic system of logic—a project that Whitehead recognized as a failure from an early date and from which he withdrew his energies. Indeed, Whitehead became one of the ablest critics of subject-predicate logic before the third volume of *Principia* was ever published and he maintained that criticism for the rest of his life.[19]

And finally, Whitehead's readers, from students to the most informed teachers of his texts, have not understood Whitehead's actual achievement in formalizing a non-metrical theory of extension, and how his other methods in *Process and Reality* have contributed to all sorts of subsequent advances in computer science, abstract mathematics (Whitehead is the true pioneer of mereotopology), and the critique of General Relativity. At the same time, Whitehead avoids the conflation of space with extension that has dogged almost all philosophy since Kant.

Pie in the Sky

In this book, we address and (we believe) *solve* each of these deficiencies in the understanding and application of Whitehead. We make the case that Whitehead's thought, properly understood, is not only *relevant* to a host of contemporary philosophical problems, but also offers both formal and critical tools for addressing serious problems in the contemporary interpretation of science, with applications to every branch of theoretical knowledge. In this book we sketch a formalization of his views of possibility and actuality in ways that may open up empirical research on questions of time and non-local energy as they relate to the idea of possibility and the physical world.[20]

Thus, the importance of this book, we would argue, lies not only in the correction of a long history of misreading, but contains guidelines for much more effective reading and application of Whitehead, which also makes it possible to *teach* his books with less confusion (see chapters 1–5). In addition to these contributions, it is hard to predict what importance may be accorded to our constructive applications of Whitehead's ideas to problems in the interpretation of science and the relations among philosophy, empirical, knowledge, and nature. Obviously we hope these ideas are found to be important and that they do some good in the world.

We hesitate to say that our ideas in the middle chapters are fundamentally "new," since the vast majority of what we say is found somewhere in Whitehead, or in his background sources, or in the history of science and the history of mathematics he assumed, as well as elsewhere, but we will take credit—or blame—for the arrangement and presentation of these ideas here, which *is* without precedent, either in Whitehead scholarship or in any other philosophical work. The *effect* of our ideas will be new, we hope, if they receive any attention. This middle part of the book culminates with a "new" account of philosophical naturalism that would actually *settle* the current naturalism debate in both analytic/pragmatic and Continental philosophy, if it were to be considered in the context of the philosophy we defend here. We realize how bold this claim is, but we have aimed high and hope to attain some of what we aim for.

Theological Questions

The final chapters of the book (chapters 11–13) complete the account of naturalism and address the important traditions of process theology and theological naturalism that have developed among intellectuals drawn to process thought as a way of re-articulating spiritual traditions and experience among humans. Much creative work in this domain has been done since Whitehead's death, although we think most of it has little *directly* to do with Whitehead's own philosophical concerns or with his actual achievements. As Stengers rightly points out in *Thinking with Whitehead*, it was Charles Hartshorne, and American philosophical theology, that preserved the strands of interpretation that have become today's revival. Auxier has addressed his development in previous writings.

Much of what we have to say in these final chapters is aimed at *extracting* Whitehead from this discussion that has now spanned some four generations, leaving, we hope, both the achievements of process theology *and* Whitehead's own thought intact. The tendency among process theologians to venture into scientific metaphysics and to associate their results with Whitehead's name has had the effect of deterring other serious interpreters of science and philosophers of science from the study of Whitehead, due to a common (although far from universal) scientific superficiality in the theological books and essays, and due to an aversion on the part of philosophers of science to pursue any kind of theological discussion. Some of the theologically oriented process literature is very good, scientifically and mathematically, such as the work of Wesley Wildman, but others are not so good. No one's misinterpretations of Whitehead are repeated more frequently than those of the theologian Lewis S. Ford. We want it understood that animus toward *his views* expressed sometimes in our main text is not indicative of any personal disdain. His viewpoint is basically theological and his method of interpretation is an adaptation of New Testament biblical criticism. It is misguided, and it has misled others who are even less well prepared than Ford to deal with the systematic and technical aspects of Whitehead's philosophy.

Due to the wide reading Ford received, many people simply start studying Whitehead wherever he sets them, with a full set of assumptions in place that, in our view, impairs a person's ability to understand much of what is plainly before him or her in Whitehead's text. That situation needs to change, and so Whitehead *as* a natural philosopher, cosmologist, and philosopher of mathematical order needs to be separated from the theological tradition he helped to start. Whitehead's concerns and achievements are not of a sort that contributes very much to theological discussion, *except through extrapolation*. These days, the interpretive tradition in process theology is so well established, with its own fine elders (John Cobb, David Ray Griffin, Marjorie Hewitt Suchocki, and others), that younger theologians and advocates of natural religion are not often aware of the extent of their own extrapolation from Whitehead, and thus, some of their work is passed off as Whitehead "scholarship." It usually is not. This needs to be corrected and situated rightly in the literature. Theological thinking in a process vein is valuable and ought to be pursued vigorously. Still, our last two chapters do the work of sorting out process theology (which is never "Whiteheadian" in any strict or even fair sense) from genuinely Whiteheadian ideas about God. We draw attention to some of the most important deviations in this tradition from Whitehead's thought and we show some of the reasons to press for a Whiteheadian naturalism that is for the most part independent of theological concerns.

Thinking with Whitehead

Isabelle Stengers' *Thinking with Whitehead* is currently, one might say, "the book to beat" in Whitehead interpretation. Stengers really is only *thinking with* Whitehead. Her work is not Whitehead scholarship, nor is it intended to be. She leaves that work to others. Whitehead interpreters are very much divided by the position they take regarding Lewis Ford's "compositional analysis," as an interpretive approach to Whitehead. Compositional analysis is, as we have said, essentially an application of the methods of New Testament higher criticism to the question of Whitehead's development (with this supposed "development" limiting and determining interpretation). We absolutely reject "compositional analysis," and we find Ford's work to be a house of cards, highly speculative, implausible, and damaging to the understanding of Whitehead. Stengers not only endorses Ford's method and his results, but does so "unequivocally" and even suggests her book wouldn't exist without Ford's results.[21]

This changes everything, even though Stengers acknowledges that her basic problems are different from Ford's. Rather than argue with Stengers at every turn, or highlight the places where we agree with her (and there are many), we will set aside her interpretation here and argue mainly along our own line. We leave it to others to decide whether Stengers' insights can be extricated from the Fordian web.[22] Our interpretation of Whitehead stresses the continuities and confluences of his thinking as part of his life as lived. We reject major breaks and sudden discoveries of the sort Ford depends on, and, unfortunately, Stengers allows to influence her understanding.

The common distinction among Whitehead interpreters, first offered by John Cobb, is, or has been, a threefold distinction of "genetic" interpreters, "systematic" interpreters, and the partisans of "compositional analysis." Our book is both "genetic" (chapters 1–5) and "systematic" (chapters 6–13), but we do not accept anyone's account of Whitehead's development (Lowe especially), and no one else has quite the same view of "system" that we defend, although we find in work by Jorge Nobo, James Bradley, F. Bradford Wallack, and several other "systematic" interpreters, much we can endorse and little with which we disagree.

Systematic Interpretations

Among the systematic interpreters, there has not been another book on Whitehead as ambitious as this one since William Christian's *An Introduction to Whitehead's Metaphysics*.[23] Christian's isn't really an introductory book at all, in spite of the title. It is a good book, but limited by the fact that Christian does not seek to coordinate Whitehead's mature thought with his earlier writings and, as a result, offers an interpretation that forms a semi-stable island in the much larger sea of Whitehead's thought and his historical context. Christian's book is the only one that could easily be compared with ours on the whole, but even it is not nearly as comprehensive.

There are several studies that combine shorter systematic treatments with the aim of "introducing" Whitehead. Ivor Leclerc's *Whitehead's Metaphysics: An Introductory Exposition*,[24] is one such, but it suffers from the same overemphasis on *Process and Reality* as Christian's book, and also fails as an introduction (being too difficult to read). It has the advantage over Christian of situating Whitehead more astutely in the history of philosophy (emphasizing only those figures Whitehead actually studied), but this virtue eventually becomes a vice due to an over-emphasis of the same. Leclerc tends to reduce Whitehead to Plato or Aristotle or Leibniz in the course of discussion and to play the great philosophers off against one another, a comparison in which Whitehead is usually the loser, according to Leclerc. Thus Leclerc fails to present Whitehead's true intellectual context as belonging to the mathematics and physics of his own day.

In terms of the actual systematic, total interpretation of Whitehead, carried out in our middle chapters, our book is most similar to Jorge Nobo's *Whitehead's Metaphysics of Extension and Solidarity*.[25] We agree with much of what Nobo has to say. He is one of the few interpreters of Whitehead careful enough to get huge chunks of the philosophy of organism right, in our view. His book is very ambitious, but not nearly as comprehensive as ours. It also suffers from being more difficult to read than Whitehead himself. When one has done the work to figure out what Nobo is actually saying, one finds his book illuminating, helpful, in places brilliant (both intellectually and from a scholarly viewpoint). But Nobo's influence has been blunted both by the difficulty of his book and by constant negative representations from Lewis Ford's powerful perch as founder and editor of *Process Studies*.

Apart from Nobo, there is no total attempt at an interpretation of Whitehead similar to ours, but there are several books that investigate vital parts of the philosophy of organism. The most important of these are John Lango's, *Whitehead's Ontology*,[26] F. Bradford Wallack's *The Epochal Nature of Process in Whitehead's Metaphysics*,[27] and Judith A. Jones, *Intensity: An Essay in Whiteheadian Ontology*.[28] These studies have a good deal to recommend them, and their results can be used to supplement ours, but they do not attempt total readings.

Genetic Interpretations

Among genetic interpreters we need only mention Victor Lowe's *Understanding Whitehead*.[29] Lowe was one of Whitehead's last students and became his reluctant biographer, although his colleague Jerome Schneewind actually had to edit Lowe's extensive notes and writing into the two-volume biography that every Whitehead scholar now reads and relies upon. But Lowe's 1962 book is actually a summary of Whitehead's development as a thinker from 1891 into the early 1940s, when Whitehead really stopped writing. Lowe does, therefore, connect the island of *Process and Reality* to the rest of Whitehead's own development, but unfortunately Lowe had no serious training in the history of science or the history of mathematics, and, in our opinion, not much ability as an intellectual historian (something Lowe repeatedly admits himself; he did all this work because *someone* needed to do it and no one was stepping up).

Lowe's 1962 book, like his biography, is misleading on many crucial points. His concern was to save Whitehead from criticisms that were popular among narrow-minded positivists and other language analysts during the time after the Second World War. He distanced Whitehead from anyone who might kill his influence among such people.[30] He essentially tried to make Whitehead respectable for analytic philosophy, to the extent he could, given the style and dogmas of analytic philosophy during that time. The context has changed in our time.

Verbal Disputes

As for guides to Whitehead, especially Whitehead's language, there are two well-known companions to *Process and Reality* by Donald Sherburne and Elizabeth Kraus, and there is John Cobb's *Whitehead Word Book*.[31] These works could be seen as competing with and complementing both our method of reading Whitehead and our systematic analysis of *Process and Reality* in the middle chapters. Indeed, our middle chapters, if correct, would largely work against Sherburne's book as a viable guide, and our account and renders Kraus's book valuable only for parts of her commentary. But neither of these books attempts the total interpretation offered in our book. We will address ourselves to Sherburne and Kraus in the notes as circumstances warrant, and along with Cobb, these sorts of interpretations can be classed as "genetic," but limited in scope to a given book.

Out on a Limb

There are many studies applying aspects of Whitehead's thought to various problems he did not himself address. There are also numerous books that attempt to "update" Whitehead to bring his ideas to bear on the problems in one discipline or another. The results are very mixed. A good deal of success has been achieved in religious philosophy (see Wesley Wildman's numerous works), and also especially where religious experience touches on social experience (for example, Catherine Keller) and the experience of nature (see Richard Weidig and Ralph Pred, for instance[32]). Our book does not address directly this sort of study, since science, mathematics, ontology, logic, and the tools of analysis and knowledge are our primary concerns. Our reading has *implications* for these sorts of books, of course, but only to the extent that it may suggest these various authors are working with ideas *they* believe to be Whitehead's, but are not Whitehead's. We are not purists and are untroubled by the good use of good ideas wherever they may be found, but putting Whitehead's name to those ideas can be misleading.

The only direct competition regarding this aspect of our book comes from those Whitehead interpreters who seek to apply Whitehead to science in light of more recent developments in physics, especially, after Whitehead's lifetime. In this domain, our work, if we are right, invalidates the approach of Michael Epperson in *Quantum Mechanics and the Philosophy of Alfred North Whitehead*.[33] This earlier work of Epperson's was especially problematic in that it slipped from employing aspects of Whitehead's metaphysics as interpretive tools in the philosophy of physics, to making more broadly exegetical claims about Whitehead's metaphysics itself, claims which do not hold up under a careful reading of Whitehead's texts. However, the more recent work by Epperson and Zafiris, *Foundations of Relational Realism: A Topological Approach to Quantum Mechanics and the Philosophy of Nature*, does not suffer from that weakness.[34] In this work, Epperson (who handled the specifically philosophical parts of the discussion) maintains a much cleaner focus on the philosophy of *physics*, and thus on *applying* Whitehead's thought rather than going into deeper interpretations *of* that thought. But precisely on that account, Epperson and Zafiris are not offering an account that competes with the argument presented here. Rather, their work supplements and complements what we are doing, or so we believe.

In particular, it is worth highlighting two salient features of Epperson's and Zafiris's work. The first of these is the "relational realism" from which the book gets its title. One of the most profoundly unfortunate ideas in contemporary physics is the "multiple universes" claim, an especially unhappy piece of nonsense for which we will have some choice words later in this book, especially in chapter 10. This sort of groundless speculation on the part of physicists is driven, at least in part, by a failure by physicists to take relations *seriously*, while treating their mathematical formalisms as somehow ultimate. Thus, we agree with Epperson and Zafiris when they say that, "when elegance of formalism . . . trumps empirical applicability, the measure of scientific progress begins to derive from the measure of its appeal, rather than its appeal deriving from its progress."[35] But Epperson's and Zafiris's insistence that relations should be treated as having a genuine standing of their own, and not as merely parasitic upon their *relata*, makes their work one of the most important contributions in the philosophy of physics in recent years.[36] The second point to highlight is the use by Epperson and Zafiris of category theoretic ideas, especially that of a "sheaf," both as an interpretive tool and a primary formal technique applied to quantum mechanics.[37] As we will be arguing throughout this book, "algebraic" modes of thinking (which includes category theory) are of vital importance in understanding Whitehead's relationalism.[38]

In other areas, we tend to agree on many points with the book by Murray Code, *Order and Organism: Steps to a Whiteheadian Philosophy of Mathematics and the Natural Sciences*,[39] but Code has revised his view on much of what he said in that book, especially the parts we think he had right. We will note our reasons for this view in our notes as the book unfolds.

Our approach tends to agree with the results of the plasma physicist Timothy E. Eastman, who has published many articles on Whitehead and physics, and has gathered together some of the principal contributors to this discussion in his *Physics and Whitehead: Quantum, Process, and Experience*,[40] but one problem we encounter repeatedly is that such interpreters do not rightly understand what Whitehead means by "quantum" and they wrongly assume it is close to what contemporary physicists mean by that word. One challenge for our book will be the displacement of this habit of thinking. What Whitehead means by "quantum" is far more radical, and far more important, than what contemporary interpreters of science suspect.

1 Reading Whitehead

> The order in which the parts should be studied will depend upon the psychology of the reader. I have placed them in the order natural to my own mind, namely, general principles, particular applications, and finally the general exposition of the mathematical theory of which special examples have occurred in the discussion of the applications.
> —Whitehead, *The Principle of Relativity* (vi)

Difficult Reading

Whitehead is difficult to read, and there can be little doubt that the challenge of his language is partly responsible for limiting his influence. The difficulty, however, is not just mastering a new vocabulary. The problems also have much to do with the unfamiliarity, the novelty of the individual thoughts and relations among the ideas Whitehead describes, and this situation is not so much linguistic as intellectual. The ideas themselves are not easy to *think*. But in addition, it is not always easy to grasp the structure of Whitehead's *exposition* because it operates simultaneously at several levels of abstraction (or generality), and Whitehead is in the habit of moving from one level to another with little warning he has done so. He kindly overestimates his readers and under-estimates the subtlety and elusiveness of his thinking. Whitehead does define his vocabulary carefully and he does clearly describe the various levels of generality among which he hops, and he is very conscientious about laying out the scope, limits, and aims of every inquiry. But he asks too much of us.

The effort to understand Whitehead is further complicated by the organic character of his thought. Each part of his philosophy, every level of generality, every term, and every goal, is immanent in and achieved through all of the others. To understand clearly any given passage in one of Whitehead's books requires an understanding of the rest of the book. Whitehead is not one to waste words, so unpacking the full meaning of even a single sentence, in light of the work that sentence does in the book, can require a long narrative. Organicity in a philosophy is familiar to readers of Hegel, Bradley, Bergson, Royce, Heidegger, and numerous others, but the problem is especially pronounced in Whitehead's most important works, since the order of exposition is often very different from the order in which the ideas would be traced if the exposition were linear. As Whitehead says, "In the presentation of a novel outlook with wide ramifications, a single line of communications from premises to conclusions is not sufficient for intelligibility. Your audience will construe whatever you say into conformity with their pre-existing outlook."[1]

In addition to these problems, there is also a great difficulty in understanding how one work by Whitehead, whether an essay or a book, relates to or bears upon others in the full corpus. Here we have to confront not only issues of the development of Whitehead's thought, but also the *unity* of it. Many incompatible interpretive theses have been defended by various scholars over the decades, but the most influential theses usually involve the idea that Whitehead's thought culminates in his 1929 work *Process and Reality*, which is by any standard the heart of Whitehead's philosophy, and that all other works are to be measured by the content and aim of that work. This prevailing view is neither correct nor helpful in learning to read Whitehead, so we will not assume it in what follows.

Whitehead's Lexicon

To begin the daunting task of "reading Whitehead," it is a good idea to start with a discussion of terminology and how it is deployed in service of systematic inquiry. Whitehead organized his inquiries in the way mathematicians do, defining terms clearly but flexibly and generally, and redefining them when he changed from one inquiry to another. Thus, for example, what he *calls* "eternal objects" in *Process and Reality* has a close relation to what he calls by the same name in *Science and the Modern World*, and both are close to what he calls "ideal forms" in *Religion in the Making*, but there is no reason to assume an identity of meaning for technical terms across different books. Indeed, the meaning of technical terminology develops and deepens more *within* the context of a single inquiry than across books, so that there is no reason to assume that a technical term, such as "ideal form," means exactly the same thing from one occurrence to the next within even the *same* book, let alone across books. For example, in *Religion in the Making*, the term "ideal forms" is used more or less interchangeably with the terms "ideal entities," and "abstract forms" and all are closely related to the term "ideal world."[2]

Whitehead says, "The fate of a word has to the historian the value of a document."[3] To be a reader of Whitehead is to be a historian of a sort. But Whitehead's writing style tempts one to treat individual occurrences of technical terms as though they might serve as fixed definitions. He has a habit of saying things like "an actual entity is the outcome of a creative synthesis, individual and passing."[4] But elsewhere he may say something like "to be an actual entity is to have a self-interest,"[5] which seems to have no clear relation to what has been said before, and also seems to have definitional force. As a matter of writing style, Whitehead makes hundreds of statements like these in his shorter works, and thousands in the longer works.⁶

Even if it were clear what force a definition has in Whitehead's thought (and that could be a long discussion), it would never be clear when he intends a given statement to have definitional force. It is better to think of the occurrence of terminology in Whitehead in hermeneutic terms, to see each occurrence of a term as a promissory note of further refinement, in terminology and in total context, over the course of a work. There is, however, no such promise of refinement when we move from one work to another.⁷ Whitehead's philosophy does develop and fill out. His numerous footnotes in later works to earlier works suggest that he thought of the various inquiries in relation to a total development and effort, but with technical terminology he makes no such assurances. In the domain of technical terminology, the safest interpretive course is to confine one's attention to a single work until the full meaning of the term *in that context* has been appreciated and understood. Only then is a reader in a good position to consider how the differing aims of various books may affect the comparison of terminology that may seem similar or even be common to both.

For example, the subject matter of *Science and the Modern World* (1925) is not too far removed from that of *Process and Reality* (1929) since the latter treats cosmology, which Whitehead understands as a subdivision of scientific thinking, while scientific thinking is the topic of the former book. We might expect very similar meanings to be associated with similar or identical terminology, and the expectation is, we think, fulfilled. But the context of usage does make a difference; the aims of the two inquiries are different, and comparisons in meaning should always resist the urge to assume premature identities of terminology. Thus, not only should we refrain from asserting simple definitions *within* a single work, we should be even more reluctant to assume identical definitions from one work to another. To use the meaning given to a term in one work as the standard by which occurrences of similar (or identical) terms must be measured (as is lamentably common among scholars of process thought) denies to each inquiry the relative independence Whitehead claims for his works and over-emphasizes the continuities among them in ways that are sure to distort one's reading and understanding of Whitehead.⁸

Whitehead's philosophy of language actually requires such caution as we advise here. Not only is *error* the ground of meaning for Whitehead, but he continually complains about the poverty and inadequacy of language.⁹ It is best to take the development of terms in each of Whitehead's works as open-ended but increasingly adequate metaphors that point to some fundamental relation—a relation that will not come fully to language.¹⁰ A definition for Whitehead is not the demonstration of an essence from first principles, or from the four causes. He does not accept the idea that predication (whether traditional or Kantian) is our best route to capturing the meaning of an experience.¹¹

The best place to begin one's study of Whitehead's usage is to read *Religion in the Making* in conjunction with *Science and the Modern World*. Because the two works were written in the same year, one can largely ignore the question of whether a time lapse has affected Whitehead's thinking. Whitehead says that these two books are the same train of thought in different applications, so variations in terminology are just as informative as similarities. To master Whitehead's usage, then, means to grasp first the fate of terms within a single work, and to begin to grasp what is involved in comparing terminology from across works, starting with the easiest. Another option is to compare *The Concept of Nature* to *An Enquiry into the Principles of Natural Knowledge*, which treat similar questions at different levels of abstraction, "companion volumes," as Whitehead terms them, but these are more demanding works and for that reason perhaps less than ideal for mastering vocabulary.[12]

In addition to what is said here, it is sometimes illuminating to trace the origins of the technical terms Whitehead chose for his system beyond his own works. For example, the term "ingression," which Whitehead uses for the way eternal objects enter into concrescence, may seem like a neologism, but it isn't. It goes back to Middle English and has a long-standing legal meaning, and has been adapted to mathematical and economic uses over the centuries. To ingress is the complementary of to egress, but one hears neither term very much, and "egress" wasn't adopted by Whitehead as the complementary; rather, he chose "to express" for that role.[13]

Similarly, sometimes the odd word "prehension" is attributed to Whitehead as a new term (and indeed, his *sense* of the term *is* new), but its use dates back at least to 1828 and also has a legal meaning. Whitehead, in spite of appearances, is wary of neologisms, but not wary of making full use of the English language and adapting existing terms to new shades of meaning. As you might expect, however, these adaptations are dynamic and developmental. The *introduction* of a new term is just the beginning of a process of getting it to do the work he has in mind for it. The cluster of essential terms becomes a mutually defining organic whole—or in Whiteheadian, the multiplicity of terms becomes a plurality.[14] That organic whole is firmly in Whitehead's mind before he writes a book, and the order of exposition in the book is independent of the process by which his thinking proceeded in developing the web of meanings. This is a mark of mathematical inquiry, where what is published has little or no relation to the process by which it was discovered or invented.[15] In short, Whitehead is presenting the *results* of his thinking, *not* the process. The arrangement is determined both by his habits as a mathematician and his (less than reliable) estimation of what will be clear to his readers.[16] Thus, fully grasping one term requires fully grasping all of the terms, and not confusing the order of their exposition in a work with their mutual relations of meaning. For this reason, it can be difficult to know where to begin.

It will be clear that among the strategies we recommend, a good one is listing occurrences of key terms and compiling near verbatim summaries of the meanings Whitehead attributes to them, which is what Auxier did in the glossary to *Religion in the Making*, while Sherburne has done something similar for *Process and Reality*. This is obviously a labor-intensive process. One might use Auxier's glossary as a substitute for doing that work, and to a lesser degree, Sherburne's *A Key to Process and Reality* may also be consulted, although its scope of the latter work is limited and it does not try to preserve the order of usage in the book (thus the internal development of terms in the exposition is usually lost). However, even doing this kind of work for oneself will not necessarily provide an adequate understanding of the reading. Some other principles have to be understood.

Whitehead's Structure of Inquiry

We will now make a suggestion as to how we should understand the structural organization of Whitehead's inquiries in various works, and how, for him, "method" is subordinate to "theory," or working hypotheses, *within* an individual work. As he says, "theory dictates method, and . . . any particular method is applicable to theories of one correlate species."[17]

The first thing to be aware of is that Whitehead is not attempting phenomenology in any of his inquiries. The type of description of experience that (in effect) denies both genuine discontinuity and continuity to the perceptual flux, substituting for that flux some sort of *Fürsichsein* or intentional condition (by some *epoche* or reduction), so as to discover discreteness and continuities in *acts of consciousness* (and structures and principles) is *not* Whitehead's aim. Thus, interpreting Whitehead as a phenomenologist is a misstep. He knows very well the importance of phenomenological description, but it is not his task. If Whitehead were to have a phenomenology, it would look very much like the one Ernst Cassirer created in *The Philosophy of Symbolic Forms* (1923–1929), since the relation of culture and nature was Whitehead's main preoccupation, as distinct from consciousness and psychology. The complementarity of Whitehead and Cassirer is instructive and worth examining because Cassirer's project brings to Whitehead the *kind* of phenomenology his thought requires.

Wilbur Urban, who contributed essays to both of the important Library of Living Philosophers volumes on Whitehead and Cassirer,[18] identifies the central issue in attempting to bring their ideas together as follows:

> If the ideal form and immanental law of all knowledge is, indeed, to be found in the mathematical-physical sciences, then it would seem the symbolism of metaphysics must also be a symbolism of relations and that a philosophy of events, such as that of Whitehead for instance, would necessarily be the resultant metaphysics. On the other hand, if

it is true, as we are told by Cassirer, that science as symbolic form has no exclusive value, but is only one way of constructing reality, and has value only from the standpoint of science, then it would appear that a metaphysics, to be adequate, must be a metaphysics of art and religion also and must have a language and symbolic form which includes these forms also—in which case it could no longer be a symbolism of relations merely, but must be a symbolism of things also.[19]

Here Urban expresses a point of view quite common not only among critics of both Whitehead and Cassirer, but even among their most sympathetic interpreters. The effort to treat science as both an independent *and* culturally embedded form of knowing is a delicate task. Urban does not grasp "relation" as a concrete metaphysical hypothesis, as what we call "real" in this book, but sees it only in its logical or mathematical sense, hence assuming that *method is prior to theory*. He has not understood that for Cassirer and Whitehead, there are no "things," in science, religion, art, or any other symbolic domain, *until* the "things" are constituted as the highly mediated relational outcome of a symbolization process. Nor has Urban appreciated that art, religion, etc., are modes of symbolization, ways of relating to and having a "world." The various symbolic forms of culture invite separate inquiries with methods and ontologies adapted to their variations.

Urban has *rightly* intuited that a metaphysics of events, like Whitehead's, is appropriate to the type of philosophizing for what Cassirer is doing, but Urban has not considered that there are "events" in more than the scientific sense of the term. He has also not considered the importance of the deeply aesthetic character of scientific events. Understanding Whitehead involves the recognition that even if, for instance, "artistic events" and "scientific events" are both "events," the price one pays for seeing them as similar, classifying them under a single term, is that one forfeits most of what makes them interesting, valuable, and even knowable (in any rich sense). To know something abstractly is not to know it very intimately. Certainly the factors that point to any event, as seen in the light of art and the factors that point to the "same" event as seen in the light of science are, in many ways, positively correlated, but there is no good empirical reason to assume the ontological identity of the factors picked out in those variant horizons.

Yet, Urban rightly recognizes that the tendency of Cassirer, in keeping with Whitehead, to endow science with the highest authority in "knowledge" is puzzling to many in light of their similar claim that science is a culturally embedded and thoroughly mediated form of knowledge. Whitehead and Cassirer are indeed both philosophers of science in the highest sense. Urban, like so many other twentieth-century realists, is assuming that science, done rightly (i.e., according to its supposed "method"), has a privileged ontological path to the way things *are*, independent of the knowing mind. But even the question of mind-independence is a holdover from substance metaphysics, with its assumption that there is some necessary relation between "knowing" and "being" (and that this relation is secured in mathematical or logical relations). In short, Urban is a sort of unconscious idealist, while Whitehead consciously insists upon the stubborn reality of particular experience, i.e., its inexhaustibility in the effort to render it symbolically for our limited and self-interested aims.

Where one assumes only *possible* relations between reality and knowing, and places necessity strictly within the internal operations of knowing, reality becomes a contingent process, an inexhaustible resource for our symbolic appropriations. On this point Cassirer and Whitehead agree, and in no way does this restriction upon the reach of science reduce its authority. The move to treating possibility as the fundamental modal category in metaphysics simply allows us to situate scientific knowing historically and empirically, and to account for the growth and alterations in our scientific claims. In short, if science wants to know nature, it has to approach nature as a growing, contingent, and dynamic system. That is a stiff demand to make upon static, categorical thinking.

But philosophy deals not with nature or religion or art directly, it deals foremost with the structures of their possibilities. As Whitehead says: "Philosophy is the ascent to the generalities, with the view to understanding their possibilities of combination. The discovery of new generalities thus adds to the fruitfulness of those already known. It lifts into view new possibilities of combination."[20] Its aim is *philosophical* knowledge. It is worthwhile at this point to give special note to something Whitehead often says near the beginning of his inquiries. Whitehead complains often about how little philosophy can really accomplish, blaming this usually upon "weakness of insight and deficiencies of language."[21] As a result, our initial descriptions remain "metaphors mutely appealing for an imaginative leap."[22] The difficulty, he continues "has its seat in the empirical side of philosophy":

> Our datum is the actual world, including ourselves; and this actual world spreads itself for observation in the guise of the topic of our immediate experience. The elucidation of immediate experience is the sole justification for any thought [note, not just *philosophic* thought]; and the starting-point for thought is the analytic observation of components of this experience. But we are not conscious of any clear-cut complete analysis of immediate experience, in terms of the various details which comprise its definiteness.[23]

This is the cost of a radical kind of empiricism.[24] With this issue understood, it is nevertheless true that Cassirer and Whitehead undertake very different projects, phenomenological and metaphysical, respectively, from the same basic convictions about process and reality. We see no fundamental disagreement between the two, but we believe that few interpreters rightly understand their full complementarity.[25]

Interpreters of Whitehead have commonly confused the issue of the *development* of his ideas, with the *meaning and purpose* of his individual works. We will take up questions associated with the external motivations for Whitehead's individual inquiries in subsequent chapters. For the moment it is sufficient to note, for example, that the "triptych"[26] was written in response to Einstein, while the development of Whitehead's metaphysics was a response to the reception of his natural philosophy in the intellectual community. With each individual work, we descry reasons for its existence that are external to its contents. Yet, one must not carry too far the speculations that are invited by supposing we have access to motives and perceptions in the distant past. It is common for interpreters (Lewis Ford is a good example[27]) to think of each book as presaging some further development of the whole philosophy of organism,[28] culminating, for most, in *Process and Reality*, with the two major books he wrote thereafter drawing out consequences. There is great value in studying Whitehead's development, but it has led most readers to overlook the common pattern in each of his books, and very often they have misunderstood his method (and how it is dictated by a different working hypothesis in each book) along with his way of structuring an inquiry. The *structure* of the inquiry is essentially the same in each book, but the *methods* vary with the hypotheses under consideration.[29]

Whitehead is always careful in his introductions to supply a clear indication of where his present inquiry fits in the general effort of human thought. He carefully qualifies the project at hand and suggests its limits. The proper "criticism of a theory," or "working hypothesis," for Whitehead, "consists in noting its scope of useful application and its failure beyond that scope."[30] In truth, each book is an individual inquiry into some important domain of experience. Each book proceeds along the same basic line: he will describe the phenomena under investigation to provide a version of them, how they *may* be seen (not how they *must* be seen), and he also subjects that description to a systematic analysis. In no instance does he claim that the initial description of the phenomena (what he calls their "genetic account" in *Process and Reality*) is the final or the only way of describing these phenomena. In every case he notes that their analytically detailed and systematic characterization (what he calls the "coordinate account" in *Process and Reality*) is presupposed by the genetic description, but the coordination of the phenomena sets the scope of the inquiry and their genetic description is limited by that scope. Coordination of the phenomena is closer to the way they are experienced than is their genetic specification.[31]

In adopting this structure Whitehead follows standard mathematical methods. One must specify the entities and rules in order to provide their systematic interrelations. There is a great deal of freedom in specifying the phenomena or entities genetically, but the test of whether it has been done well is whether interesting systematic relations, "new generalities," and "possibilities of combination" do come to light in the coordinate phase of the inquiry (the order of genetic and coordinate exposition can vary, as we will explain, but regardless of what order these phases appear in a given book, the genetic specification presupposes the coordinate whole). And the whole analysis, both genetic and coordinate, is measured against the ways in which it illuminates *experience.* In answer to the question "why should we accept this genetic account?" Whitehead holds that its warrant looks back to experience, first, and attempts as accurately as possible to carve out the phenomena under discussion in ways that accord *with* our experience. (This is the closest Whitehead ever comes to doing phenomenology—a simple and accurate description of experience, but the coordinate whole is not treated as "given," and thus the genetic specification does not claim to get at "the things themselves," as phenomenologists usually want to do.) Adjustments in any genetic account are inevitable in light of our further investigation of the phenomena. In spite of that, we are justified in specifying the phenomena to the best of our ability for the purposes of a more thorough speculative understanding of their interrelations. The adjustments of *terminology*, along the way, as discussed above, are justified by the way that the initial specification of phenomena is always too abstract, too stipulative, and needs concretization. Looming in the background of this activity is the genuine result of prior thinking, i.e., the coordinate account without which we would have no idea of how to specify phenomena.

Thus, the genetic phase of an inquiry is an initial narrativizing of the phenomena, under descriptive principles determined by the *purpose* of the investigation at hand, more specifically, the working hypothesis or "theory," the common character of which is to sequence and specify those phenomena, i.e., to limit what we *mean* in referring to them *in that context* so that we can discuss them systematically. The sequencing does not have to be historical or chronological, but it *can* be, as in *Religion in the Making* and *Adventures of Ideas*. In every case, however, the aim is to get the phenomena to "stand still," to make possible a conceptualization which counters their ephemeral and changing character, so as to provide us with something we can trust from one thought to the next.

With the initial genetic description in place, and the flux controlled, we can move to the presentation of a coordinate analysis (bearing in mind that the coordinate analysis of our inquiry was the prior condition for the genetic specification in our exposition of our results). This is a second order narrativization, providing a bridge to what may fairly be called a "spatialization" of the problem, and this procedure allows us to examine and bring to light systematic features that were immanent in the initial genetic specification of the phenomena. This is, of course, the trick that makes mathematical exposition seem magical. The mathematician certainly knows where the exposition is going, but the auditors or readers do not. When it is shown that immanent relations in the seemingly innocent genetic specification were "loaded," it is as if the lady sawed in half appears again whole. The truth of the matter is that the "coordinate lady" *logically* precedes the "analyzed lady," and the analysis was virtual, not actual.

The exhibition of these relations presumably enables us to see how our own activity of characterizing phenomena genetically will *necessitate* certain conclusions about their relations and *exclude* other conclusions. To alter the necessitated results of a coordinate account, one must back up and alter the genetic account, and in so doing attend again to the ways in which experience presents those phenomena *together* that we had singled out and specified for an initial genetic description (bearing in mind that, in terms of the method, the coordinate account remains primary). Experience comes together, genetic description specifies, and coordinate description restores some portion of the togetherness we lost in specifying, but does so according to some (narratively informed) *purpose*. The effect of the process is to produce a coordinated description that approximates our original experience, but as adapted to the aims for which the inquiry was undertaken. The coordinate description should exhibit just those characters in experience which are relevant to the defined purposes that produced the inquiry.

The entire undertaking is thoroughly hypothetical.[32] No conclusion about reality in itself is forced upon us by any result in the genetic or coordinate accounts, because the option always remains, in Whitehead's structure of inquiry, of altering the purpose or scope of the inquiry itself, or of describing the phenomena genetically using different principles of specification. No ontological knowledge is claimed for any result, by which we mean, no knowledge of any necessary relation obtaining between the way things are in themselves, our logic, and our knowledge. It is taken for granted that the relationship between *experiencing* and philosophical *understanding* is thoroughly contingent, yet there are still pressing reasons to attempt to gain philosophical knowledge, even if that knowledge is only descriptive. Philosophical knowledge of a subject allows one to recognize important possibilities within the adventure of living, possibilities that will be overlooked without it, to the great diminishment of life.

Thus, Whitehead's books are individual inquiries into well-circumscribed aspects of experience, submitted to a genetic specification and to a coordinate analysis (both are descriptions), the results of which are limited by the way in which the initial inquiry and its purposes were defined. For example, when Whitehead confronts the phenomena of religious experience in *Religion in the Making*, he is not immediately interested in their implications for modern cosmology. He does not describe these phenomena in *Religion in the Making* for the purpose of seeing their relation to cosmology, or science of any other kind; rather, he describes them for the purposes stated:

> to give a concise analysis of the various factors in human nature which go to form a religion, to exhibit the inevitable transformation of religion with the transformation of knowledge, and more especially to direct attention to the foundation of religion on our apprehension of those permanent elements of reason of which there is a stable order in the world, permanent elements apart from which there could be no changing world.[33]

When similar concepts come up, describing similar phenomena in *Process and Reality*, they have been described for a quite different purpose, that of investigating "a phase of philosophic thought which began with Descartes and ended with Hume."[34]

In reading the interpretive literature, one gets the impression that Whitehead's followers simply do not *believe* him when he states explicitly what he is doing and why, in spite of the fact that he goes on to explain in detail what limits he places upon each inquiry. The inquiry in *Process and Reality* is about the way philosophers thought between Descartes and Hume, a certain turn of thinking, novel in human history, in which the algebraic and geometrical modes of thought came together. *Process and Reality* makes no attempt at a full assessment of philosophic thought from Kant forward or prior to Descartes. Whitehead thinks the modern philosophers had some things right that became obscured after Kant. He wants to draw out what they had right. That is the "theory," to use the term as Whitehead does.[35] The *method* (genetic and coordinate analysis) is post-Kantian, but not the subject of the inquiry, i.e., *modern* cosmology. Thus, Whitehead's major work attempts to come to terms with the fundamental insights of *modern* philosophy.

Another clear example of the ordering of "genetic" and "coordinate" analysis is in *Adventures of Ideas*. Part I of that book specifies genetically, in the mode of the history of ideas, "the influence exerted by the Platonic and Christian doctrines of the human soul upon the sociological development of the European races." Part II specifies genetically "the influence of scientific ideas upon European culture."[36] The transition to the coordinate account is in Part II, where Whitehead breaks with his earlier narrative style, numbers the paragraphs into a categoreal scheme of the essential ideas that have been genetically specified, and then proceeds in subsequent chapters to analyze them in coordination with one another. The book's subsequent structure requires that we pay close attention. The break between the coordinate analysis and the *results* of it does come when Whitehead moves to Part IV, but he does not there explain it. For whatever expository reason, he saves the explanation for the beginning of the final chapter, but he is perfectly explicit about what he has done and how. He says:

> Our discussions have concerned themselves with specializations in History, of seven Platonic generalities, namely, The Ideas, The Physical Elements, The Psyche, The Eros, The Harmony, The Mathematical Relations, The Receptacle. The historical references have been selected and grouped with the purpose of illustrating the energizing of specializations of these seven general notions among the peoples of Western Europe, driving them towards their civilization.[37]

The *selection* was done in the genetic accounts of Parts I and II, and the *grouping* was accomplished in the coordinate account of Part II. These are simply other names for genetic and coordinate analysis, although we may assume that the method of "selection and grouping" here corresponds to the "theory" that Platonic and Christian doctrines of the soul can tell us much about the sociological development of Europeans. To recur to our earlier point about Cassirer's cultural phenomenology, one can see that in this "theory," these Platonic ideas function similarly in the domain of a philosophy of Western civilization as "symbolic forms" function for Cassirer's much broader "philosophy of culture."

Whitehead goes on: "Finally, in this fourth and last Part of the book, those essential qualities, whose joint realization in social life constitutes civilization, are being considered. Four such qualities have, so far, been examined:—Truth, Beauty, Adventure, Art."[38] Such structural statements as these are the ones that reveal Whitehead's habits of inquiry, and they cannot be overlooked. For Whitehead, "Truth," for example, does not mean the same thing from one inquiry to another, for scientific truth and historical truth are related but separate ideas, and they require different genetic phenomena and coordinate analyses to show their respective meanings.

To assume, for example that the account of "Truth" in *Adventures of Ideas* trumps the account in *Enquiry into the Principles of Natural Knowledge*, or *Science and the Modern World*, or *Process and Reality* is simply to miss what Whitehead does in a book. This error generalizes the results of one inquiry into or out of the results of another in which the phenomena have been genetically specified for other and different purposes, and coordinately analyzed differently, with different results presupposed. It cannot be assumed that later books reveal a development or change in Whitehead's basic ideas. Perhaps there is development, but it is not shown to be the case simply because he says different things about, for example, Truth, in later books.

What Whitehead says about Truth in *Adventures of Ideas* is tied intimately to the genetic specification and coordinate analysis (or more accurately, the "selection and grouping") of truth in *that* book, and for *its* purposes. The project there has to do with the way the *idea* of truth contributed to the growth of Western civilization, and not more than that (at least not directly more). The results of any inquiry are dependent upon this general way of approaching problems. Results are revisable even *within* a single inquiry. One can genetically specify different phenomena, or ostensibly the "same" phenomena in other ways. The coordinate analysis is more restricted, because it is logically prior to and therefore answerable to the genetic specifications, but from any accomplished genetic specification, many other coordinate analyses may be undertaken. This is to say, results from one inquiry can be extracted for other purposes, of course, but they must be adapted to those new purposes.

The criterion by which we adjudge the entire process, as having been done either well or poorly, is the criterion of experience. This criterion holds for every inquiry, although one always needs to say what aspects of experience are most relevant to forming the sub-criteria. In general Whitehead does adhere to "adequacy," and "logical rigor," and "exemplification" as functional criteria. These are the criteria made explicit in *Process and Reality*.[39] All of these criteria are experiential, but in different ways (a point that will be taken up in subsequent chapters). It would not distort what Whitehead has done in every book to say that

(1) "adequacy" applies most directly to whether the genetic specification is both in agreement with the phenomena it selects from experience (that these phenomena do occur and are appropriately specified in light of the coordination that precedes them), and whether those phenomena are representative of experience more generally, not simply within the domain of the inquiry at hand;[40]
(2) "logical rigor" (or coherence or necessity, which come to the same thing) deals with whether the phenomena, as genetically specified, are analyzed in accordance with all and only the logical principles required and implied by the prior coordination, and without illicit inferences, contradictions and/or undischarged assumptions when we move from one level of generality to another;
(3) "exemplification" (and/or applicability) deals with whether the outcome of (1) and (2) does indeed illuminate our actual experience as *had*, while no part of our experience as *had* fails to be exemplified in the "theory."

These are ambitious criteria and they set the bar very high for successful inquiry. We would expect nothing less. There is nothing in this procedure that is in the least out of line with radical empiricism as William James defined it (as taken up in chapter 2 and subsequent chapters). It is simply more worked out and systematic than anything James ever managed.

To use Whitehead's results from one inquiry in another context requires the results to be correspondingly modified for the new purpose. The analogies from one of Whitehead's inquiries to another are rich and tempting. And there is much in Whitehead's earlier works, especially on the developments in science, and several explicit discussions of Kant and transcendental idealism, which would aid us in a post-critical framing of the phenomena of post-Kantian philosophical thought, including cosmology, if we wished to undertake the inquiry. Whitehead's interpreters have habitually treated *Process and Reality* as if it were a straightforward, post-Kantian piece of ontology. That is a mistake. Whitehead explicitly distances himself from the entire Kantian project and insists that the Moderns were closer to being correct in their efforts at speculative cosmology. To treat Whitehead's metaphysics as a post-critical contribution to the twentieth-century conversation misses what is so radical and so interesting about his viewpoint. *Process and Reality* is a book about Modern philosophy and the way it was done from Descartes to Hume—and what it had right. *Adventures of Ideas* is about Platonic philosophy, seven of its "generalities," and how they influenced European civilization. That is what the inquiry is. It is not some other. And so on.

Simply to assume that Whitehead's results in any of his inquiries have automatic application to problems that lie beyond their clearly specified scope is a great mistake. But it is a lamentably common one. To assume that the results of Whitehead's inquiries include ontological knowledge, necessitated by the order of being and the order of knowing, is to misunderstand completely Whitehead's structure of inquiry, and how "theory" determines method. Philosophical knowledge is not *necessary* knowledge, for Whitehead, although the means by which it is attained will employ "necessity," as a tool and concept, internal to the terms of the inquiry itself.

Levels of Generality in Process and Reality

Another obstacle to understanding Whitehead's thought has to do with the difficulty many readers have in sorting out the various levels of generality he simultaneously employs in the course of a single inquiry. Apart from the common strategy of genetic and coordinate analysis (or selection and grouping, or something similar,) a Whiteheadian inquiry will be framed by a pair of "bookends," that is, the *most* general level of description and the *least* general (or most particular) level. These levels are adapted to the inquiry at hand, and as the inquiries differ, so do the levels of generality. For the purposes of this chapter, we will make explicit the levels of generality and particularity in *Process and Reality*, but the pattern is the same in every book and even in the essays. It is almost impossible to do any philosophical thinking without moving among various levels of generality. Since *Process and Reality* is probably Whitehead's most important book, and perhaps the most difficult, we will use it as an example.

The most general or abstract level of description in *Process and Reality* is the "theory of cosmic epochs," featuring an account of how "the extensive continuum" is divided *in our own* cosmic epoch. The extensive continuum itself is conceived as an irreducible undivided divisibility, or a divisible continuum as yet undivided. On a few occasions, Whitehead allows himself to speculate about whether there might be any common rules or characters that would necessarily belong to the extensive continuum in cosmic epochs *other* than our own, but for the most part the theory of cosmic epochs is intended as a construct that limits the highest applicable level of generality for any human thinking as it inquires in the domain of cosmology. Beyond the constraints and presuppositions found in our own cosmic epoch, we possess no method of inquiry that can be counted on to avoid the pitfalls of vacuous abstractionism. Thus, the "cosmic epoch" is the whole to which all parts belong in the course of *Process and Reality*. The whole is, of course, characterized differently in other books (European civilization in *Adventures of Ideas*, for example, or what goes toward making a religion in *Religion in the Making*, and so on). But the cosmic epoch, being the *functional* limit of inquiry in *Process and Reality*, is the "universal" to which all particulars belong *in that inquiry*.

The first division of the extensive continuum in *our* cosmic epoch is the *temporal* division of the continuum, which may be specified in any number of ways, but what is revealed in each *mode* of temporal division of the continuum is the genetic *divisibility* (note the modality) of the continuum itself.[41] In whatever way the continuum is temporally *divided* (note the modality), the cosmic epoch is designated as the whole to which such temporality belongs, the temporal divisions, no matter how they are specified, being the ultimate character of all *parts* in the epoch. This, then, is the most general level of description in *Process and Reality*—the theory of cosmic epochs. Whenever Whitehead uses the term "cosmic epoch," he is speaking at this level of generality. Other verbal clues have to do with language about "the whole."

On the other side, the most concrete (or least general or most particular) level of description in *Process and Reality* is the theory of prehensions.[42] Prehensions are "entities" whose mode of existence is described as "concrete facts of relatedness." Their "entitative-way-of-being" is exhibited in the analysis of actual occasions, but they are not found existing independently of an actual entity in its actual world, nor are prehensions independent of one another. In their concrete existing they *are* the togetherness of the cosmic epoch, its "solidarity"; prehensions may be *treated* individually only by an act of abstraction through which real continuities are treated as disjuncts.[43] Because prehensions are the most concrete particulars in Whitehead's inquiry into modern cosmology, they coordinate all the levels of generality in the whole theory. Thus, prehension stands to cosmic epoch as ultimate part to ultimate whole, hypothetically framing the inquiry. There is no claim that cosmic epochs or prehensions *are* the ultimate units of the universe *itself*, only that they will serve the roles of ultimate whole and ultimate parts for the inquiry that is *Process and Reality*. Every inquiry has to make clear what whole it takes as its subject matter, and with what concrete parts it will coordinate that whole.

Between the theory of cosmic epochs and the theory of prehensions in *Process and Reality* are three major levels of generality: (1) the theory of concrescence and transition, or more simply, the theory of actual occasions; (2) the theory of causal efficacy and presentational immediacy, or more simply, the theory of perception; and (3) the theory of propositions and generic contrasts, or more simply the theory of symbolic reference (or even of judgment). These are presented in order, from more general to less general, as the intermediate levels in *Process and Reality*.[44] The theory of judgment is almost as concrete as the theory of prehensions, while the theory of actual occasions is almost as general as the theory of cosmic epochs.

If some of you find yourselves saying that the theory of actual occasions is like Peircean Firstness, the theory of perception is Secondness, and the theory of judgment is Thirdness, we can only say that we would not discourage that kind of creative thinking. But this is not Peircean phenomenology. It definitely is metaphysics, and the relations of part and whole are what interest us here. Each of these levels of generality has a *view* of the whole and the whole is analyzed into "entities" appropriate to each level of generality. Is the "whole," when analyzed into prehensions, the *same* whole as the whole when analyzed into perceptions? Only by hypothesis, but yes, we do *assume* it is, even though there is no need to assert its identity or unity in any strict sense. It is more accurate to think of the levels of generality into which the whole may be analyzed as *analogous* levels. This is how Whitehead speaks, which is another way of saying that we gain in determinateness when we move from more concrete to more general levels, but we lose definiteness.[45]

Even the whole takes on a different character at various levels of generality. For example, the "whole" as articulated by the theory of perception is called "nature," and its parts are perceptions, their objects, and the forms of relation between these. Taken together, the (symbolically) coordinate result of these parts is called "science," when it is derived according to certain kinds of genetic specification (mainly measurement), while that same whole is called something like "beauty" when derived according to certain intensities of organization, best exemplified in art. The point is that the levels of generality in other of Whitehead's inquiries, apart from *Process and Reality*, are similar but not identical with those in *Process and Reality*. Past philosophies have made too much of the problem of perception and have tried to get this level of generalization to do more work than it can credibly bear (these are the famous "dogmas of empiricism")—a problem which persists into the present. This treatment of the structure and levels of generality in *Process and Reality* is brief, but it will be explained more thoroughly in chapter 2 and throughout this book.

2 Whitehead's Radical Empiricism

> In order to discover some of the major categories under which we can classify the infinitely various components of experience, we must appeal to evidence relating to every variety of occasion. Nothing can be omitted, experience drunk and experience sober, experience sleeping and experience waking, experience drowsy and experience wide awake, experience self-conscious and experience self-forgetful, experience intellectual and experience physical, experience religious and experience sceptical, experience anxious and experience care-free, experience anticipatory and experience retrospective, experience happy and experience grieving, experience dominated by emotion and experience under self-restraint, experience in the light and experience in the dark, experience normal and experience abnormal.
>
> —Whitehead, *Adventures of Ideas* (226).

In this chapter, we are arguing against the mainstream of Whitehead scholarship. There is a standard story about the development of Whitehead's metaphysics which is not altogether incorrect, but it rests on a misunderstanding of his orientation on the scope and limits of philosophy. In the previous chapter we argued that the right way to read Whitehead respects both the development of his thought and the individuality of his inquiries. We stressed the mathematical habits of inquiry found in Whitehead's philosophy, even though these habits are unfamiliar these days among conventional philosophical procedures. Now we go further.

Whitehead was a radical empiricist, following William James, Henri Bergson and John Dewey.[1] One can be a radical empiricist without being a pragmatist, as we will explain.[2] However, good pragmatists tend toward radical empiricism because one of the keys to pragmatism is its commitment to philosophizing all of experience and only experience. Yet, if pragmatism has an Achilles heel, it is in assuming that *possible* experience need not be robustly philosophized. Radical empiricists who do not profess pragmatism, such as Whitehead and Bergson, will tend to give possible experience greater weight than do their pragmatist counterparts.[3]

The Unholy Trinity

As we have argued earlier, Whitehead always held that metaphysics is descriptive and hypothetical in character, and that philosophical hypotheses do not yield ontological knowledge—knowledge of necessary relations among knowing, being and thinking. For example, Whitehead says that "speculative philosophy embodies the method of the 'working hypothesis'." He continues:

> The purpose of this working hypothesis for philosophy is to coordinate the current expressions of human experience, in common speech, in social institutions, in actions, in the principles of the various special sciences, elucidating harmony and exposing discrepancies. No systematic thought has made progress apart from some adequately general working hypothesis, *adapted to its special topic*.[4]

Please note the use of the term "coordinate" in this passage. This statement, along with others, is the basis for our repeated assertion of the *primacy* of the "coordinate account" (over the "genetic account") in Whitehead's method. His assertion here is unqualified. *All* progress in systematic *thought*—not just philosophy—depends upon the prior coordination of an "adequately general" working hypothesis. In this chapter, we will more thoroughly discuss how working hypotheses are adapted to their special topics in Whitehead's own inquiries. As we mentioned above, Whitehead uses the word "theory" to describe a working hypothesis when it is adjusted to the proper level of generality, and coordinated with the needed concepts for the purposes of a given inquiry.[5] But for the moment, we ask, do "working hypotheses" or "theories" in philosophy result in ontological knowledge? Whitehead says they do not: "Philosophy has been afflicted by the dogmatic fallacy, which is the belief that the principles of its working hypotheses are clear, obvious, and irreformable."[6] That is, philosophers think that their hypotheses, when taken as conclusions, provide them with ontological knowledge.

By "ontological knowledge" we mean basically the set of characteristics Dewey associated with "the quest for certainty," which we would summarize as an unholy trinity in which the necessitarian structure of logic is used to close the gap between knowing or *experiencing* something, on one side, and the being or *existence* of the thing known or experienced on the other side.[7] The exact character of that gap will be addressed in the chapters 4 and 10, when we defend Whitehead's "radical realism." For now we are interested only in Whitehead's rejection of ontological (or necessary) knowledge.

The tradition of associating logical and epistemological with ontological *necessity*, the "unholy trinity," was the basis for claiming ontological knowledge from Aristotle up to the mid-nineteenth century.[8] As the last

150 years have unfolded, the old assumptions about all types of necessity have come unraveled. All philosophy and most science that depends upon the iron association of logical, cognitive and ontological necessity have become untenable. Many scientists simply will not speak of "laws of nature" these days, except as approximate descriptions of regularities that can be profitably generalized, and even then they often balk at the word "law," for fear it will be taken in a universal sense.[9] And as a negative example (of what not to do), the conception of "rationality" used by mainstream analytic philosophies is dependent upon at least the association of logical and ontological necessity (this is the very meaning of "rationality" in the way the term is used), and to the extent that scientific conceptions of knowledge have moved beyond this outdated idea of necessity, science rightly ignores the demands made by mainstream philosophy that science should be or must be "rational."[10] "Rationality," in common and even intellectual discourse, today, is based on the structure of possibility and probability, not on the structure of necessity and old-style, essentialist actuality; and knowledge today does not require necessity of any kind.

Even the movements, following C.I. Lewis, to reinvent logic so as to deal effectively with possibility, have been rendered largely pointless because they have clung to various modes of necessity to *account for* possibility, with each version of modal logic interpreting possibility according to a slightly varied sense of necessity and its corresponding notion of validity.[11] So long as necessity is taken to be the guarantor of rationality, the conception of rationality advocated will be as useless to science as it is to practical life. We will not here go into David Lewis's "modal realism."[12] We will only make the point that realism (of any kind), when it adopts a narrow conception of rationality and necessity, has been left behind by any philosophy that has a genuine chance of surviving in the present century and beyond.[13] Whitehead's ontology presents "living ideas" precisely because it does not employ necessity of the indefensible kind for its ground.[14] Whitehead's ontology is grounded in the assumed equiprimordiality of possibility and actuality, and whatever necessity may be, it is subordinate to possibility.[15]

Thus, Whitehead never subscribed to the unholy trinity.[16] When he dealt with questions of the broader aesthetic order, it was grasped by imagination and was prior to the enlivening exercise of reason as brought to bear upon the possibilities we imagine.[17] But since there is an entrenched habit of reading Whitehead as though he embraced ontological necessity and associated it with logical necessity, it is needful that we begin to modify that habit.

Radical Empiricism

We want to make it clear that by "radical empiricism," we do not mean a "method" in philosophy. It is a broad orientation on the scope and limits of philosophy itself, within which any number of methods might be adopted, including but not limited to those favored by pragmatism. Let us begin at the source. William James wrote:

> Radical empiricism consists first of a postulate, next of a statement of fact, and finally of a generalized conclusion. The postulate is that the only things that shall be debatable among philosophers shall be things definable in terms from experience. The statement of fact is that relations between things, conjunctive as well as disjunctive, *are just matters of direct particular experience*, neither more nor less so than the things themselves. The generalized conclusion is that therefore the parts of experience hold together from next to next by relations that are themselves parts of experience. The directly apprehended universe needs, in short, no extraneous trans-empirical connective support, but possesses in its own right a concatenated or continuous structure.[18]

It is not crucial here to pin down every historical detail to show that Whitehead was a radical empiricist in *precisely* James's sense. We think he was.[19] What Whitehead called "the fallacy of misplaced concreteness" (and James called "the philosopher's fallacy") is not avoidable unless one is a radical empiricist. Perhaps the greatest barrier to clear philosophical thinking is our habit of substituting the abstractions produced by our simplification and analysis of experience for the fullness of experience itself. Whitehead certainly used abstractions, so what prevents an abstraction-dealer from misplacing a few of them? It is really simple: never claim ontological knowledge, and never sell experience short.

The place where we depart from many other readers of Whitehead, has to do with whether Whitehead claims for his view "ontological knowledge," or whether the logical rigor and rational necessity in his metaphysics entitles *us* to claim ontological knowledge. Many readers assume that Whitehead *took himself* to be claiming ontological knowledge. This is incorrect. A close reading of Whitehead's defense of speculative metaphysics at the opening of *Process and Reality* shows that Whitehead never claims any such thing there, and his discourse on the scope and limits of philosophy in *Adventures of Ideas*[20] is only one of several places one can check to confirm it—indeed he never makes a claim of ontological knowledge anywhere.

One might approach a defense of Whitehead's radical empiricism relative to any of a number of topics, but we will restrict our demonstration to the issue of what Ford calls Whitehead's "discovery" of "temporal atomicity" (a term Whitehead himself never uses). As Ford reads the text, in works prior to 1925, Whitehead defended a view of overlapping temporal events, or an overlapping hierarchy of durational epochs, but by 1929, Whitehead was defending "temporal atomicity," the view that "continuity concerns what is potential whereas actuality is incurably atomic,"[21] and that there is a "becoming of continuity but no continuity of becoming."[22]

The term "atomicity" occurs twice in *Process and Reality*. The first is the reference mentioned above. At page 292, Whitehead says: "In this theory the notion of the atomicity of the actual entities, each with its concrescent privacy, has been entirely eliminated. We are left with the theory of extensive connection, of whole and part, of points, lines, and surfaces, and of straightness and flatness." Please note the use of the word "theory" in this passage, in light of what we argued in chapter 1. If atomicity can be "entirely eliminated" in a theory that is not itself guilty of misplaced concreteness, then how could it be ontologically necessary that actual entities are incurably atomic? The incurable atomicity is a *function* of the "theory," the generalized working hypothesis, applicable in *Process and Reality*, and is an instance of the quantum of explanation. Understanding more thoroughly what temporal atomicity is, in the context of Whitehead's radical empiricism, is an important illustration of our larger claims about how he conceives of philosophy, how methods are subordinate to working hypotheses and "theories," and how this result provides some instruction for how to read Whitehead. We will take up in detail later the proper interpretation of these passages, especially with respect to the relation between the theory of extension and the concept of physical space.

Mementos of a Timequake

We will approach the question of Whitehead's radical empiricism in relation to "temporal atomicity" using a pair of artworks, the film *Memento*, written by Jonathan and Christopher Nolan, and directed by the latter, and Kurt Vonnegut's whimsical 1996 memoir *Timequake*. These two works point to the central problem of temporality *in experiential terms*, and will nicely orient us upon the issue of framing any sort of empirically adequate philosophy about time. We begin by confronting the problem of discontinuity—which is the very problem posed by "temporal atomicity." After all, the temporal atom is characterized in the first instance by its indivisibility and in the second instance by its disruption of the continuity of the temporal flow.[23]

Memento teaches us the problem of discontinuity "from the inside out," as it were. The story assumes that time is continuous and orderly in the *objective* world, but our central character, Leonard Shelby, has suffered a head injury and can no longer move his short-term memories into long-term memory.[24] Thus his world effectively begins anew every five minutes or so. This is a fundamental *internal* disruption of the temporal flow. In order to live life, Leonard has to find ways of arranging the objective world so that he can interpret it, and most importantly, he must discern the difference between the truth and deception. Is there an adequate method of arranging the objective world so as to compensate for radical discontinuity in our internal temporality, our concrescence?

Leonard combines many methods, but the most poignant among them is tattooing on his body "true propositions," certainties. These get him into trouble, eventually, since it turns out that they may not all be true. Nevermind that. It's a different problem than the one we are interested in. Our point is a small one. How Leonard *remembers* that the things on his body are "truths" is one of many problems the Nolans cannot really solve, since such memory requires the continuity, in Leonard's experience, of the proposition "the propositions on my body are true," to be both true and to hold from one radical break to the next in Leonard's experience. This is how, having posited radical internal discontinuity, or subjective temporal atomicity, the Nolans smuggle temporal continuity back into Leonard's inner experience.

Kurt Vonnegut's *Timequake* teaches us the problem of discontinuity "from the outside in," as it were. In this fanciful memoir, the universe suffers a tiny crisis of confidence regarding its own "decision" to expand. It pauses, hiccoughs, contracts for just a moment, and then recovers its confidence and expands again. The "moment" of contraction turns out to be almost exactly ten years on earth, which is nothing much from the standpoint of the universe, but it's a significant durational epoch for the little creatures on earth, as they have to live the years 1991 to 2001 all over again, a repetition of what Whitehead calls "transition". Disappointingly, they quickly discover that they are not free to change anything they did—which suggests that in fact they weren't really free the first time either, they just didn't know what would happen, so their actions and decisions *felt* like "free will." When the rerun of those ten years ends, and "free will" (i.e., ignorance of the future) kicks in again, everyone is caught off guard, having become bored with the rerun, and all sorts of accidents occur because people have to start actively paying attention to things like steering their cars and watching where they are going.

Vonnegut is toying with Nietzsche's eternal recurrence, of course, but he would have no story at all if he didn't assume that people were *conscious* of the fact that the rerun *was in fact* a *re*run. This is to say, they all *remember* the timequake and retain their memories of the events that have happened to them before, and will again, and they remain conscious of the radical objective discontinuity in time, the timequake, even though they are not free to study it or do anything to prevent it, or make any different decisions while reliving the ten years. Vonnegut thus compensates for the discontinuity of the objective space/time "continuum" by smuggling in the *continuity* of *particular* conscious experience (concrescence). Otherwise timequakes could be occurring all the time and it wouldn't matter, since no one would be aware that reruns were reruns—and this idea is actually closer to Nietzsche's eternal return of the same, but it doesn't frame a story in any interesting way, and it's a purely imaginative construction, not an experiential possibility.[25]

When we mess around with the idea of time's continuity, whether it is the continuity (or at least the accumulative tendency) of internal time consciousness, or the assumed continuity of objective temporality, things get strange in a hurry. We humans can't make phenomenological or narrative sense of radical temporal discontinuity, or as one might just as well call it, "temporal atomicity." In order to make sense of the supposition of radical discontinuity, we must cheat, and reintroduce continuity at some other level of supposition. But we need not go to the world of film or literature to find versions of this problem. Whitehead was a mathematician. He understood the problem of discontinuity very well indeed.

Getting from any natural number to the next poses the same difficulty of continuity. If one thinks of the natural numbers as embedded in a continuum, so as to overcome their discreteness, that continuum is subject, then, to the same problems time suffers in the hands of the Nolans and Vonnegut. Internally it proves nothing to count one's way from 1 to 9 unless the person counting can retain 1 and 2 and all the way to arriving at 9; this is a feat of continuous memory. Similarly, if one wishes to proceed from a square on the sidewalk numbered 1 and arrive by successive steps at a square numbered nine, the continuity of the walker as a continuous body against which the discreteness of the squares is overmatched, is a requirement. Thus, the discontinuity of the natural numbers cannot be treated as radical, or ontologically atomic.

How many ways can we pose the problem of atomicity? Well, how many ways can we introduce discreteness between, say, 3 and 4 (however defined)? Infinitely many ways; for example, in the case of Dedekind cuts.[26] But is there genuine, *ontological* discreteness between three and four? We don't really know; the Pythagoreans thought there was, but for more modern number theory the situation is vague. In the case of 0 and 1, for example, in certain mathematical inquiries we do suppose, hypothetically, an absolute break in the continuum, in order to commit to binary coding the situation we observe when either a circuit is complete or not complete (even a value of 0 draws current). We know that empirically, the situation is not absolutely discontinuous. We know that *some* energy moves between and through an incomplete circuit, but we discount that energy because it's "nothing" compared to what happens when a circuit is complete. In fact, if the circuit were not at least potentially complete-able, there would be no point in describing it at all.[27]

Yet, we have no compunction about calling the completed situation "1" or "the switch is on" and the incomplete situation "0" or "the switch is off." In binary, the discontinuity between 1 and 0 is, abstractly at least, radical discreteness—we choose either 1 or 0 to represent the situation, and there is nothing in between allowed by the description, considered abstractly (i.e., the code strings are composed entirely of 1 and 0). The description of the situation for the sake of binary is idealized (i.e., it is a "theory" prescribing a method); we begin with a continuous empirical situation—the continuity of a potentially complete-able circuit, and encode it with discrete characters for certain purposes. If we take numbers to be our imaginative descriptions, then binary is sort of like Leonard Shelby's efforts to tattoo words on his body to arrange the world in a reliable way. There is a sense in which binary is far more reliable, and to that extent "truer," than actual circuitry. It would be like offering Leonard some phrases he could *really* count on, more than he can count on the empirical world to live up to them.

In calculus, of course, the situation is reversed. We begin with a discrete empirical situation, say, where the moon is observed to be now, and where it will be a year from now—two discrete observed places, and we attempt to find symbols to integrate the two. We still use 0 and 1, but now as extremes or continuous poles of what we seek to measure. The whole effort is to find ways of overcoming the discontinuity (very real) between 0 and 1, on the one hand, and between the position and motion of a body at one time, and its position and motion at another time, on the other hand. Things approaching one another asymptotically meet at 1, in infinity of course, for calculus, but empirically, well, the moon gets where it is going, just as Achilles outruns the tortoise and his arrow arrives at the target.[28] Numerical continuity is created by a recursive function so as to approximate the continuous change in an empirical body, at least when we make no errors in the calculation and constrain our calculation within sufficiently narrow durational epochs. This is not greatly different from the *timequake*, if we think of the moon's motion as the first run-through and our calculus as the rerun—mechanical, predictable, and not quite real.

And this is to say nothing of the way we have to scratch our heads about physics, when it comes to quantum discontinuity in relation to relativity (which asserts continuity on a grand scale). It is not as if Whitehead lacks awareness of this situation. His writings from the first make it abundantly evident that he is interested in dealing with *both* sorts of discontinuity. Do his interpreters believe he would simply grow impatient with this when it comes to metaphysics and suddenly make a final decision about how things are, choosing discontinuity as the reality and continuity as an appearance? The text simply does not support any such reading. Whitehead does not believe temporal atomicity is "just the way things are." He believes that *metaphysical explanation* needs, in order to give a description that is logically rigorous and empirically adequate, to privilege atomicity, and all he means by this is precisely what James meant in saying that "direct particular experience" is the stated fact of radical empiricism. Particular experience imposes atomicity on metaphysical explanation, not vice-versa.[29]

If we insist or assert that there is any sort of thing as "experience in general," which is neither yours, nor mine, nor anyone's, then we commit the fallacy of misplaced concreteness. Experience is always, as far as we know, experience from some particular perspective, and since we do not know much about the "experience" of non-human beings, we cannot safely postulate that all experience is *like* human experience. Nor can we easily be certain, on the one hand, what characteristics of human experience are shared by other beings, and, on the other hand, which are accidents of our kind of existing. To make *our* way of experiencing the measuring rod of all that exists is the mistake Protagoras made. The idealists exacerbated this error by making up an abstract (i.e., fallacious) God and putting all time, and hence all experience, into that being's "perspective" (in truth, this is no perspective at all).

So Whitehead postulates the "actual entity" as a metaphysical version of a geometrical point, calls it a "drop of experience" (James's phrase), and denies to it any alteration or change as a matter of definition. It is surprising how powerful this strategy of explanation turns out to be (as we will take up in chapter 4). The actual entity is a spatialization, by extensive abstraction, of time, for the purposes of metaphysical description. The idea may be employed beyond metaphysics, but the strategy for explanation is the more important characteristic in this scenario. In the end, we will have no choice but to restore its continuity with other unanalyzed events, and that is what the coordinate phase of inquiry does.[30] Coordinate analysis proceeds algebraically by emphasizing what is structurally common and internal to the relatedness of the whole to any actual entity (as distinct from the self-externalizing achievement of satisfied concrescence, which is "narrative" in structure). The structural description is co-present throughout the whole; in the coordinate phase, no traditional narrative can be constructed, but values that satisfy the requirements of a coordinate description of any given actual entity are structurally continuous with all other values. A working general hypothesis is *not* a likely story, structurally or otherwise.[31]

Whitehead is simply counterbalancing James in this, when James says: "... my 'radical' empiricism denies the flux's discontinuity, making conjunctive relations essential members of it as given, and charging the conceptual function with being the creator of factitious incoherencies."[32] But James could only get good phenomenology this way, not good metaphysics. When it comes to the arrangement of the objective world in accordance with this kind of continuity, James is about as badly off as Leonard Shelby, and cannot distinguish a true proposition from one that merely seems to work. The other side of radical empiricism denies to the flux its continuity of becoming, in order to make it yield to spatialization. That gets us good metaphysics, but its descriptions have a kind of necessity that is untrue to experience. But no one would say Whitehead is a determinist, so the necessity in the logical rigor required of good metaphysics has to submit to some limitation.

Yet, in order to be a consistent "temporal atomist," if Whitehead really believed this were a bit of ontological knowledge, determinism *would be his only option*, since he would be allowing logical necessity to become ontological necessity in governing whole-part relations (as Vonnegut does in *Timequake*). But radical empiricism, so long as it treats experience as *particular* experience, *can* have a good metaphysics. In fact, it does have a good metaphysics. It is in *Process and Reality*, and there Whitehead's Ontological Principle is what guarantees that misplaced concreteness will not occur, i.e., we allow only one actual entity per actual world. But he does not take the principles of his working hypothesis to be "clear, obvious and irreformable," any more than he thinks it merely a likely story. He adopts the principle for this one inquiry, and different versions of it for other inquiries.

Let us review and summarize what we have outlined in this chapter:

First, what we have presented here is perhaps less of an argument, in the "strict" philosophical sense of the term, than a narrative framework for interpreting Whitehead's philosophical methodology (the method of genetic and coordinate analysis and division, upon which we will expand greatly in chapters to come). As we pointed out in the Introduction, narrative intelligibility is an essential component in the business of "making sense" in philosophy, and especially important due to the general absence of correctives that would be sticks and goads in other disciplines.

But we do not wish to suggest that this story in the first part of this book—*by itself*—composes a compelling demonstration. Nevertheless, our story does help to establish in no small part: (1) *how* we are reading Whitehead, with the *why* part of the argument making its many appearances in the chapters that follow; (2) *why*, among other problems with the existing interpretations that are so deeply problematic, the assertion of "temporal atomism" is the one we chose for critique; and (3) the *nature* of Whitehead's philosophy (including the point that it isn't phenomenology).

As for radical empiricism, which, James insists, is composed of a postulate, a statement of fact, and a generalized conclusion: the *postulate* (that everything must come back to experience) is so obviously present at every stage of Whitehead's thought (and, we hope, *ours* as well) that one would scarcely know how to challenge it. That said, every aspect of this book returns to this point again and again. The central thesis of our argument, that the concrete explains the abstract, rather than *vice versa*, drives home this postulate of radical empiricism. Yet, even without our argument here, we submit that the burden of proof rests upon anyone who would dispute this postulate, or the fact that Whitehead accepts it. His repetitions of it as an orientation on all philosophical questions is constant from his earliest works to his last ones.

With respect to *the statement of fact* (that relations are real, and real constitutive elements of experience), we will be showing in the following pages that not only did Whitehead accept this point, but also, as a mathematician he carried it forward well beyond what James, the psychologist, could readily imagine. When James thought of relations, he thought of feelings that could be interpreted in the higher phases of experience. Whitehead's view was far more radical, reaching down into every concrescence, however simple. There is radical empiricism of the sort that complex biological organisms with cultures and civilizations adopt for their own improvement, and then there is the radical empiricism that describes the real cosmos. We do not set aside the importance of the higher phases of experience, but we insist upon the priority of the order exemplified in the physical universe in our interpretations of that experience.

Finally, as to the *generalized conclusion* (that experience holds together by virtue of the relations that are already in it), anyone who has considered Whitehead's criticisms of Hume in *Process and Reality* can hardly entertain any lingering doubts of Whitehead's thoroughly reasoned conviction on this point. Relations are not simply to be included among the real things, there are no things apart from the more or less stable relations that constitute them as societies of occasions with physical and, in Whitehead's special sense, "mental" histories. Thus, where many in the secondary literature have lost themselves on the garden path is in their failing to see *all three* of these points of radical empiricism as operating together in Whitehead's thought and philosophical methodology. We intend to take significant steps to correct this failure in this book.

3   The Logic in Metaphysics

> The idea of equivalence requires some explanation. Two things are equivalent when for some purpose they can be used indifferently. Thus the equivalence of distinct things implies a certain defined purpose in view, a certain limitation of thought or of action. Then within this limited field no distinction of property exists between the two things.
>
> —Whitehead, *A Treatise on Universal Algebra* (5)

Introduction

At first blush it might seem odd, even unreasonable, to speak of a mathematician as a radical empiricist. After all, a mathematician trades in high abstractions, while radical empiricism is about the concrete fullness of experience. However, this way of phrasing matters rushes past essential aspects of that very experience which we are supposed to be attending to. Specifically, experience in its concrete fullness includes multiple layers and dimensions of robust relational connections as an ineliminable component of that concreteness. James spoke of the immediate experience of such relations as "and," "or," and "but." However, these terms represent only the thinnest and most tepid of the relational possibilities to be found in experience. Other relations include such things as rotational and translational forms of symmetry, congruence (an example that Whitehead frequently uses), continuity, and discontinuity, whole and part. When we look at such relations in their purest forms—and this is true of "and," "or," and "but" as well—our inquiry has centered itself solidly within the field of mathematics. Absent a thorough grasp of mathematical reasoning, the true wealth of relational connections that infuses our experiences might easily be overlooked for its subtlety.

Indeed, so central is the field of mathematics in schematizing and understanding relational structures that perhaps the surprise is not that a mathematician could be a radical empiricist but rather that a radical empiricist should ever fail to be a mathematician. But just as one does not need to be a jazz musician to develop an appreciation for jazz itself, we can, and, in the case of Whitehead's thought, *must* develop an informed appreciation for the nature and role of mathematical thought and inquiry in philosophical contexts. So in the first part of this chapter we will examine Whitehead's ideas about mathematics both from a general, pedagogical perspective and more specifically as an instrument of philosophical investigation. The second part of this chapter will involve some direct application of these ideas as we examine the role of one particular thread of mathematical/relational thinking as a "bridge" between logic, broadly construed, and metaphysics. As such, this is not about the philosophy *of* mathematics, nor even of philosophy *and* mathematics. Rather, we wish to draw attention to the nature of mathematical *thinking* as Whitehead argued for it, and its central role as an instrument of philosophical advance that is not limited to any one "tradition" of philosophical inquiry. Our discussion in this part will be divided into two major sections, the first focusing on pedagogy, the second on philosophy *per se*—a distinction of degree and emphasis rather than fundamental kind. For focus, we will draw *mainly* on two sources: Whitehead's 1917 essay "The Mathematical Curriculum,"[1] and the 1941 "Mathematics and the Good."[2]

Mathematical Thinking: Preliminary

So, as a first step in cultivating our sensitivity to the character and function of mathematics, let us set the stage with a pair of quotes from a couple of Whitehead's works. In the first of these he states:

> It certainly is a nuisance for philosophers to be worried with applied mathematics, and for mathematicians to be saddled with philosophy.[3]

In the second, without any hint of apology, he says:

> Philosophy is akin to poetry, and both of them seek to express that ultimate good sense which we term civilization. In each case there is reference to form beyond the direct meanings of words. Poetry allies itself to metre, philosophy to mathematic pattern.[4]

The first quote came from the latter-middle part of Whitehead's career when he was still making the transition from mathematician to philosopher; the second is from the very end of his last major publication. Despite the evident sop at apology in the first quote, Whitehead never backed away from his commitment to mathematical thought as an essential component of a liberal, well-educated mind, in general, and philosophical thinking in particular. These two are, of course, related, as we have tried to demonstrate in the first two chapters.

Mathematical Thinking: Pedagogy

It has been observed on a number of occasions that there is considerable sympathy between Dewey and Whitehead on the subject of education. And indeed, while Whitehead never stated matters so succinctly, it seems reasonable that he would agree with Dewey's claim that, "If we are willing to conceive education as the process of forming fundamental dispositions . . . toward nature and fellow men, philosophy may even be defined as the *general theory of education*."[5] Certainly with respect to mathematics, its *purpose* in education—whether liberal or technical—is for Whitehead very much of a piece with its place in philosophical thought itself. Thus a solid first step toward understanding the importance of mathematical thinking in philosophy is to grasp its general pedagogical role.

Stating a general characterization of mathematics—we deliberately avoid the term "definition" here—would not be very useful by itself, but it will provide a helpful framework that the following discussion will flesh out. Two more quotes will provide us with this starting place:

> Mathematics is merely an apparatus for analysing the deductions which can be drawn from any particular premises, supplied by commonsense, or by more refined scientific observation, so far as these deductions depend on the forms of the propositions.[6]

> (M)athematics is nothing else than the more complicated parts of the art of deductive reasoning, especially where it concerns number, quantity, and space.[7]

Both of these quotes were written at a time when there was still a reasoned belief that mathematics might, in some *formal* sense, be "reducible" to *symbolic* logic.[8] Yet the operative terms in the forgoing are "formal" and "symbolic." The co-author of the *Principia* shows little interest here (or anywhere else, for that matter) in taking such "reduction" to logic as naively as some of his contemporaries, including his counterpart; he remains firmly settled in the conviction that the *forms* of reasoning he is describing are decidedly *mathematical* in character.

Three of those forms Whitehead holds out for special attention: number, quantity, and space. Mathematical thinking is not exclusively limited to these forms—it is, in fact, a far more general science than just these subjects.[9] But these forms are of special interest in the *concrete* pedagogy of mathematics. "Concrete" in this instance does not mean an avoidance of abstractness, but rather a commitment to *relevance*. "Education to be living and effective must be directed to informing pupils with those ideas, and to creating for them those capacities which will enable them to appreciate the current thought of their epoch."[10]

The thing to be avoided in this process, so that one might have living and effective education, is "reconditeness." Whitehead says ideas are recondite when they are, "of highly special application and rarely influence thought."[11] He continues: "The goal to be aimed at is that the pupil should acquire familiarity with abstract thought, should realize how it applies to particular concrete circumstances, and should know how to apply general methods to its logical investigation."[12] To this purpose, the above named classifications of study are brought to bear. However, in this context Whitehead refines and specifies things from just the naming of those classifications, and states instead that study should concentrate on, "the *relations* of number, the *relations* of quantity, the *relations* of space."[13]

We should bear in mind that this pedagogical theory is not directed at persons who will be requiring specialized mathematical skills only, but even and especially toward those engaged in a purely *liberal* course of education.[14] At the same time, this course is not to be a "fluff" exercise, and one is to eschew all "pretty divigations."[15] Whitehead insisted on emphasizing these basic notions of functions, of relations, and their rigorous connections with concrete examples, such as the laws of mechanics.

In addition, one of the primary tools that Whitehead highlighted was the study of mathematics from the perspective of its *history*. Whitehead emphatically does not mean the "mere assemblage of dates and names of men," which is as tediously recondite an exercise in pointless memorization as one might care to identify. Rather, historical study is "the exposition of the general current of thought which occasioned the subjects to be objects of interest at the time of their first elaboration."[16] In other words, what Whitehead is advocating as a minimal component of any liberal education is a thorough facility with the basic forms of mathematical *inquiry*: Why were these particular problems interesting? How do they get applied in specific instances? How do these forms illuminate the "general methods" of "logical investigation"? It is worth noting that since Whitehead's time, excellent historical works of just the sort he was wishing for are readily available, yet still largely ignored by both mathematical and humanistic educators.[17]

Whitehead's focus becomes even clearer as we look more closely at the details surrounding the pedagogy of the above relations. The first such relational fact that Whitehead draws our attention to is that these three areas of interest (number, quantity, and space) are *interconnected*—densely so.[18] Focusing for the moment on the relations of number, Whitehead means something far beyond just and only arithmetic. Whitehead himself observed that, "the amount of arithmetic performed by mathematicians is extremely small. Many mathematicians dislike all numerical computation and are not particularly expert at it."[19]

The *relations* of number offer an entrance into that most general logic of relations known as algebra. In such study our ability emerges to formulate mechanical laws accurately, to illustrate functions, and to draw out relational structures that take on an interest of their own. But again, this study is not done in isolation. Rather, these instruments are developed as tools in the discipline of thought and inquiry; we avoid reconditeness by developing their *real* connections with real issues. One prescient instance Whitehead gives in this regard is the early introduction to elementary statistical methods in the social sciences, since this "affords one of the simplest examples of the application of algebraic ideas."[20] As Whitehead foresaw, s grasp of basic statistical methods is today an essential requirement for any functional member of a modern democratic society. Yet, there is almost no comprehension of statistical tools except among those who specialize in the topic.[21]

Another notion that wants some clarification is that of "quantity" and its relations. One immediate reaction to the term "quantity" is that this is just another term for number—unless, of course, one is an Aristotelian, in which case what will come to mind is much closer to Whitehead's meaning: quantity, in this classical sense, is about relations of *more or less*. Now there is certainly this aspect in Whitehead's meaning, but he takes it much further as the general theory of *measurement*.[22] Among other things, the use of quantity means here that the relations involved must take care to be ones comparing *like to like*. Linear measurement cannot properly be compared to volume measures, regardless of whether those quantities have been assigned "the same" numbers or not.[23] (This issue of "quantity" will be developed in detail in later chapters.) The field of "quantity"—i.e., measurement—is one of those places where the dense interconnectedness of the relations of number, quantity, and space stands out with striking clarity. The theory of measurement, for example, is at the heart of Whitehead's critique of Einstein's general theory of relativity.[24] More recently, Krantz *et al.*, published a massive three-volume study on the subject of measurement that touches on every major area of applied mathematics.[25]

This interconnectedness and wide-ranging applicability serve to highlight one of the central pedagogical features that the study of mathematical inquiry brings out. It is, Whitehead insists, "the chief instrument for discipline in logical method."[26] While all of the three special foci of mathematical study that Whitehead noted are useful in this regard, he believed the study of the relations of space to be especially important. In addition, even as "such training should lie at the basis of all philosophical thought"—that is, the relations of number, of quantity, and of space—the relations of space will not only best highlight logical method, but those relations do so by surveying some of the *functional* aspects of "identity" that the undisciplined use of language tends to gloss over.[27] This point provides us with our segue to more philosophical concerns.

Mathematical Thinking: Philosophy

The elephant in the room here is Plato. First off, we have that grand piece of apocrypha, the declaration over the entrance to the Academy: "Let no one ignorant of geometry enter." Geometry was the first formal system for developing "discipline in logical method." There was a tremendous amount of formal geometry in Plato's time, and this was his basis for rigorous thinking in philosophy. Quoting Whitehead:

> Plato, throughout his life, maintained his sense of the importance of mathematical thought in relation to the search for the ideal. In one of

his latest writings he terms such ignorance "swinish." That is how he would characterize the bulk of Platonic scholars of the last century. The epithet is his, not mine.[28]

Now, obviously "discipline in logical method" is a matter of central concern to all philosophers, regardless of their thematic orientation. But the structures of mathematical thinking cut even deeper than this, according to Whitehead, and, on Whitehead's reading, Plato agrees. We will return to those relations of space that best exemplify and cultivate that discipline, but first, Whitehead finds an even deeper source in the connections between mathematical thinking and philosophy in his home field of algebra. "What is the fundamental notion at the base of algebra?" Whitehead asks.

> It is the notion of "Any example of a given sort, in abstraction from some particular exemplification of the example or of the sort." *The first animal on this Earth, who even for a moment entertained this notion, was the first rational creature.*[29]

Whitehead has here taken the extraordinary step of identifying algebraic thinking with the very foundation of rationality itself.

Pedagogically—and, in terms of actual formulation, historically—geometry claims first place amongst the mathematical disciplines. But *logically*, it is the algebraic mode of thinking that is *the* distinguishing characteristic of rational thought.[30] This logical primacy can be attributed to the fact that, "You can observe animals choosing between this thing or that thing. But animal intelligence requires concrete exemplification. Human intelligence can conceive of a type of things in abstraction from exemplification."[31] This abstraction from the concrete—which we should recall is different from being recondite—this taking of types *as types* without the crutch of specific examples, is the principal move of algebraic thinking. And it is the first instance of genuinely rational, as opposed to narrowly biological and immediately practical, thought. (This is not a bifurcation of nature, it is a development and essential qualification *from* nature, a matter we address in detail in chapters 10 and 11.)

With the onset of algebraic reasoning the study of pattern for pattern's sake—"and mathematics is the study of pattern"—begins in earnest; and so too does the beginning of philosophical thought.[32] Philosophical thought begins because, "[t]he history of the science of algebra is the story of the growth of a technique for representing finite patterns."[33] Any pattern must be an exemplification of a type or kind, specifically a relation that connects one thing with another (actually, many others). Failure of such relatedness—which would amount to absolute metaphysical individualism—would also be totally outside of the possibility of thought. If it is not possible to *identify* some subject-predicate complex "X" in which subject "A" is an instance of that complex, then that subject "A" genuinely fails to qualify as any-*thing* at all. Yet any such "X" is a kind of *boundedness* on "A"; in other words, it is a way or form of *finitude*, and that finitude is of the very essence of "A's" relatedness, algebraically speaking (which includes the high abstraction of subject-predicate forms). In contrast, that which is without bound—the infinite—is also without the kind of "generic specificity" (such as "A is an X") that is the characteristic of an identifiable thing. "The superstitious awe of infinitude has been the bane of philosophy. The infinite has no properties. All value is the gift of finitude."[34]

The previous quote references "value" because it is part of Whitehead's larger argument in "Mathematics and the *Good*," in which Whitehead is reconnecting contemporary philosophy's project with Plato's. The value that is given is a result of the *algebraic* character of relatedness that is fundamental to rational thought and inquiry. All finite things are variously related to other finite things: they are material or ideational, simple or complex, subtle or brute. But they are *related*; and this is both what makes them finite and what makes them intelligible. Yet finitude itself cannot be had *sui generis*. Infinity may have no properties of its own, yet infinity is that which makes finitude possible; the two do not stand in opposition so much as in constructive tension. Glossing over the details as Whitehead develops them, he concludes that, "the notion of 'understanding' requires some grasp of how the finitude of the entity in question requires infinity, and also some notion of how infinity requires finitude. This search for such understanding is the *definition* of philosophy."[35]

Prior to turning our attention to the promised examination of "identity," a few more words situating mathematical thought are in order. Whitehead was quite emphatic that, "mathematics is the most powerful technique for the understanding of pattern, and for the analysis of the relationships of patterns."[36] Algebra is thus a logically—but, recall, not pedagogically—primary tool in this task of developing pattern and understanding thereof. Completing his circle with Plato, Whitehead states, "if civilization continues to advance, in the next two thousand years the overwhelming novelty in human thought will be the dominance of mathematical understanding."[37]

Mathematical Thinking: Identity

We come at last to the mathematical analysis of identity. Despite the fact that Whitehead develops his examples in the context of specifically spatial reasoning—which, we will recall, is particularly suited to the "discipline of logical method"—the tools he brings to bear are themselves highly algebraic in character. This should not be a surprise. Quite aside from the fact that the relations of number, of quantity, and of space are highly interconnected; beyond even the fact that mathematics itself is a much more general science than just these relational studies; Whitehead's first major work, *A Treatise on Universal Algebra*, was an *algebraic* generalization of abstract *spatial* relations.

So what is it about identity that makes it even *possibly* suitable for mathematical analysis? Let us start with the fact that when we speak of identity, we are speaking in shorthand for the identity *relation*. After all, some "thing" that retains its identity as the "thing-it-is" does so across some background of changes which leave the "thing" in some stably functional state of its own "self-hood." That is, we have the functional identification and *re*-identification of some "thing" as the *same* thing, across a more or less fluid multi-dimensional context of transformations. Mathematics offers a vast selection of tools for comparing and contrasting these modes of "sameness" within densely related systems of changes.

Whitehead develops three especially relevant mathematical relations in spatial reasoning. While he does not specifically focus on their roles in the analysis of identity, by the very nature of the case they serve just such a function. Each of the relationships evinces certain classes of transformations and the sorts of "things" that remain the "same" throughout those changes. The three mathematical ideas he describes are congruence, similarity, and the relations of trigonometry.

Whitehead first notes that, "our perception of congruence is in practice dependent on our judgments of the *invariability* of the *intrinsic properties* of bodies when their external circumstances are varying."[38] The mathematical study of congruence gives logical rigor to the idea of a physical thing being the *same thing* even when its position and orientation in space, or other "trivial"—that is, non-identity affecting—changes are made. With similarity one has, "an enlargement of the idea of congruence"[39] However, the enlargement goes beyond just sameness of the individual thing to the *simple* identity of type, a sameness of relational *kind*.

But similarity comes to genuine fruition in trigonometry, which is, "the study of the *periodicity* introduced by rotation and of properties preserved in a correlation of similar figures."[40] While the second half of the last quote clearly defines trigonometry as a study of relational and typic sameness (that is yet another advance from mere similarity), the comment about "periodicity" is independently interesting because it opens us up to the possibility of *rhythm*.

Rhythm is an avenue into identity, and one that Whitehead identifies as of supreme metaphysical importance. Rhythm, Whitehead suggests, is the "causal counterpart of life: . . . wherever there is some rhythm, there is some life."[41] Periodicity by itself—and hence the analysis that trigonometry provides us—is not the same as what Whitehead calls "rhythm," since periodicity is just the empty repetition of pattern. But this is still an essential component of rhythm: "A rhythm involves a pattern and to that extent is always self-identical."[42]

Mere algebraic pattern does not take us all the way to the rhythmic fullness of *organic* identity, but it does provide us with an essential tool of analysis that philosophers have under-utilized. And upon that note—a conscious pun for any musicians—we will change keys as we segue to the second movement of this chapter.

Logic and Inquiry

Before going any further, a few words need to be said about the nature of logic itself. The term has been used quite freely in the preceding, and the attentive reader might well and justifiably come away from such readings with the sense that it is being used in a way that goes beyond that particular adjunct that is variously qualified as formal, symbolic, or mathematical logic. And, indeed, the guiding notion of logic that is at work here is very much akin to that which Dewey articulated in his 1938 *Logic*[43] and is very similar to ideas recently presented by Jaakko Hintikka.[44] This approach is also similar to Whitehead's method. This last claim might come as a surprise to those for whom the *Principia* occupies a dominant place in their image of Whitehead's work and thought. However, such imagined dominance is a result of the tremendous influence that the *Principia* exercised on others, Russell most particularly. For Whitehead, that work was never more than an interesting formal inquiry, a *façon de parler* with respect to a certain style of mathematics.

This last is an inference rather than an explicit claim Whitehead makes in any of his published works. Nevertheless, as should be evident from the preceding examination of Whitehead's pedagogical theories regarding formal reasoning (all of which were written *after* the *Principia* was published), the *Principia* simply did not loom in Whitehead's mind in the way it did Russell's or that emerging methodology that would come to be known as analytical philosophy.[45] As we will see below, the one part of the *Principia* that Whitehead retained any particular interest in was the theory of types, although his interest was in formal and structural relations—class logic and propositions—and not the logical atomism and the metaphysics Russell built from this part of the *Principia*. And even here this conclusion must be taken in a purely hypothetical mode.

None of this can be construed in any way as a criticism, much less a dismissal, of *formal* logic or its importance.[46] Aside from their purely mathematical interests, such symbolic methods share with general mathematics the virtue of cultivating disciplined habits of thought. In addition, formal logic is directly relevant to the process of inquiry, since formal logic highlights possibilities so as to illuminate explicitly the *next* constellation of questions we ought to ask. Thus, formal logic is not so much concerned with what we know as with what we want to find out. By precisely displaying the ways in which our hypotheses connect, we can generate concrete methods of testing those ideas and eliminating those ideas which do not stand up when confronted with further facts that we *now* know to look for.

However, as Hintikka observes, the question of what we want to find out divides into two primary trends: one "tactical" and the other "strategic"; the place where traditional pedagogy of logic fails is with its excessive focus upon issues that are purely *tactical* in nature.[47] Such tactical questions include matters of consistency and validity, the avoidance of fallacies, and so on. These tactical considerations are on a par with knowing what the allowable moves are for the various pieces in a chess game. However, knowing that you can or cannot move a chess piece in a certain way tells you nothing about whether you *ought* to do so in any particular situation. Strategic issues come into play with that "ought."

Within the traditional divisions of formal logic, such strategic approaches do in fact make themselves felt. In model theory,[48] for example, game theoretic methods are explicitly applied in the building of models and the application of quantifiers.[49] Hintikka highlights the methods born of the work of Gerhard Gentzen, which originated in proof theory, specifically the sequent calculi.[50] Sequent calculi are proving especially useful in computer science in their applications developing *strategic* rules and methods needed in automating the proof of theorems.[51]

The tie-ins for us are several, and we will develop these in greater detail in chapters 9 and 10. But for now, we can say that, first, the above simply makes evident the connection between logic and inquiry by emphasizing the fact that this is not some vague arm-waving by persons allergic to formal methods. A rather more indirect connection, yet one which will be coming up throughout our discussions, is the numerous and unexpected points of contact between Whitehead's ideas and methods and results in computer science[52]—none of which will be referencing the *Principia*. Finally, when it comes to making strategic, rather than merely tactical, decisions about inquiry, the most strategic choice of moves is metaphysics itself. Once we begin to appreciate the *bridge* that spatial reasoning provides between logic and metaphysics, we will be better situated to understand that this bridge is not merely a formal, but a strategic one which will help illuminate what we are calling "the quantum of explanation."[53]

With that in mind, Whitehead's theory of extension has been generating an increasing amount of interest amongst logicians, computer scientists, and mathematicians in recent years, yet arguably it still has not received the measure of attention it deserves from Whiteheadians themselves.[54] Whitehead's mature theory of extension represents the culmination of a lifetime of work on what might be called "the problem of space," a problem that Whitehead always approached as an *algebraist*. The significance of the algebraic approach is that it focuses squarely upon systems of relations, a focus that is exemplified throughout Whitehead's metaphysics.

But the same algebraic approach to spatial reasoning, the center of Whitehead's metaphysics, is also of paramount importance in the study of formal logic itself. The connection is that spatial reasoning is a *bridge* between logic and metaphysics. This point argues for a much broader range of application and importance for Whitehead's theory of extension than just process thought, or the formal developments mentioned above.

We want to eschew any attempt to *reduce* metaphysical thought either to logic or spatial reasoning. We find Whitehead's own refusal to make such a move quite compelling. It is not an accident that *Process and Reality* is built around *two* relational structures, developed as the coordinate *and* the genetic account. And while Part IV of *Process and Reality* can be read as a treatise on mereotopology, the irreducibly *narrative* structure of Part III requires that it be read more in the way of *Finnegans Wake*—especially since the holistic character of the theory of prehensions means that the "end of the story" brings us back to the "beginning of the story" and is already presupposed in it.[55] Thus, the relationship we are arguing for here is a *bridge* only: the gap between the two sides is not eliminated simply because a connection is made.

Spatial Reasoning and Logic

It is important to understand that the algebraic approach to problems—whether in the direct application to mathematics or as an analogical technique in other areas—is particularly *relational*. The manner in which such relational characteristics are highlighted is by way of "structure preserving" transformations—where the "structures" *are* the relations—that connect disparate systems to one another in such way that the interesting relations/structures are shown to have an *equivalent* role within the disparate systems. This emphasis on structural relationships is nicely illuminated in algebraic logic. Paul Halmos notes that, "it turns out that from the point of view of algebraic logic the most useful approach . . . is not to ask 'What is a proof?' or 'How does one prove something?' but to ask about the *structure* of the set of provable propositions."[56] Structure is revealed in this and other instances by the *equivalencies* of relations among different systems. This relational focus is particularly noteworthy since Whitehead's approach to logic in both his earliest and later works was explicitly algebraic in character.[57]

The algebraic approach is especially fruitful in the study of various kinds of spatial entities because it enables investigators to classify spaces according to their relational structures, thus capturing their essential characteristics while paring away details of lesser relevance. For example, this was the focus of Felix Klein's *Erlangen Program*: the classification of geometries by the use of the algebraic theory of groups, where a "group" is a system of *reversible* structure preserving transformations.[58] These group-theoretic ideas are also important in developing such applied spatial concepts as, for example, the nature and degree of uniformity of space (an essential component in the development of Einstein's general theory of relativity and Whitehead's criticisms thereof).[59]

While there are many parts to this tale, the one that is of greatest interest to us here comes with the development in the 1940's of "category theory." Category theory emerged from inquiries into common algebraic features of various geometric/spatial systems. In particular, it was discovered that structure-preserving transformations in the field of topology bore interesting connections to a much wider array of mathematical relations.[60] Topology might be loosely described as a kind of "spatialized set theory" (our phrase), in which the non-metrical relations of connections and neighborhoods of a point or region are developed; these connections and neighborhoods are then studied for their algebraic properties and their common structural relationships with otherwise heterogeneous spaces.[61] As we shall see, topological notions (although Whitehead does not name them such) came to play a vital role in Whitehead's mature theory of extension.

Category theory's connection with spatial reasoning was never lost, but it rapidly became apparent that it generalized the algebraic and structural approach to the entire spectrum of mathematics.[62] Among other things, it became clear that a category can be viewed as a kind of deductive system, and it was shown that category theory with its focus upon relation-preserving transformations offered a competing "foundation" for mathematics to that of set theory, as well as offering a robust alternative to the orthodox concept of logical structure itself.[63]

The first of the two central categorical ideas in developing this connection between category theory and deductive systems revolves around the concept of a "topos."[64] Robert Goldblatt notes that the sources of the idea include algebraic geometry, category theory itself, as well as logic and model theory.[65] Skipping over the technical details of its definition, it turns out that the spatial construct of a topos can function as a generic structure with which to model or interpret other structures with such universality and effectiveness that it can serve as the above noted "foundation." The second aspect of this formulation of the topos is the primary tool used in building those models. Rather than falling back on the traditional notion of "set," the fundamental building block here is the neo-Russellian idea of the "type." We will have a bit more to say on this matter in the final section of this chapter. But it is worth noting here that type theory is a way of concentrating one's attention upon *functional and relational* aspects of the units of one's operations—that is, on the structural transformations—as opposed to the merely *collective* aspects that are characteristic of sets.[66]

A couple of points can be made to bring the above together: First, since model theory is the attempt to connect relational structures to formal languages,[67] the fact that the formal languages of spatial reasoning as manifested in category theory are such an effective instrument for understanding relational systems in general seems scarcely the sort of thing that can be dismissed as mere coincidence. Furthermore, the structures of spatial reasoning have a fundamentally *empirical and experiential* character that differentiates them from the abstractness of set theory and thus offers potential insights into ontology that set theory does not.

Finally, it is worth observing that the category theoretic systems of *topoi* are directly related to what are known as "complete Heyting algebras," which in turn are a generalization of *typed* intuitionistic logics. As these are algebraic generalizations of spatial structures, the emphasis is once again on forms of transformations of relationally *equivalent* systems. Also, it is suggestive to note that Michael Dummett turned to intuitionistic logics for his work *The Logical Basis of Metaphysics*.[68] Dummett did not associate this turn with spatial reasoning, but that association is already built in to the algebraic and relational structures of the intuitionistic approach via the above mentioned Heyting algebras.

Spatial Reasoning and Metaphysics

Let us recall that Whitehead was quite explicit that the theory of extension was every bit as central to his metaphysical schema as the theory of prehension; thus, prioritizing one *over* the other has the general effect of falsifying both.[69] The mature theory of extension as presented in *Process and Reality* is a kind of "mereotopology," which means that it is a *topological* theory of part and whole. Just from the relevant terminology, one can begin to sense suggestive connections between the foundational studies in category theory glossed above and Whitehead's "coordinate account," which will be discussed in detail throughout the remainder of this book.

Whitehead's extensive structures—what we call "extensa" for convenience—are essentially topologies in a *point-free relational system*. We use this latter, somewhat awkward phrase rather than just saying a "space" to emphasize that the relational structure of Whiteheadian extensa is *metaphysically* prior to any system that could properly be called "spatial" in the sense of physical nature or experience. *Mathematically*, of course, extensa still qualify as "spatial," and it will be in this latter sense that we will continue to speak of these structures as fundamental forms of "spatial reasoning." Again, mathematics is one of the cardinal forms of relational thinking, and the one that Whitehead argued is of greatest importance to philosophy.[70] Within Whitehead's metaphysics the system of extensa is logically *and* metaphysically prior to both time and space as these express themselves in nature.[71]

As topological relational systems, extensa carry within themselves most if not all of the previously discussed functional corollaries with category theory and general logic. We qualify the foregoing only because the point-free nature of Whiteheadian extensa introduces some complications into the standard theory of topoi. But these complications do not seem to be essential differences, particularly since "points" can be recovered in Whitehead's system as ideal elements, while most of the principal logical relations of topoi and their related intuitionistic Heyting algebras depend on the topological and typic relations rather than any of the specifically set-theoretic and/or intrinsically punctiform aspects that are introduced for formal simplicity.

But Whiteheadian extensa and their concomitant interpretive framework carry an even deeper *metaphysical* meaning than suggested by their purely formal relations. For Whitehead recognized that the part/whole relation is the fundamental form of *internal* relatedness itself.[72] For our purposes we may differentiate the notions of internal and external relatedness by how they reveal/form (the language here is tricky) the nature of identity. For external forms of relatedness, a "thing's" identity is the first, analytically given fact; for internal forms of relatedness "identity" is the final, synthetically achieved result. Part/whole relationships are the image of internal relatedness because there is no actual part until the whole is given. Yet by the same token, the whole is itself presupposed, but vague—that is, undefined if one wants to hint at some Aristotelian metaphysics—until the *parts* are definite. The *identities* of part and of whole are thus synthetically correlative to one another and not initially given, independent facts; it is Whitehead's argument that it is logically impossible for things to be otherwise.

The theory of internal and external relatedness is an essential piece in Whitehead's solution to the problem of the one and the many. Internal relatedness is an expression of the holism of reality, of the world manifesting itself as the One. Thus the theory of extension is the "story" of the world becoming One.[73] Our earlier discussion of the contrasts between transition and concrescence at one level of generality, between causal efficacy and presentational immediacy (i.e., perception), at another level of generality, and between physical and conceptual feelings (i.e., prehension) at the most concrete level of generality, is but an application of these mereotopological structures to various experiential domains.

The other half of the story is of course that of the genetic account, or what might be called the external*ization* of relatedness in the concrescence of an occasion as its superject. This, again, is why spatial reasoning is only a *bridge*—one cannot reduce external relations to internal ones, or vice versa. Each has an essential—and essentially different—role to play.

But already we are seeing signs of a larger project than even Whitehead's theory of extension. His mereotopological system of part and whole relates in one direction to the formal and logical structures characteristic of relational thinking that connect to category theoretic notions of topoi and logical structures of typed intuitionistic calculi. In the other direction his system connects to foundational aspects of metaphysics, especially as these emerge in the forms of internal relatedness and the central problem of the One and the Many. These issues obviously run beyond "just" process metaphysics, grand as that project is. Furthermore, neither direction of these connections can be dismissed as an accident or a merely coincidental artifact of Whitehead's work.

What might be called "the problem of space" composed the earliest part of Whitehead's work as a professional mathematician. The *logic* of spatial relations was the driving interest in *A Treatise on Universal Algebra*, the theory of extension was to be his primary original contribution to the *Principia*; it underscored all of his work in the philosophy of nature and science as the basis for his systematic criticism of Einstein's general theory of relativity. The development of Whitehead's metaphysical work was a direct outgrowth of his epistemologically oriented work on nature and science, due to the need for a more fundamental *logical* grounding of that *historically* earlier work. Thus the fact that his most mature development of pure spatial reasoning is presented in his metaphysical *magnum opus* is anything but whimsical.

Spatial Reasoning and Ontology

A few additional points deserve to be made here. Rather than speak directly to the issue of spatial reasoning and *metaphysics*, we want to close this chapter with a few words on ontology. Clearly metaphysics and ontology are related; we tend to view the connection as one between relational structures (metaphysics) and their various *relata* (ontology).

One of the most interesting of the non-speculative matters that emerges here is the fact that Whitehead's mature theory of extension is frequently identified as foundational in studies of robotic vision and spatial reasoning in artificial intelligence. (See our notes on Randell and Pratt). Of particular interest to us here is the motivation of the researchers involved in these developments. It is not a matter of abstract mathematical consideration nor speculative philosophy; rather, their interest is driven by both the logical and the practical constraints of programming machines as they *actually* attempt to "see." Quite independently of any philosophical concerns, these researchers have offered a strong *empirical* argument that Whitehead's logic of extension is a kind of "formal ontology of vision." Obviously this line of research in computer science opens up tantalizing avenues of inquiry along a third axis from those of pure logic or metaphysics, an axis that runs well beyond any of Whitehead's explicitly stated projects.

A second point worth wondering about, although this one is entirely at the speculative level, is the connections among type theory, spatial reasoning, logic, and metaphysics. The fact that there are such profound connections among typed intuitionistic calculus, category theory, and spatial reasoning does seem to beg the question of what else might be lurking in the shadows. Our own sense is that either we have stumbled onto some profoundly deep connections between logical structure and metaphysics, or we are looking at a pure coincidence of genuinely comic—that's "*comic*" not "cosmic"—proportions. In either case, it is also worth wondering if there is any connection between type theory and Whitehead's broader philosophical concerns. First, as important an innovation as it was in its day, the type theory of the *Principia* is less sophisticated than its latter progeny in typed intuitionistic calculus. Also, while hints of these later developments are in the original, Whitehead does not tell us much about even that early version of type theory, apart from what has cited above. He gives some additional discussion of it in his essay "Mathematics and the Good,"[74] but his comments there are painfully brief, amounting to little more than the *possibility* of an argument for moving beyond formal theory to a sound metaphysics of relations.

Last, given the widespread interest in relating Whitehead's ideas to quantum mechanics, efforts that (we will treat in the next chapter) are often badly misdirected, it seems reasonable to suggest that Whitehead scholars ought to explore the possible relations between the theory of extension and another point-free geometry that was specifically formulated with an eye toward quantum theory. We are referring here to von Neumann's *Continuous Geometry*.[75] Among his many achievements, von Neumann demonstrated the deep formal equivalence between the Schroedinger and Heisenberg conceptions of quantum mechanics. But by some reports, at least, von Neumann came to be dissatisfied with this work, and sought to find a deeper unity in these theories. Such concerns were evidently at least part of the motivation behind von Neumann's work with mathematical rings of operators and projective geometry that led to his *continuous* geometry, a point-free projective geometry of non-finite dimension. Whitehead's own background in projective geometry was quite significant, and this work certainly informed his thinking with respect to his inquiries into the foundations of spatial relations in general.[76]

One significant point of *formal* difference between Whitehead's and von Neumann's work is that the latter was explicitly framed as an *infinite* dimensional geometry. Whitehead did insist that *nature* was essentially four-dimensional, but these dimensions are of the structure of *events*, not of spaces and times. We deliberately pluralize these latter terms because for Whitehead *spaces* and *times* are particular manifestations of the relational structures of events and not ontologically prior facts about the world. The "multi-threaded" character of these spaces and times (this being our phrase, not Whitehead's) leaves them unbounded in their formal dimensions.[77] So formally, both theories exhibit exceptionally high levels of compatibility; ontologically, both theories seem particularly suited to the interpretation of quantum mechanics; yet historically, neither theory's relations to the other appears to have been explored in any depth, much less substantially.[78] This is an issue that merits some further attention.

4   The Quantum of Explanation

> In mathematics, the greatest degree of self-evidence is usually not to be found quite at the beginning, but at some later point; hence the early deductions, until they reach this point, give reasons rather for believing the premisses [sic] because true consequences follow from them, than for believing the consequences because they follow from the premises.
>
> —Whitehead (and Russell) *Principia Mathematica* (vol. 1, v)

The Quantum Principle

As we have seen, Whitehead's theory of extension is intimately and consciously connected with both the basic structures of spatial reasoning and fundamental ideas of formal logic itself. We will continue to emphasize that the theory of extension is not a subject Whitehead casually tacked onto his metaphysics; rather, it is an ineliminable part of the very structure of reasoning internal *to* that metaphysics. However, there are two types of interpretive failure that can occur with this aspect of Whitehead's speculative scheme: the first is to fail to appreciate the importance of extensive relatedness; the other is to assign the wrong kinds and qualities of importance to it.[1] This latter will be a primary focus of this chapter, for it is a source of, or contributor to, an array of errors in interpreting Whitehead.

Among the problems that can arise is an inappropriate concentration upon Whitehead's seeming shift from the continuities of extensive forms of relatedness in the Triptych—with its emphasis on nature and science—to the more thorough-going "atomism" of the later, metaphysical works.[2] Taken to an extreme, this misreading of Whitehead's "atomism" becomes the misguided idea of a fundamental shift, if not a complete disruption, in Whitehead's philosophy between these two general sets of works, occurring in the years between 1922 and 1929. This view simply cannot be supported by a careful reading of Whitehead's texts (see chapter 5).

Of immediate interest to us here is the error of stampeding over the Triptych entirely by leaping into the metaphysics and attempting to convert that part of Whitehead's thought into a philosophy of nature of its own. Following such an ill-advised path, the theory of extension is then taken in a superficial way as glossing basic natural spatial relationships, though even in this case its role is typically given short shrift. This error seems most likely to happen when the above noted emergence of the "atomic" character of reality in Whitehead's thought is taken up in a naïve form, often enough because the *word* "quantum" has made its arresting and seductive appearance upon the stage.

Issues arising from the quantized behaviors of phenomena at the microphysical level were not new in the 1920s: Clerk Maxwell and Boltzmann had both encountered them in their theories on the behaviors of gasses in the 1860s-70s, although they did not know at the time what to make of their observations; Planck's constant on the minimal quantum of energy was formulated in 1899; and Einstein's work on the photo-electric effect—demonstrating the corpuscular nature of light—was published in 1905. But the seminal works of Heisenberg and Schrödinger appeared in the years 1925–1926, after the work that became *Science and the Modern World* had been done, and seeing these as important (if not dominating) influences in the system of *Process and Reality* has tempted many persons with backgrounds in physics to leap to conclusions. Nevertheless, it is a temptation resolutely to be avoided, as it results in a profound misdirection in the effort to grasp Whitehead's thought.

The unfortunate part of the above error is that both generic and specifically physical ideas of the quantum had already appeared in the 1919 *Enquiry into the Principles of Natural Knowledge*[3] and the 1920 *The Concept of Nature*.[4] Whitehead explicitly eschewed metaphysics in the Triptych, and indeed there is nothing particularly *metaphysical* about the notion of the quantum as it appears in those works.[5] So when the quantum makes its appearance in his metaphysical works, it is essential that the idea of "the quantum" be handled both in its *properly* metaphysical context and with a view to its relevant history; one must address oneself to individual instances and the situationally relevant circumstances of the particular text and inquiry that Whitehead is engaged in. As we have noted, this interpretive care is needed because at all times and in all places Whitehead adjusts the details and peculiarities of his methodology so as to address the topic of his present concern; exactly, one might add, as any good mathematician always does.[6] So imposing external interpretations upon the idea of the quantum, such as occurs when quantum *physics* is taken as the model of Whitehead's *meta*physics, is not the road to understanding but the path to misrepresentation. Whitehead had already warned us of the dangers of mistaking an exemplification of a principle for the principle itself in *The Principle of Relativity*; and, for better or worse, he was not in the habit of repeating himself once he had made a point that he considered settled.[7]

For example, amongst Whitehead's earliest arguments in *Process and Reality* we find that, to the oft-quoted claim that "the ultimate metaphysical truth is atomism," is then added the caveat that, "atomism does not exclude complexity and universal relativity. Each atom is a system of all things."[8] Whitehead then immediately contrasts *metaphysical* atomism with the wave/particle duality of light, distinguishing the latter as "corpuscular." Whitehead's choice of examples here seems rather telling since, as previously noted, it was precisely such behavior of light that Einstein highlighted in his 1905 paper, a paper which served as an important step in the quantum revolution in *physics*. The fact that Whitehead is so careful to distinguish the two can hardly be an accident. "A corpuscle," Whitehead states, "and an advancing element of a wave front, are merely a *permanent form* propagated from *atomic* creature to *atomic* creature. A corpuscle is in fact an 'enduring object.'"[9]

Thus, the idea of the quantum as it appears in the Triptych is an expression of matters of *nature* as these are investigated in natural philosophy. But the quantum in *Process and Reality* is of a much more general character than that which is revealed in nature, especially nature as it manifests itself in this cosmic epoch.[10] And by "general" it must be understood that the subject matter of Whitehead's work here is not limited to a mere illumination of this or that theory of *physics*. Such work remains squarely a part of *natural philosophy*, and carries with it none of the issues relating to the full range of metaphysical interpretation. Moreover, Whitehead was not one given over to racing after the latest *façon de parler* in physical theories and mathematics.[11] The puzzles that commanded his attention were the same ones that had stayed with him from his days as an undergraduate at Cambridge.[12]

This aspect of Whitehead's professional focus is what we have called "the problem of space." The physical-scientific source of this issue predates both relativity and quantum theories; the problem of space came thundering into physics with the publication of Clerk Maxwell's seminal works on electromagnetic theory.[13] Indeed, Maxwell's electromagnetic theory is the very real source of the crisis in physics that led to relativity itself. It was Maxwell's theory that motivated Michaelson and Morley in their attempts experimentally to measure the earth's movement through the æther that was postulated as the medium for electromagnetic waves. And while Einstein glosses over *all* of the experimental evidence that supposedly led up to his 1905 paper on special relativity (SR), he gave explicit pride of place to the *formal* problems of space that Maxwell's theory introduced into physics, and the mathematical asymmetries between it and Newtonian mechanics.[14]

This problem of space is the most visible thread of continuity in all of Whitehead's work, both as a mathematician and as a philosopher. Whitehead's thesis on Maxwell's theory earned him his Trinity Fellowship at Cambridge. The effort to find the most logically basic relational formulation for the characters of space was the primary motivation behind his *Universal Algebra*. His final statement on the problem of space does not occur until Part IV of *Process and Reality*. So it strains credulity to argue that he abandoned this line of inquiry—a focus of the previous fifty-plus years of his scholarly life—in favor of the "latest-greatest" topic in physics.[15]

As we have seen, the developed results of Whitehead's work on the problem of space display a fundamental connection to basic principles of logic itself, relations which Whitehead was aware of from the earliest stages of his professional career. He realized that this work was more fundamental than projective geometry—and more logically basic than affine and/or metrical geometries: The theory of extension is the instrument from which these others are constructed. This focus upon logically fundamental characteristics is, by the same token, our Ariadne's thread for finding the role of the quantum in his metaphysical thought. However, the path we need to follow is not in the emerging studies in physics in the latter part of Whitehead's career, but in the *logical structure of explanation* that was already manifesting itself in his earliest publication.

As such, it is important that we maintain a sharp distinction between the order of *explanation* and the order of *reality*. Here we begin to make good on the promissory note in the previous chapter regarding the strategic rules of logic. Care is needed to avoid conflating the order of *reality* with the order of *nature*. Whitehead's naturalism (see chapters 10 and 11) is nonreductive, and it is most assuredly *not* the grand total of all that is real. It is with the latter that Whitehead's metaphysical speculations are concerned (and even then, the *discussion* of reality is almost exclusively limited to the present cosmic epoch), while it is to a single structural feature of those speculations that we must here attend. And precisely because we are primarily concerned with such structural features, our attention must be drawn to what we call the "quantum of explanation."[16] Given the close connection between Whitehead's logic of extensive relatedness and his metaphysics, the insights we gain into the quantum structure of explanation, our toe-hold on the real (in our own cosmic epoch) is similarly reinforced.

We can already see from the quotation a few pages above, where the "atom" (the undivided) is a system of all things, that, since the metaphysical *atomicity* of the real does not exclude *complexity*, the order of the real is not an order of simple parts contributing to the making of a complicated whole. Indeed, given that the "atoms" of which Whitehead speaks are "systems of all things," we already have before ourselves rather telling evidence that Whitehead means his "atoms" to be taken in the original Greek sense of "uncut"; that is, as irreducible wholes. This holistic sense of reality was not driven by exotic physical effects such as "quantum entanglement."[17]

While there are sources of inspiration for Whitehead's "atoms" in the physical literature, the deeper sense of it comes from his educational background, which includes his mathematician's appreciation for systematic totalities, his understanding of Clerk Maxwell's *field* theory, and his poetic ear for the holism and rhythm native to human experience.[18] But the holism at stake is, once again, as much a matter of the order of *explanation* as it is the order of reality. In any discussion of a systematically interconnected whole such as Whitehead attempts, there is a kind of "falsification" of the real that occurs the instant one chooses to emphasize one aspect of that whole in one's descriptive explication.[19] Yet such selective attention is not optional; it is the *sine qua non* of engaging in any manner of rational inquiry whatsoever, regardless of whether one is engaging in high-energy physics, speculative philosophy, abstract mathematics, or making dinner with what is currently in the refrigerator.[20]

Ultimately it is not falsification but abstraction that is occurring here.[21] And as long as one does not confuse or conflate the order of explanation with the order of reality, then just as with a great work of literature, such an abstract metaphor—that poetic "little lie"—becomes the instrument of telling a much greater truth.

The above point was not in any way lost upon Whitehead himself:

> Philosophers can never hope finally to formulate these metaphysical first principles. Weakness of insight and deficiencies of language stand in the way inexorably. *Words and phrases must be stretched* towards a generality foreign to their ordinary usage; and however such elements of language be stabilized as technicalities, *they remain metaphors mutely appealing for an imaginative leap.*[22]

The italicized portions of this passage are particularly striking, since they are precisely the kinds of tropes required to engage a *literary* work. Quoting Whitehead again, "but no language can be anything but elliptical, requiring a leap of the imagination to understand its meaning in its relevance to immediate experience."[23] Whitehead had already emphasized in the preface that, "the unity of treatment is to be looked for in the *gradual development* of the scheme, in meaning and in relevance, and not in the *successive* treatment of particular topics."[24] This highlights once again the focal role of the structure of *explanation* in *Process and Reality* as opposed to just and only a dogmatic claim about the constitution of reality. As is the case with any thoughtful piece of writing, one cannot say what the real "story" is if one is content to shatter it into pieces and simply gaze upon the disconnected heap.

It seems at least a little ironic that it should be a mathematician reminding us that comprehensive explanation depends not only upon empirical adequacy and logical coherence, but also upon the "narrative intelligibility" of an account as well. That is to say, no small part of the explanatory force of an idea comes from the kind of appreciation that only emerges from a pan-synoptic grasp of the whole that is both cognitive and *felt*, this latter term being chosen quite deliberately.

Extensive Relatedness

It is appropriate that a consideration of some of the more literary aspects of Whitehead's explanatory schema provide us with our segue into his most mathematical discussion in *Process and Reality*. But the logic of this step is two-fold: first, it is Whitehead himself who insists that the intelligibility of the whole can be found only *in the whole*, and to leave out any part of it is, in effect, to falsify that totality by destroying its *a*tomicity (un-cut-ability). Secondly, the theory of extension is of immediate relevance to the misreading of the quantum as somehow derivative from its physical exemplification. So, while the fullest discussion of extension must wait for chapter 5, we can no more ignore it entirely until that time than Whitehead could meaningfully quarantine his discussion of it until just and only Part IV of *Process and Reality*.

We must always keep in mind the distinction Whitehead draws between atomic characters of reality with their "system-of-all-things" holism, and the corpuscular societies of occasions that are tagged in non-metaphysical parlance as physical "particles." There is simply no reducing the former to the latter without completely losing sight of Whitehead's real meaning. Whitehead's "atoms" are uncut in a sense that would still be the prevailing one at the time Whitehead was coming up through University. Physical corpuscles, on the other hand, are not "atomic" at all in Whitehead's sense; rather, they are *societies*—ordered and organized serial collections—of occasions, "objects" in the non-metaphysical sense, propagating themselves in *physical* space and time, under a coordinate analysis.[25]

Second, extension (or more appropriately, "extensive relatedness," a term favored throughout this book) is one of those aspects of structure that is native to any actual entity in its actual world. For instance, Whitehead notes that:

> Thus, the [extensive] continuum is present in each actual entity, and each actual entity pervades the continuum. . . . Thus, an act of experience has an objective scheme of extensive order by reason of the double fact that its own perspective standpoint has extensive content, and that the other actual entities are objectified with the retention of their extensive relationships.[26]

The term "extensive," it must be recalled, is used in reference to Whitehead's scheme of mereology—that is, logic of part and whole—which morphed into the mereotopology of *Process and Reality*. This is not simply a perspective of *natural* space or time, but of the logical and coherent (in Whitehead's sense of mutually interdependent) mereotopological sense that lies at the foundation of our conceiving the very possibility of natural spaces and times.

In the third place, "this extensive continuum is 'real,' because it expresses a fact derived from the actual world and concerning the contemporary actual world."[27] This is part and parcel of Whitehead's "radical realism" (a term we will explain in chapter 10). This relational system is not *abstracted* from concrete experience but rather infuses every vital, innermost part of it. Thus, extensive relatedness is not a set of concepts that is just casually superadded to actuality in its naively "atomic" ultimacy, but is an inherent fact expressed *by* that actuality; it is an irreducible "part" of the "system of all things."

Extensive relatedness was something that Whitehead argued was of immediate relevance to the issue of the quantum; indeed, the quantum is at the center of his opening discussion of the technicalities relating to the coordinate division.[28] But in coupling these ideas as closely as Whitehead does, he also significantly undermines any hope of identifying his quantum of *explanation* with quantum *mechanics*. The latter is expressly a matter of *microphysical* reality while the theory of extension—as Whitehead explicitly tells us—is logically prior to any *metrical* considerations whatsoever. Yet, such metrical properties are absolutely essential in order to distinguish something as "small" or "large." Whitehead reminds us of this absence of intrinsic metrical content in *Process and Reality* at various places: "the properties of this continuum are very few and do not include the relationships of metrical geometry."[29] Indeed, "these extensive relationships are more fundamental than their more special spatial and temporal relationships. Extension is the most general scheme of real potentiality, providing the background for all other organic relations."[30] On the other hand, quantum mechanical relationships and physical *corpuscles*, in physical theory, are inherently metrical in the sense that ideas of measurement completely permeate physical theory—even when such measurements prove rather paradoxical in nature.[31] As such, these metrical concepts are not, and cannot be, fundamental in any relational system that is itself *logically* prior to, and grounds the very possibility of, measurement.[32]

Indeed, metrical relationships are, in comparison to other structures within a generalized "spatial" logic, scarcely general at all. Rather, they are highly specialized developments in their own right. At the point at which one has a metric, one has proceeded through many successive layers of differentiation and particularization, such as those that generate topology, projective relations, affinities (similarities of angles but not size), until one has finally developed such detailed and narrowly determined relationships ("congruences") that will finally allow one to speak meaningfully of metrical forms of relatedness. One of the most obvious such forms is size itself (see our chapter 7). All such measuring is far from Whitehead's concerns in *Process and Reality*.

If one's logical machinery is so general that it has yet to include metrical functions, then considerations of size, including differences as gross as that of "micro" and "macro" physical scales, are logically *meaningless* because they can *not* be expressed. However, caution must be exercised (here as everywhere) because as Whitehead quite carefully illustrates, there are other senses of "micro" and "macro" that one can employ without presupposing a reductive collapse to mere physical size. "In the philosophy of organism," Whitehead tells us,

> it is held that the notion of 'organism' has two meanings, interconnected but intellectually separable, namely the microscopic meaning

and the macroscopic meaning. The microscopic meaning is concerned with the formal constitution of an actual occasion, considered as a process of realizing an individual unity of experience. The macroscopic meaning is concerned with the givenness of the actual world considered as the stubborn fact which at once limits and provides opportunity for the actual occasion.[33]

What is striking about this initial definition of these ideas is that the "givenness of the actual world" that "limits and provides opportunity," which Whitehead identifies with the "macroscopic," is precisely the sort of *real potentiality* that is part of the province of *extensive relatedness*.

The macroscopic cannot be *identified* with extensive relatedness, of course. Indeed, one possible criticism of the above reading is, on the surface, provided by Whitehead himself some pages on. There he says,

There are two species of process, macroscopic process and microscopic process. The macroscopic process is the transition from attained actuality to actuality in attainment; while the microscopic process is the conversion of conditions which are merely real into determinate actuality."[34]

In some ways, this might seem like the opposite of the previous characterization. However, Whitehead immediately refines this last quote, stating that, "The former [macroscopic] process effects the transition from the 'actual' to the 'merely real'; and the latter [microscopic] process effects the growth from the real to the actual. The former process is efficient; the latter process is teleological."[35] The fact that Whitehead so carefully associated the macroscopic with efficient causation is, again, a very thorough-going connection of the macroscopic with extensive forms of relatedness, because it is here that the overtly causal forms of interaction—in their more commonly understood *physical* sense—play out most directly. This becomes particularly obvious when one casts an eye to the nature of presentational immediacy and the theory of perception. (See chapter 8.)

Once again we must emphasize that the connection between macroscopic forms of relatedness and extensive connections is emphatically *not* an identity. What this brief examination shows is the focus within Whitehead's *magnum opus* upon the structures of *explanation*. The uses to which Whitehead puts the concepts of the "macroscopic" and the "microscopic" have explicitly eschewed such trivialities as *metrical* size. Whitehead addresses these concepts in their generically cosmological sense; that is, in their "specific" metaphysical relevance to *this* cosmic epoch. And it is within the scope of *this* relevance that we find the *real* meaning of the quantum.

As previously noted, Whitehead's concern with extensive forms of relatedness initially emerged from purely physical, scientific connections, but those connections are not seated in *micro*-physics. Rather, the scientific breakthroughs that inspired Whitehead and started him on his lifetime inquiry into the logical foundations of space, time, and reality came from Clerk Maxwell's work in electromagnetism. It was Maxwell who unified electromagnetic phenomena into a conceptually integrated whole. Only on account of such a unification were the respective developments of special and general relativity on the one hand, and quantum mechanics on the other, even possible in physical theory.

Maxwell's revolutionary theory of electromagnetism systematized what had previously seemed to be disparate phenomena under the aegis of a single, comprehensive explanatory framework. It also brought to light a fundamental problem in nineteenth-century physics, illuminating a fundamental asymmetry between the theory of electromagnetism on the one hand and Newtonian mechanics on the other. Maxwell's theory set a fixed, finite speed to the propagation of light waves, a speed which did not seem to vary with differing frames of reference or the relative motions of those frames, which starkly contrasted with the way motion behaved in Newtonian physics. By the evidence, this irreconcilable difference between the *mathematics* of Maxwell's and Newton's physical theories, rather than any issues relating to specifically empirical evidence, was the real motivation behind Einstein's development of relativity.[36]

With regard to the micro-*physical* theory of quantum mechanics—and here it is essential that we be clear about the distinction between physical and metaphysical principles,—it is only because Maxwell so meticulously unified electricity, magnetism, and light in a single, relational whole that the puzzles of quantum mechanics could manifest themselves *as* puzzles. It was Maxwell's work that demonstrated the unity of electro-magnetism as an undulatory phenomenon. But at a later date—1905, to be precise—Einstein also experimentally demonstrated the corpuscular nature of light with his paper on the photo-electric effect. This contradiction of the wave and corpuscular properties of light, in conjunction with Planck's work on the transitional energy states of physical systems, defined the research program for micro-physics for the foreseeable future by creating the equally irreducible and seemingly irreconcilable wave/particle tension in physical theory (raised by the theories now known as quantum *mechanics*). But physical corpuscles are not the minimal units of explanation. This is not because there are corpuscles that are smaller still, but because there are *logical* quanta that are more basic in the process of fundamental explanation.

The term "basic" here is not tossed about casually. The logical/explanatory primitives Whitehead is working with are prior to space and time. Referring to the development of extensive relatedness in Whitehead's *corpus* as "the problem of space" is simply a metaphor. We justify ourselves here in the use of this half-truth on the Whiteheadian grounds that it creates an effective path of narrative intelligibility. But that justification does not change the fact that an imaginative leap is required, a leap that can only be made from the precipice of algebraic reasoning.

For example, Whitehead tells us that, "other actual entities are objectified with the retention of their extensive relationships. These extensive relationships are more fundamental than their more special spatial and temporal relationships. Extension is the most general scheme of real potentiality, providing the background for all other organic relations."[37] He expands upon this notion by saying, "Physical time makes its appearance in the 'coordinate' analysis of the 'satisfaction.' The actual entity is the enjoyment of a certain quantum of physical time. But the genetic process is not the temporal succession: such a view is exactly what is denied by the epochal theory of time. Each phase in the genetic process presupposes the entire quantum, and so does each feeling in each phase."[38] And further on, Whitehead says, "physical time or physical space . . . are notions which presuppose the more general relationship of extension. . . . The extensiveness of space is really the spatialization of extension; and the extensiveness of time is really the temporalization of extension. Physical time expresses the reflection of genetic divisibility into coordinate divisibility. . . . The immediately relevant point to notice is that time and space are characteristics of nature which presuppose the scheme of extension."[39]

The serial, narrative structure of language makes it impossible to talk about anything—even purely logical/algebraic relations—except in terms of "before" and "after." But as the above quotes from Whitehead make clear, the sense of priority that appears at the foundational basis of extensive relatedness is not one of temporal "earlierness," but of logical—which is to say, *inquirential*—presupposition. The order, once again, in not an expression of nature, space or time, but of the quantum of *explanation*.

Propagation

Whitehead's metaphysics includes a detailed development of a theory of internal and external relations or, more accurately, a systematic account of the mutually referencing internal*ization* and external*ization* of relatedness. Whitehead's quantum is a quantum of explanation precisely because it coherently outlines the salient structural features of these forms of relatedness within the linguistic and algebraic frames of expression available to him at the time. He did this while consciously pressing the boundaries of those frames to say more than they were accustomed to saying. To follow through with this task, we must go well beyond simply trying to characterize the quantum, as it stands by itself—as though it were even possible to stand by itself(!)—in a manner that other philosophers might describe as "in-itself." We must also trace an outline of the paths through which the quantum expresses *its*-self as one atom in the many creating the one. Which is to say, we must gloss the ways in which the quantum propagates *its*-self in the systematic and relational totality of the real. Such paths will illuminate the *logical* function of the quantum, and help further distance us from the unfortunate physicalistic reductions that have dogged so much of contemporary Whitehead commentary.[40]

It is worth reminding ourselves here of some of the details of the categoreal scheme. Perhaps the most important of these is that "philosophy is explanatory of abstraction, and not of concreteness."[41] There are two salient points here. The prioritizing of the concrete over the abstract is the obvious one, and must be understood in the context of Whitehead's radical empiricism. However, of greater importance for us is the easily neglected claim that philosophy is *explanatory*, and not merely descriptive, even if the term "explanation" is being used in an unfamiliar way. We are going well beyond a mere analysis of sentences or a phenomenological description of experience. We are stretching our cognitive "legs" in an effort to reach the real and the true.

The direction of explanation is always from the more abstract back to its grounding in the more concrete. But the nature of that grounding is never in the simplistic form of "one thing::one explanation." To begin with, the "principle of relativity"—which might be called the "principle of relationalism" to avoid confusion with the far more narrowly construed physical theory of Einstein's—requires that, "every item in its [the actual entity's] universe is involved in each concrescence. In other words, it belongs to the nature of a 'being' that it is a potential for every 'becoming'."[42] We must not suppose that the use of the term "potential" in the preceding is somehow to be viewed as nugatory, a "mere" potential. The *pure* potentials—the eternal objects—together with actual entities stand out from the other categories of existence with what Whitehead characterizes as having "a certain extreme finality."[43] So Whitehead is not one to treat potentiality as a second-class citizen.

Indeed, it is this "potency" (a term we use guardedly for its common root with "potential") that makes each metaphysical atom a "system of all things." Whitehead characterizes the meaning of potentiality in the sixth Category of Explanation as "indetermination rendered determinate in the real concrescence."[44] The indetermination is rendered determinate because while "each entity in the universe . . . can . . . be implicated in that concrescence in one or other of many modes . . . in *fact* it is implicated only in *one* mode."[45] It is quite determinately *one* mode, and not merely "possibly one" mode. The entire universe is expressing itself in some determinate manner in each concrescence. That expression—which is to say, its prehension by the concrescing entity—might be negligible or even negative. But it will be that determinate *one* way which contributes its mode of potency to the systematic totality of *that* concrescence.

What stands out for us is the enormous difference between these richly textured modal forms of explanatory propagation in Whitehead's metaphysics versus the extremely narrow possibilities of a merely physical set of relations.[46] Physical waves and physical particles are not the *paradigms* of relatedness, but merely their simplest *examples*. It is worth noting how Whitehead reinterprets the wave/particle aspects of microphysics in the light of his logical Categories of Explanation. Microphysics is an abstraction *from* the fullness of nature, and it must find its explanation in the completeness of the categoreal scheme as a model.[47]

Thus, Whitehead notes that, "both a corpuscle and an advancing element of a wave front are merely a permanent form propagated from atomic creature to atomic creature."[48] Notice that neither the corpuscle nor the wave *is* itself such an atomic creature, and hence *can not* be identified with the fundamentals to which all explanations must be referred. Furthermore, while a train of waves will always exhibit social order, at the earliest stages of its propagation this order takes on the character of a more or less loosely organized *personal* one.[49] At the same time, a corpuscle is the type of society that manifests itself as an "enduring object." But Whitehead goes on to remind us that, "The notion of an 'enduring object' is capable of more or less completeness of realization." On account of this, a wave can be more or less corpuscular; indeed, "the train of waves starts as a corpuscular society, and ends as a society which is not corpuscular."[50]

There are two points in the above that are worthy of note. The first is that Whitehead completely inverts the standard interpretation of quantum mechanics: for Whitehead, there is no "collapse of the wave-packet." Rather, there is a dissipation of the "corpuscularity" in the initiating social order. Thus Whitehead states that:

> [W]henever we are concerned with occupied space, we are dealing with this restricted type of corpuscular societies; and whenever we are thinking of the physical field in empty space, we are dealing with societies of the wider type. It seems as if the careers of waves of light illustrate the transition from the more restricted type to the wider type.[51]

The second point is that the relational structures that are involved in this discussion are highly specific instances of the far more generic quantum of explanation. For example, immediately preceding the quote above, Whitehead highlighted the fact that, "in speaking of a society . . . 'membership' will always refer to the actual occasions, and not to *subordinate enduring objects* composed of actual occasions such as the life of an electron or of a man."[52] Whitehead quite explicitly sets an electron and a human being upon equal—and equally *derivative*—footing. These kinds of high end "entities" are the conjunctive *abstractions* from the genuinely *concrete* actual occasions that go into their extensively constituted natures. So, while, on the one hand, the "general principles of physics" emerge from these purely *logical* considerations exactly as one would require them to from any metaphysics with even a marginal claim upon our attention, it remains the case that these are purely logical considerations.[53] Physical instantiations are merely exemplars, and not paradigms, of the holistically organic approach that Whitehead is working from. The densely structured societies that compose the abstract "enduring objects" required by our habits of subject/predicate speech are not the foundations of explanation, but merely the cognitively identifiable abstract constructs of a deeper and more immediately present experience.[54]

So when we speak of "propagation" we cannot limit ourselves to just and only a physicalistic propagation of mere abstract corpuscles and/or waves.[55] We must dig deeper for the concreteness of the objectifications made by actual occasions in their variegated processes of becoming that go into making all manner of such structured societies. It is here that we find the logical order, which is to say, the quantum of explanation.

This quantum is, by logical—if not altogether "formal"—necessity, sensitive to context. The explanatory quanta relevant to a structured society of electronic occasions will vary in significant details from those applicable to a structured society of human occasions. Yet, as was noted above, both such structured societies are abstract derivations of a deeper layer of concreteness than that to which our lazy usages of language have for long ages habituated us.

One way to look at this is through the lens of reductionism. Reductionists say that, ultimately, there is only one type of explanation, and all forms of explanation are, by logical and/or metaphysical fact—or even necessity!—reducible to examples of this *one* privileged type. This approach to explanation is different from monism, which says that all that is real can be traced back to a single kind of "stuff." However, while every hybrid entity will be rooted in that stuff in a unique way, the explanatory path that traces those roots might be different from other entities in fundamental ways. Moreover, there may be even more than one such path for any given entity, depending upon the context and the inquiry. Thus, not only is "is" said in many ways for Aristotle, but things variously participate in all four "causes," a term that might be better translated as "reasons" to avoid confusion with the contemporary scientific senses of "cause."

It is worth recalling Whitehead's frequent use of the ideas of both efficient and final "cause." And if we fully embrace the sense of "cause" as *reason* noted above, then we can find a sense of formal "cause" occasionally used.[56] An immediately notable point about these techniques of explanation is that they are all highly *generic* in themselves, and as such none of them can be *reduced* to a specific type. Whitehead's sensitivity to the multidimensional demands of explanation, exemplified throughout *Process and Reality*, is part and parcel of his holistic, *organic* approach. The analysis of any part is itself only a *part* that must wait upon the development of the *whole* for the fullness of even that partial, analytical meaning, while the *whole* has no meaning in the absence of its parts.

Death and Dissipation

The requirements of explanation cannot be limited to just and only those positive factors that capture our attention in any given instance, any more than a part can take on its meaningful place in the scheme of things in the absence of the whole of which it *is* a part. The whole, of course, is also *just the whole it is* because of the parts of which it is composed. This is the *a*-tomic, undivided character of the quantum of explanation: the most concretely real "elements" of "the world" are those undivided totalities of experience which can only be analyzed through abstraction, but which are the explanatory basis of that which has been abstracted.

The concrete—and, hence, the atomic whole—is the center for relevance. Or, more correctly, any particular "atomic occasion" (by which we mean any actual entity in its actual world) is centered upon its unique, perspectivally informed universe. However, the *relevance* centered upon this perspective is not (in general) a Pascalian God whose center is everywhere. One need not slip into the fallacy of simple location to realize that there is a fringe of diminishing relevance to any particular atomic occasion. Relevance is undoubtedly the primary requirement for a quantum of explanation. But for any finite entity, the scope of that relevance is bounded. This boundedness of relevance works both upon the relevance *for* the particular atomic occasion and also upon the relevance *of* that atomic occasion.

For how such an occasion—typically as a member of a society of such occasions—slips beyond the horizon of relevance for most if not all other entities is itself an essential component in the quantum of explanation.[57] In other words, in order fully to appreciate the vividness of explanatory relevance we must also appreciate the evaporation of that relevance in dissipation and death. We will make a major theme of this dissipation in the final chapter of this book, but for now, a preview is needed.

As always, the concrete—and hence, the *a*-tomic—is the basis of explanation. How then does an atomic occasion lose its purchase upon other occasions such that the traces of relevance of the first occasion cease to manifest themselves in others? The answer can be sketched only in a generic way (to believe otherwise would be to endorse a form of reductionism), but the central idea is that as an ordered society begins to lose its cohesion, its atomic occasions cease to exercise the collective potency to propagate their objectified selves with any but negligible intensity. The model for this is not the death of an ordinary living creature, since biological creatures are such astonishingly complex societies that one can too easily lose track of the more generic patterns amidst the chaos of details. Nevertheless, the simpler examples we will turn to below are still possessed of significant organic interest.

The first of these examples is less obviously so: the Cosmological Microwave Background Radiation, or "CMBR." In what respects is the CMBR still relevant to human existence? It has only been in the past few generations that the CMBR has even been detected, and then only with the most exotic of high-tech equipment. It would seem as if the entire history of life on Earth has been blithely ignored by the cosmological context of its existence and evolution. Or is this way of framing the question perhaps missing something? The answer to this question is a straight-forward, "Yes."

It is a matter of comparatively little interest whether or not beings like ourselves are consciously aware of the fact that we are awash in a sea of electromagnetic fields—what Whitehead tellingly and repeatedly refers to as "electronic occasions." The simple fact is that everything on Earth has been inundated by the CMBR since the inception of life and, indeed, much earlier than that. The CMBR has been an ever-present fact with which all actualities on Earth have been interacting since such interaction was even possible. But this physical influence is quite small, so very small that it is arguably the case that its relevance to this planet had all but effectively died. Such loss of relevance is, however, a thing very much in need of qualification. After the construction of highly specialized technological enterprises, humans were able to revivify the CMBR in their own conscious thought processes, surprising some small number of us in our practical activities by the discovery of what had always been there (e.g., radio waves in general, but the CMBR more specifically).

To put this point in Whitehead's vocabulary, elements of the CMBR concresced into our field of relevance long past the point of any substantive physical effects it might have had upon simpler occasions. But herein lies both the strength and tragedy of the CMBR. Its tenacity of endurance is reflective of its grotesque simplicity and absence of organizationally coherent nexūs. The CMBR endures because, from almost the very moment of its inception, it was already so irrelevant to most concrescences that its ongoing propagation entailed no meaningful costs to the occasions propagating it. Meanwhile, its numbing simplicity closed the doors almost entirely to any possibility of "error" such as might lead to creative advance.[58]

As another example, consider the influences of childhood memories and how these variously propagate and dissipate in individual human lives. Events that seem sharply etched on our experiential fabric at the time of their occurrence dissipate into a kind of shadow realm of our own ongoing concrescence. These earlier actualities can propagate themselves in ways that range from a kind of potency so thoroughly established in our actuality that we scarcely notice them. For example, consider when we first began to walk. This was a tectonic change in our lives, and one whose echoes have not subsided insofar as our walking improves for many years, and stays with us as long as our knees and hips last. But no one actually remembers the event itself. What a pity that so momentous an event is dead to us forever even as its consequences so immediately permeate our every concrescent actualization.

Alternatively, consider a memory from childhood that actually can return in force many years later. Such memories will often be painful or shameful ones. There is a good chance that our mere mention of such things already will have caused one such memory to leap to the reader's mind. Two things are worthy of note here: first, the original event and its revivification in memory occurred as discrete *pulses,* as quanta. This is also true of the walking example: The first time one walked more than two steps bipedally was exactly that, the *first* time, a Zenonian burst of origination that never occurred before or after, since all subsequent trials were no longer first and were fundamentally informed by what had come before. The mature experience of walking seems like a continuous stretch from the period of very young childhood, but even that impression misses the significantly distinct and *a*-tomic act that constitutes each step. Moreover, each one of those steps is itself informed by the quanta of past experiences that led up to it, no matter how specialized the present "two steps" may have become in the interim. For example, if one is fencing, the change in foot movement that became habituated when one "learned how to fence" comes to be internalized in one's walk. One can identify the walking gait of a dancer if one is attuned to such movements. Similarly, we detect the compensations in the hips as the cartilage in the knees degraded from too many years of fencing, or too much stress from dancing, or too little exercise, or a thousand other variations that descend from the first two steps a person took.

The overarching purpose of these last examples is to emphasize that the *logical* relationships of propagation and dissipation, of origination and loss, of creation and death, are *forms* that manifest, themselves, the variegated quanta of explanation. Each quantum of this relational fabric might trace out forms that share analogies with other quanta: the painful memory from childhood vanishes into the background of actualization like the CMBR until some unexpected event (a casual sentence, a radio-telescope) suddenly focuses it back into a relevance that seemed—up until the immediately preceding "moment," which is to say, appropriately *a*-tomic duration—to have been lost entirely to any *real* potency. Instead, and seemingly against all reason, the past event bursts like a nova upon the current scene of actualization. Conversely, a childhood memory can cross the boundaries of the normal into the traumatic, and hang on with an operational force as definite as that of walking, while yet being pathological to its core. Some such memories are genuine horrors, not to be minimized, yet others—seemingly against all reason—hang on with nearly equal force and dog a person's life at every new step.

Yet it is certainly the case that none of these (and an indefinite variety of examples that could be offered) is ever against *all* reason. However, the reasons—despite their analogies and even because of them—are uniquely the *quanta* of explanation, irreducible and *atomic*. If we lack complete clarity in regard to the whole, we need not lack it relative to a perspectival part. It is in many ways here, in the realm of death and dissipation, that the proper disanalogies of Whitehead's quantum of explanation stand out with sharpest clarity against the scientistic reductionism that is often applied to his metaphysics. These quanta have reality *only* as analogies, and not simplistic identities.

5   Extensive Connectedness (The Metaphysics in Logic)

> If civilization continues to advance, in the next two thousand years the overwhelming novelty in human thought will be the dominance of mathematical understanding.
>
> —Whitehead, *The Interpretation of Science* (199)

Modes of Thought

As we have argued, Whitehead was a mathematician to the core, and if we ever hope to understand his philosophy we will have to understand the particular nature*s* of his modes of thought—and we very deliberately pluralize both. We have been pressing the idea that an insufficient amount of attention has been paid to Whitehead the algebraist, leading to a frequent failure to appreciate some important functional characteristics of his philosophy. We have emphasized the unity of Whitehead's thought as against the prominent interpretations that see a fundamental disruption between the natural philosophy of the triptych and the later metaphysical works. We will contend here that no small part of the reason for this failure to take proper account of the unity of Whitehead's thought is the absence of a sufficiently nuanced appreciation of the nature of algebraic thinking.

Whitehead was no ordinary mathematician, and no ordinary grasp of mathematical thinking by itself will serve to interpret the body of his work. Whitehead frequently demonstrated a poetic, even lyrical, appreciation for language—a rare gift amongst publishing mathematicians. Nevertheless, the nature of our topic here is such that we will, of necessity, set aside Whitehead's oft demonstrated sensitivity to poetic forms and focus on his other central modes of relational thinking, on the mathematical mind that framed his "pre-geometric" spatial theory of extension. But this is not the most important difference between Whitehead and other mathematicians. Rather, the stand-out fact for us is that Whitehead was a radical empiricist. Evidence for this claim is already present in Whitehead's first major publication, his *Treatise on Universal Algebra*, and we will be visiting some further suggestive passages in the second half of this chapter. It is a matter of comparatively little importance for us here whether Whitehead was already a full-blooded radical empiricist in 1898, or if he was "merely" so natively sympathetic to the position that he (more or less seamlessly) made the transition at a somewhat later date.

This chapter will be completing a "local" constellation of ideas within the broader context of this book. In chapter 3 we examined the "logic in the metaphysics," while chapter 4 provided a focused examination of the quantum of explanation both as a logical "atom" and a functional unit in Whitehead's metaphysics. This chapter will be the first "ribbon on the box" as we look now at the "metaphysics in the logic."

The first part of this chapter will be critical, directed at that thread of interpretation that tends to downplay (if not outright dismiss) the role of the theory of extension in Whitehead's thought. This opening critical move is intimately connected to the later constructive work, in that the dismissal of the theory of extension serves—or rather, *dis*-serves—both to shatter the unity of Whitehead's thought and to lose sight of the truly radical form of empiricism that is to be found in that thought. We will be using Lewis Ford's influential work, *The Emergence of Whitehead's Metaphysics*, as our foil in this initial stage of the chapter. We have great respect for Ford's scholarship and subtlety of thought, and we dislike commenting upon another commentator's work in an exclusively negative manner. But the constraints of this chapter do not permit us the space to treat of Ford's principal theses as anything but mistaken.

Skip to Page 337 [512]

The title of this section is a reference to the opening page of Part V of *Process and Reality*, and a swipe at that portion of the Whitehead community that so recklessly skips the preceding part. We are, of course, referring to Whitehead's mature and final contribution to his mereotopological theory of extension. We emphasize that this theory has *not* received the attention from many of Whitehead's interpreters that it deserves—either in terms of quality or quantity of that attention.[1] There is a substantial irony in this neglect. But for the moment we have a different issue to focus on: specifically, *why* have so many scholars chosen to skip to page 337?

It is perhaps uncharitable to answer a question that requires speculating upon other people's psychological processes. But it seems inescapably likely that no small part of this problem stems from a general "math-phobia" that we find in almost all areas of human activity, and against which Whitehead scholars have seldom shown an especially high resistance. And there is no denying that the coordinate account of *Process and Reality* reads much like a textbook on topology and the logic of spatial relations. Few philosophers are trained in these subjects. However, such training is a vital resource in following one of the primary modes of Whiteheadian thought.

This is not the place for a comprehensive exegesis of Ford's analyses of the (by his estimation) non-role of extension in Whitehead's mature metaphysics, or of Ford's claims of a fundamental discontinuity between Whitehead's earlier and later projects. Rather, it is sufficient to highlight just a few of the issues we wish to argue against. These issues are so central to Ford's reading that if these foundational premises are undermined, the entire edifice collapses. The very first of these premises is found in the opening few sentences of his book.

Ford begins his book with the statement that, "Most readers have found a vast difference between Whitehead's earlier works in the philosophy of nature, published 1919–1922, and his later metaphysical writings, starting with *Science and the Modern World* (1925)."[2] By itself, this is not problematic. But Ford next asks, "Why is this so?" and then proceeds to treat its being so as not only an essentially correct estimation of "most readers" responses, but as an altogether accurate and appropriate evaluation of Whitehead's texts.

Let us be clear here: This move by Ford—which establishes his entire thesis and provides the foundational support for the principal ideas that inform the rest of his book—this move, as it stands, is an *argumentum ad populum* fallacy. Ford apparently takes it for granted that "of course" there *is* this fundamental difference between the earlier and the later works. But insofar as this is offered as an argument, it is a *fallacy* pure and simple. And insofar as it is not offered as an argument, we should be justifiably suspicious that it is present at all. The remainder of Ford's book is intended to justify this move, but it does so largely on the basis of a significant measure of evident confirmation bias. There are very important aspects of *Process and Reality* that directly refute Ford's primary thesis.

But returning to Ford's opening move, we should ask: what if there is no "of course" to be found where Ford is pointing? Even assuming Ford's estimation of the majority of readers is correct, what if this reading of Whitehead is a mischaracterization of the relational continuities between Whitehead's earlier philosophy of nature—in particular, the three volumes we have been calling "the triptych"—and his later metaphysical works? Even if we granted the *factual* accuracy of Ford's claim (which we do not), that factual accuracy would not (by itself) carry any *logical* force. This lack of logical force becomes even more evident once we wonder whether that reading majority is sufficiently informed in the methods and structures of formal algebraic reasoning to adjudicate the matter properly of the appropriate place of Whitehead's theory of extension, its role in *Process and Reality*, and the presence or absence of real discontinuities in Whitehead's methods and topics.

We are arguing that, among other things, such a background acquaintance with algebraic thought is an important ingredient in even a basic understanding of Whitehead's philosophy. We would all be deeply suspicious of a scholar who stood before us and announced, "I am an expert on Aristotle and Plato; but I've never learned any Greek language or culture." Why should we be any less concerned about a Whitehead scholar who never learned the mathematical language and culture? The only hope we have to "get" Whitehead is to *think* like him, which means thinking with and through the tools that defined his life as a professional *mathematician* for some forty years.

One important step toward seeing the unity in Whitehead's works is to understand how mathematicians engage their problems—a methodologically vital aspect of mathematical culture. As we have already pointed out, this methodology underlying Whitehead's texts is recognizable to mathematicians. It is a pattern of setting out the aims and limitations of one's program, then developing and deploying the tools that will be brought to bear upon the proposed analysis. The specification of the topic will presuppose a comprehensive grasp of the background materials and previous developments of the topic—often without explicitly saying as much. This is a procedure that frequently leaves the uninitiated feeling that they have entered into the middle of an ongoing conversation, because in reality they have.

Nevertheless, even as the specific topic might shift away from or build upon a previous inquiry, the techniques of setting out one's agenda—and then sticking to it with meticulous care—remains constant. On the other hand, the structure of such a methodology is not a mechanically applied formula that is indistinguishable from one thinker to the next. The *nature* of the topics chosen as well as the specific techniques of analysis employed will regularly display the intrinsic thought patterns of their authors. In this regard, the methodological approach—or perhaps one should say, the "meta-methodology"—taken by Whitehead is recognizably continuous from before and after (as well as between) such works as his *Enquiry*, his *Science and the Modern World*, and of course, *Process and Reality*.

We would add that a mathematician does not, in general, abandon a previous project without comment. This is especially true when one's current project continues to reference previous thematic development. If there is a problem with the earlier work, then one articulates the issues and discusses directions for their potential resolution. On the other hand, if there is *not* a problem then one takes the previous work as *given* and moves on, referencing it only in so far as the earlier conclusions are appropriate to the topic at hand. This is exactly what Whitehead does: On only two substantive technical points does he correct a major component of the triptych: the first of these is the shift from a pure mereology of part and whole to the mereotopology of extensive connection, and the second is the introduction of strains into the technical machinery of his natural philosophy. Otherwise, many of his numerous references to his earlier philosophy of nature found in his later works are wholesale citations of entire volumes from the triptych.[3]

It is equally worth observing the wholesale and directly topical continuities between the triptych and the later metaphysical discussions. We have already spoken of the two principal thematic threads that run through Whitehead's work as "the problem of space" and "the problem of the accretion of value."[4] We are concentrating here on the first thematic structure, the "problem of space," because it is precisely here that the substantive *continuity* between the triptych and the later works stands out most sharply. Quite aside from the fact that, here, Whitehead's natural philosophy finds its vector into his metaphysics, Whitehead's entire theory of perception collapses without the fully articulated theory of extension presented in Part IV. Moreover, absent the theory of extension Whitehead would neither be able to discuss—much less resolve—the questions of the one and the many nor the nature and connections between internal and external relations.

The roots of Whitehead's evolving theory of extension first appear in his *Universal Algebra*, which was supposed to have a second volume. That second volume evolved in time to become the intended fourth volume of the *Principia Mathematica*, which shares with its earlier planned manifestation the ontological distinction of having never appeared. But through all of its various incarnations, even into its ultimate form in Part IV of *Process and Reality*, Whitehead's theory of extension was driven by the desire to develop a "pre-geometric" logic of spatial relations. With the emergence of Einstein's general theory of relativity, this project took on the added urgency of explicitly developing an empirically meaningful logic of spatial measurement as well. All of these ideas, in their various thematic developments, are clearly present in the triptych and the later works such as *Science and the Modern World* and *Process and Reality*. So, in lighting upon the supposed "vast difference" between the triptych and *Science and the Modern World*, Ford has pushed the putative dis-analogy too far.

As one last item on this point, we find Ford's opening assertions to be themselves quite surprising. Until Ford announced them as such, the thought that there was a "vast difference" between the triptych and the metaphysical works had never occurred to *us*. The nature of our experience and study—which, admittedly, has included a moderate amount of logic and abstract algebra—emphasized to our way of reading the enormous *continuities* that we found in Whitehead's texts. Now, having criticized Ford for his *ad populum* we would do well not to slip into our own *ad vericundiam* argument here. But it is important to note that the weight we are asking you to place upon our personal authority is only that of possibility: if an alternative reaction than Ford's *ad populum* is possible, then obviously there is nothing *necessary* to this "common reaction." In accordance with what we like to call the "first law of metaphysics"—what is actual is possible.

The next constellation of problems we wish to note in the Fordian interpretation of Whitehead's theory of extension, turns upon the status of internal and external relations on the one hand, and the status of extension versus prehension on the other. Ford cites (on at least five occasions) Whitehead's note on page 202 of the 1924 reprint of *Enquiry into the Principles of Natural Knowledge*, where Whitehead says that, "the true doctrine, that 'process' is the fundamental idea, was not in my mind with sufficient emphasis. Extension is derivative from process, and is required by it."[5] Based upon this note, and his reading of *Science and the Modern World*,

Ford argues that extension is relegated by Whitehead to a secondary status. "Prehension," Ford asserts, "now replaces extension as the primary relation between events"[6] Indeed, Ford goes on to assert that, "Extension can be interpreted strictly in terms of external relations, while prehension is precisely a spatiotemporal internal relation"[7] Given Ford's otherwise detailed scholarship, this is a peculiar claim to make. We cannot see how one could hope to justify such a position in the light of Whitehead's actual statements on the subject.

To begin, Whitehead makes relatively little use of the notions of internal and external relations prior to *Science and the Modern World*, and even in that book their appearance is limited. Nevertheless the continuity of Whitehead's thought from the triptych to *Process and Reality* is readily discernible. Let us look again at the quote above, from Whitehead's *Enquiry*: "Extension is derivative from process, and is *required* by it" (our emphasis). Ford apparently only looks at the first half of that quote, that extension is derivative from process. But Whitehead concludes by noting that *extension is "required" by process*! We are hard-pressed to imagine Whitehead ever being so egregiously ungrammatical as to invert the above relation by accident, but this is evidently what Ford is taking for granted in his blanket dismissal of the extensive forms of relatedness.

Moreover, while it seems obvious that prehension adds a dimension to the structure of process, nevertheless there is nothing in *Science and the Modern World*, or any of Whitehead's other works, to suggest that extension has been relegated to a mere sidebar of his overall scheme. In addition, the notion that extension provides only external characters of relatedness is not credible on the face of it—extension is nothing if not the basic form of pre-spatial relations, as we have seen here and throughout.

Furthermore, Whitehead explicitly argues that extension is an essential, irreducible character of temporal epochs, even in the initial development of these notions in *Science and the Modern World*. As Whitehead states in that work:

> Realization is the becoming of time in the field of extension. Extension is the complex of events, qua their potentialities. In realization the potentiality becomes actuality. But the potential pattern requires a duration; and the duration must be exhibited as an epochal whole, by the realization of the pattern. Thus time is the succession of elements in themselves divisible and contiguous. A duration, in becoming temporal, thereby incurs realization in respect to some enduring object. Temporalisation is realization. Temporalization is not another continuous process. It is an atomic succession. Thus time is atomic (i.e., epochal), though what is temporalized is divisible.[8]

Extension is not an accidental or secondary adjunct to the above; it is an essential quality and principal qualification of duration and realization.

(We would note that the above quote also makes it evident that Whitehead's use of the term "atomic" does not mean microscopic, but "irreducible"— i.e. "ατομώσ" or "uncut" in the original Greek sense. It is a divis*ible*— which is to say, extensive—actual whole.)

In *The Principle of Relativity*, Whitehead had already noted that, "It is evident that essential and contingent relationships correspond closely to internal and external relations."[9] It is also worth recalling that as early as *Enquiry into the Principles of Natural Knowledge*, Whitehead was already breaking out the relatedness of extension along both internal and external lines.[10] Thus, the duality Ford notes above—setting out extension as both derivative from and required by process—is not some comet that just miraculously appeared in the Whiteheadian sky. Along the same lines, we should observe the continuities that run from *Enquiry*, where Whitehead says that "Time and space both spring from the relation of extension,"[11] to the more mature claim found in *Process and Reality* that:

> Physical time or physical space . . . are notions which presuppose the more general relationship of extension. . . . The extensiveness of space is really the spatialization of extension; and the extensiveness of time is really the temporalization of extension. Physical time expresses the reflection of genetic divisibility into coordinate divisibility.[12]

Note that the later quote has acknowledged the need to bring prehension (genetic divis*ibility*) into the picture. But note, more importantly, that space and time remain *derivative* characters of the essential relation of extension. There is nothing in *Science and the Modern World* which breaks this unity of thought, even as (for the technical purposes of that specific inquiry) different aspects of those relationships as given, are varyingly emphasized.

Since extension and prehension are prior to space and time, they are also prior to the *metrical relations of space and time*. Hence, in speaking of the atomic nature of an occasion, especially in its extensive character, it is logically incoherent to regard that "atomic" nature as in any sense referring to the *size* of the occasion. While "large" and "small" can be dealt with qualitatively in purely ordinal terms, this should not be supposed to translate into any absolute quantitative measure, a point Whitehead emphasized in chapter 3 of *Principle of Relativity*, a work he quite explicitly continued to endorse throughout *Process and Reality*.

Finally, to cast the internality of spatiotemporal relations entirely upon prehension, as Ford does, flies directly in the face of Whitehead's discussion in *Process and Reality*, where he *explicitly* associates internal relations with extension.[13] "The 'extensive' scheme is nothing else than the generic morphology of the *internal relations* which bind the actual occasions into a nexus, and which bind the prehensions of any one actual occasion into a unity, coordinately divisible." Whitehead immediately goes on to observe the internally related nature of, "the *mutual implication* of extensive whole and extensive part. If you abolish the whole, you abolish its parts; and if you abolish any part, then that whole is abolished"[14] This is a refinement of, rather than a breach with, the discussions that had come before. As one can see in chapter VII of *Science and the Modern World*, Whitehead continued to endorse the philosophy of nature that is found in the triptych, including the central and essential role of extension in the processes of nature. While there is a great deal more that can be said on this topic, we hope the above is sufficient to motivate us to return to page 283 of *Process and Reality* and keep reading, because our story does not end there.[15]

How Is a Computer Scientist Like a Radical Empiricist?

It is difficult to understand how exactly the "skip to page 337" readers of *Process and Reality* hope to deal with Whitehead's theory of perception—that level of generality where causal efficacy becomes presentational immediacy, or, in a sense, time becomes space. Surely no one is prepared to argue that that theory of perception is a disposable part of Whitehead's metaphysical program, or that the structure of presentational immediacy is in any way an otiose appendix to the theory of perception. Yet presentational immediacy scarcely ever appears in *Process and Realty*, except in more or less direct conjunction with extensive relatedness, not only in Part IV itself but throughout the entire book. In Part IV, presentational immediacy is discussed directly in connection with both strain loci and measurement. It is worth repeating the point for emphasis: how is one supposed to look upon Part IV as a mere adjunct or a tangential addendum when one of the two fundamental characteristics of perception relies entirely upon the nature of extensive relatedness even to be possible, let alone actual?

One "moderate" version of the "skip to page 337" approach to *Process and Reality* might be this. These interpreters might say: "Granted that extensive relatedness is a vital aspect of Whitehead's theory of perception, even perhaps his entire metaphysics, still the technical development of Part IV does not really advance the primary argument of that book, and can be effectively addressed by persons with more narrowly focused technical interests." We are not aware of anyone who has stated the above argument. The avoidance of such technical details has consequences both inside and beyond the Whitehead community. For one thing, it remained for Robert Palter to notice that there was a *formal* inconsistency in Whitehead's development of extensive connectedness in Part IV.[16] However, Palter himself did nothing about this issue, and it was not until the early 1980's and a pair of articles by Bowman Clarke that a formally consistent set of axioms was developed, which completed that part of Whitehead's project.[17] Clarke published these two articles in the *Notre Dame Journal of Formal Logic*. This is significant, because computer scientists do not, as a rule, browse the pages of *Process Studies*. And it is to computer scientists that we must now turn.

One notable group of computer scientists is working on the formal and practical characteristics of the artificial intelligence of visual processing. A less technical phrase for that would be "the folks trying to teach robots to see." It is a widely acknowledged fact in this sub-discipline that Alfred North Whitehead's work on extension is *foundational* for their enterprise.[18] Our experience has been that Whitehead scholars are simply astounded to learn of this fact. Yet we should have expected and even predicted such a connection. After all, Whitehead was digging into the fundamental structures of perception as these connect directly to nature and natural philosophy. Meanwhile, the only way a computer scientist will ever teach a robot genuinely to see is by becoming a radical empiricist about visual and spatial relationships.

We will return to this last claim. First, we want to cover some of the recent history of post-Whiteheadian theories of spatial relations, because these are topics of independent philosophical interest. We will not be engaging directly in any formal analysis here; rather, we will content ourselves with briefly placing some of these formal aspects of Whitehead's thought into context and attempting to situate them in the general universe of contemporary thought on the topics of logic and spatial reasoning. Much of this discussion will hearken back to chapter three. But recall that while that chapter was examining the logic in metaphysics, we are ultimately looking at the metaphysics in logic.[19] The change in perspective is important, and we will say more about how such changes function in both logic and metaphysics in the final section of this chapter.

Anecdotes are not the strongest form of argument, but one is appropriate here. It will place our topic in a more intelligible narrative frame, while at the same time emphasizing the nature of discovery rendered more accidental and almost miraculous because of the extent of the aforementioned neglect.

In the summer of 2003, one of us (Herstein) was still a graduate student. Being bored, unemployed and penurious, Herstein was naturally shifting around for ways to spend money he did not have. He discovered what was—for him—the perfect opportunity: the 2003 meeting at Indiana University of the North American Summer School in Language Logic and Information. Herstein registered for this conference, drove up on the appointed days, and joyously filled his time there with numerous things logical.

One of the sessions was a special focus group on computers and spatial reasoning. Because spatial reasoning was a topic of interest to Herstein from his work in Whitehead's theory of extension, he attended. One can appreciate his astonishment when not one but *two* leading researchers in the field got up and independently acknowledged that the originating work in their discipline was Part IV of *Process and Reality*. Herstein made a special effort to catch up with both presenters afterwards in order to get references and further information, all with an eye toward tracking down this development of Whitehead's thought in greater detail. Along the way, Herstein had a particularly amusing conversation with Ian Pratt-Hartmann:[20] "I understand all that stuff about extension," he told Herstein, speaking of *Process and Reality*, "but what is the *rest* of that book about?"

In any event, here is some of the front-and back-story on Whitehead's theory of extension:

The theory of extension that appears in Part IV of PR is a type of "mereotopology"; which is to say that it is a *mereological* theory of part and whole coupled with a *topological* (and therefore, non-metrical) theory of neighborhood and connection. The significance of Whitehead's shift from the pure mereology of the triptych to the mereotopology of Part IV is substantial (we will have more to say on that in the final section of this chapter). Whitehead's presentation in *Process and Reality* was largely informal and he never noticed that there was the previously mentioned inconsistency in his basic axioms. Again as noted, it was only with the pair of articles by Bowman Clarke that these inconsistencies were resolved in a formally acceptable manner that retained the philosophical intentions of Whitehead's theory.

Clarke's formalization had the disadvantage of being presented in second-order logic, but because these articles were published in a respectable venue, they succeeded in bringing Whitehead's system of extensive connectedness to the attention of researchers in computer science. People such as Ian Pratt-Hartmann and David Randell reformulated the system into a first-order theory, which could then be employed in computer and robotic systems. It is also worth mentioning that Whitehead's mature theory of extension includes a non-metrical theory of convexity—his theory of flat-loci—that has yet to receive adequate attention from either a philosophical *or* a mathematical perspective.

The earlier theory of extension found in the triptych is a purely mereological system. Alfred Tarski, partly inspired by Whitehead's discussion in *The Concept of Nature*, wrote his famous paper "Foundations of the Geometry of Solids,"[21] which brought mereological concepts back to the forefront of work in formal logic. It is perhaps not an accident that Tarski was also responsible for reviving interest in algebraic logic, which (as we have seen in chapter 3) is intimately connected to broadly spatial forms of reasoning. Ultimately, Whitehead abandoned the strictly mereological approach after a series of articles in the *Journal of Philosophy* by Theodore de Laguna which pointed out a logically more parsimonious method of achieving Whitehead's intended results.[22]

But the story neither begins nor ends with the triptych. As already noted, Whitehead was exploring the idea of extensive relations embedded in an algebraic logic more than thirty years earlier, the ultimate result of which was his universal algebra. At the time Whitehead was doing this work—indeed, from before he'd first begun his University studies at Cambridge—philosophical arguments were raging as to the nature and reality of such mathematical entities as the infinitesimal point and, later, given the work of Georg Cantor, the nature and reality of the "element" of a set. Many of these discussions explicitly hearkened back to Aristotle's notion of the "undivided divisible," and appealed to nascent mereological concepts as an alternative to set theory.[23]

Formal developments were slow in coming; still, Whitehead's theory in the triptych was not the first to appear. That honor goes to Lesniewski, who published his work in 1916. But since that publication was in Polish, it exercised no influence on the English-speaking world for many years.[24] However, Lesniewski's motivations in developing his mereology were not that different from Whitehead's in the development of the theory of extension: Lesniewski was an empiricist. He was an "old/new fashioned" empiricist, very much of the positivist stripe that was becoming so popular in the 1920's, which is perhaps a contributing reason for his lack of lasting philosophical influence. Still, he was powerfully motivated by a deep-rooted suspicion of set theory and its foundational pretensions. So while Whitehead was largely indifferent to the foundational enterprise in mathematics (once the *Principia* was published, he never troubled himself with any of the philosophical motivations that drove Russell), both Whitehead and Lesniewski were concerned with the *empirical* content of the ideas they espoused.

Whitehead, on the other hand, was a *radical* empiricist. We are beginning to see avenues of change with regard to Whitehead's contemporary significance, emerging in the multi-dimensional and multi-modal disciplines of computer science. And as noted above, this recent influence should not come as a surprise. Computer scientists working in the field of robotic vision (whether at a theoretical or an applied level) are obliged to deal with the world *as it is genuinely delivered to experience*. Geometries of infinitesimal points and breadthless lines are the abstracted arcana of mathematical inquiry, not the presentations of experience. Complexly involved extensa are the genuine and atomic—that is to say, *irreducible*—constituents of our presentationally immediate encounters with spatialized reality.

Indeed, any functional system with finite visual resolution—such as a computer or a human being—attempting to engage or assemble a cognitively meaningful picture of the world, while employing limited processing capacities, is not only logically and mathematically, but *metaphysically* required to do so in a manner that takes extension, mereological relations, and elementary topology as its primitive elements. Finite eyes and finite minds do not encounter infinitesimal points or the geometries built from them. The reason a computer scientist is like a radical empiricist is because she is dealing with a world of spatial relations composed of atomically real, projectively related extensa—which is to say, the mathematical structures around which Whitehead built the first forty years of his professional life. We, as Whitehead scholars, need to relearn this lesson that Whitehead never forgot.

One final point to note here: we referred to the "*projectively* related extensa" that are the cognitively meaningful spatial relations that present themselves to experience and are open to analysis by a radical empiricist. We wish to re-emphasize the *projective* aspect of that observation. Such relational structures as are formally represented in projective geometry are precisely the sorts of structures that any radical empiricist would—or at least ought to—recognize as being immediately present in the field of experience.

And we have the testimony of a particularly fine set of radical empiricists in our possession to justify this claim. Those include the artistic geniuses of the Renaissance. It is useful to recall that projective geometry did not originate in the dingy studies of mathematical researchers. Rather, it emerged as a concrete development of the well-lit studios of those painters and artisans of the post-medieval world who were struggling to render upon canvas the basic spatial relationships of ordinary perception. These relationships were not the products of abstracting intelligences unanchored to the radically real world of empirical fact; they were the direct presentations *of* experience itself. This is to say, the very relational matter that any good radical empiricist would find within the elementary structures of experience includes those complexly and projectively related extensa of our spatial experience. These are the self-same mathematical fundaments of Whitehead's thought that have been ignored in the philosophical literature.

Perhaps we might fault Whitehead for not emphasizing this aspect of spatial relatedness more than he did. However, as we have noted, Whitehead was not one for repeating himself. To include the relevant sections from his *Universal Algebra* and *Axioms of Projective Geometry* would have easily added another hundred pages to *Process and Reality* without ever once saying anything that he hadn't already said.

The Metaphysics in Logic

In chapter 3, we speculated about some potential lines of research leading from Whitehead's theory of extension into other areas of inquiry. A few reminders of and additions to those speculations will be useful here. We begin with a quote, this time from a decidedly non-Whiteheadian source:

> The eyes of the scientist are directed upon those phenomena which are accessible to observation, upon their apperception and conceptual formulation. *In the attempt to achieve a conceptual formulation of the confusingly immense body of observational data, the scientist makes use of a whole arsenal of concepts which he imbibed practically with his mother's milk; and seldom if ever is he aware of the eternally problematic character of his concepts.*[25]

These words are not those of Thomas Kuhn as he was manning the barricades of a revolution in the philosophy of science. Rather, they were written by Albert Einstein in 1953, even as Einstein himself continued his lifelong habit of blissfully disregarding the implications of these sage observations in his own work. And while Einstein's comments are certainly ironic, there is no irony *per se* in what Einstein says; he was himself largely unaware of the eternally problematic character of the concepts with which he laid the foundations of contemporary physical cosmology. Science, insofar as it hopes to be scientific, must deal with the deliverances of experience as *these are genuinely delivered*. In addition to the critiques regarding the issues of the orthodox Einsteinian formulations of the nature of space and measurement, we will have more to say about some of the consequences which accrue when scientists leap beyond this responsibility into a realm of ill- or uninformed metaphysical speculation (in chapter 10).

The fact that such poorly conducted metaphysical speculation should be running riot in physics is a bit ironic. Meanwhile, so abstract a discipline as formal computer science has trumped even our best philosophical intentions, and arrived at a radically empirical formulation of space and cognition ahead of almost everyone, other than Whitehead himself. It is past due that we in the philosophical community catch up with our more mathematical cousins. Because it is not enough that we explain to "them" what "the rest of that book is about"; we need to need to come to a rigorously formulated answer to that question ourselves. For we, as philosophers, are no less habitually blinded by the concepts that we have imbibed with our "mother's milk," which we may all happily understand as a metaphor for our too readily accepted forms and frames of interpretation.

We have previously touched upon the interesting connections between algebraic logic and spatial reasoning, as a "bridge" between logic and metaphysics. We mentioned that Paul Halmos, in his classic on the subject, observes that with regard to algebraic logic, "the most useful approach to that theory is not to ask 'What is a proof?' or 'How does one prove something?' but to ask about the *structure* of the set of provable propositions."[26] We have seen how Whitehead says that "Geometry, in the widest sense in which it is used by modern mathematicians, is a department of what in a certain sense may be called the general science of classification," where "classification" means such *structural* relations as order, betweenness, connection, containment, and so forth.[27] We brought up fruitful connections among intuitionistic logic, topology, and structures known as "Heyting algebras." Category theory—which has been significantly cited by Leo Corry as the *key* element to the very notion of algebraic structure—arose from a consideration of relations in topological and projective spaces.[28] The spatial concept of "topos" was held up as a significant alternative to the traditional approach to the foundations of mathematics.[29]

Good theories fructify in unexpected ways. We have already mentioned John von Neumann and his "continuous" projective geometry, with no "points" whatsoever. "Points" for Von Neumann were treated as limiting abstractions, while the primary structures that one works with in continuous geometry were *extensive* sub-spaces. The formal analogies here between Von Neumann's system and Whitehead's mereological/mereotopological theory of extension are more than a little striking. Thus, for example, the computer scientist Giangiacomo Gerla has observed the familial relation between Whitehead's and Von Neumann's point-free approaches to geometry.[30]

Another example worthy of mention is modal logic. This is particularly appropriate here rather than in chapter 3 because modal logics are so much more intrinsically metaphysical in nature. While modal operators are popularly treated as dealing with possibility and necessity within the philosophical literature, computer scientists are more inclined to treat them as "essentially a simple way of accessing the information contained in relational structures."[31] This connects with spatial reasoning on several levels. In the first place, the basic modal operators of C. I. Lewis's S4 are readily interpreted as *topological closure* operators—which is to say, as distinctively *spatial* functions. Furthermore, the "frames"—as in "Kripke frames"—which provide the relational connections of relative possibility, are themselves "essentially a set-theoretic representation of Boolean algebras with operators," again a distinctively spatial structure.[32]

An especially interesting—and unexplored—connection with these modal ideas is Whitehead's theory of measurement. Measurement is a central concern of Whitehead's natural philosophy, and receives an entire chapter in Part IV of *Process and Reality*. The discussion in that chapter of strain-loci and projective relations is a considerable refinement of that which appears in chapter 3 of *Principle of Relativity*. Nevertheless, it is a refinement *only*, as is made clear by his citation of the earlier book, where he advises his audience to read the entire volume.[33]

For simplicity of comparison, then, we would draw the reader's attention back to chapter 3 of *The Principle of Relativity,* where Whitehead discusses the concept of equality.[34] Whitehead makes the case that there is no such thing as equality *simpliciter*; rather one always and only has equality with respect to some contextualizing characteristic. Whitehead represents this as "A = B → γ", where "γ" is the determining common character that makes the comparison of "A" and "B" possible. In other words, this relevance function "γ" is a *modal* operator, with some pretty striking formal similarities to the structural relations of a *Kripke frame*. Stated in another form, "A = B" means that "A" is commensurably *possible* with respect to "B" within the frame "γ." The notion of this commensurable possibility is expanded in *Process and Reality* with the subdivision of the frame into strain-loci, which establish the physical anchor for the measurement, and presentational immediacy which provides the mathematical-relational content. The presence of these spatialized, modal concepts is unsurprising, since clearly the *possibility* of measurement must be implicated in the *fact* of measurement. This idea is yet another at the heart of Whitehead's theory of space and measurement, and it provides yet another example of what we have been calling "the metaphysics in logic."

One cautionary note here: the philosophical significance of extensive relatedness must not be buried under the guise of merely "formal" developments. Gerla himself commented that, "surely there are philosophical motivations on the basis of Whitehead's passage from the inclusion-based approach proposed in the books PNK and CN to the connection-based approach proposed in PR."[35] While Gerla demurs comment on those *philosophical* reasons, he goes on to point out the significant mathematical reasons for making just that change (from inclusion to connection, or straight mereology to mereotopology). Among the issues that Gerla especially attends to is the relative ease with which a mathematical point can be defined, a factor which Whitehead himself specifically highlighted.[36] But as Gerla observes, it would be a mistake to think that the *formal* advantages exhaust the reasoning behind the change.

We can find a hint of this reasoning in the computer scientists themselves. We have mentioned Tarski's development based on pure mereology. Yet, despite the fact that Tarski is such a significant luminary in the literature of logic, his approach was not the one chosen by people like Pratt-Hartmann and Randell to build their AI visual systems. Instead it was the approach of a relatively obscure (from the point of view of formal logic) process theologian (i.e., Clarke) who formalized Whitehead's mereotopology. This is only an outline of the steps that led from Part IV of *Process and Reality* to their modern applications in computer science, but it should suffice for our purposes.

More needs to be said regarding formal topics. Even when one mathematical system is *formally* identical to another, the two need not be the same *heuristically*. The principles that guide inquiry—and we should just call these *metaphysical* principles—do not always lead in the same directions, even on a formal level. An example here is the difference between Gentzen-forms of sequent calculi and Hilbert-style axiomatic systems. "In a Hilbert-style system, formulas rather than sequents are derived, starting with instances of axioms and using in the propositional case only the one rule of inference, modus ponens." However, "These [Hilbert] systems are next to impossible to use for the actual derivation of formulas because of the difficulty of locating the substitution instances." This difficulty comes despite the fact that the completeness and consistency of both types of system guarantees that they can prove exactly the same set of formulae. Moreover, "There seems to be very little relation between the simplicity of the conclusion and the complexity of its derivation."[37] As a result, automated proof systems used in computer science invariably light upon techniques derived from the work of Gentzen, even as mathematics departments continue to teach the formally "equivalent" Hilbert-style of logic.

Philosophically there are other reasons for turning away from Hilbert axiomatics. As Jaakko Hintikka has observed, if formal logic is to have any genuine interest (beyond what Whitehead might characterize as a merely *recondite* exercise) then it needs actively to inform the real process of inquiry. Just as Gentzen style tools provide a computationally effective means for deriving proofs (when such proofs are derivable), these Gentzen techniques also structure the deductive organization of inquiry in a more natural way. Hintikka further observes that the so-called "cut" rule, while formally eliminable, is so heuristically illuminating that as a technique for informing and structuring inquiry, Gentzen logics are superior tools when that rule is included.[38]

These examples bear directly on Whitehead's mature theory of extension and his methodological approach. Whitehead's theory was not an accident in either form or intent. Thus, methodologically, Whitehead frequently emphasized heuristic and pedagogical clarity (in his selection of axioms) over mathematical parsimony. We see this in works as diverse as his little *Axioms* of geometry tracts from 1906–1907, to his final development of extensive connection in Part IV of *Process and Reality*. We see this also in his choice of categoreal principles, selected for their comprehensive strengths and mutual coherences rather than their formal minimalism. This, indeed, is the method of the quantum of explanation, both for Whitehead and for inquiry in general. This is why the theory of extensive connection, built to encompass presentational immediacy *and* the nature of nature, seamlessly integrates with computer science and robotics. The metaphysical principles illuminated in the very heart of logic itself express the quantum in its purest form.

Conclusion

The above examples of formal and scientific developments all bear strikingly interesting connections with Whitehead's theory of extension and spatial relations. But this latter is a product of Whitehead's commitment to a radical empirical interpretation of experience. The formal structures of the above are representative developments of the "direct deposit" of relational characteristics that truly *radical* empiricists will find in experience, should they be meticulous enough to analyze such deposits with care and with an eye for mathematical detail.

As for Whitehead's theory of extension, we note these following senses in which that theory is patently irreducible:

1. The theory is clearly *not* eliminable from the greater *corpus* of Whitehead's work as somehow a purely secondary and ultimately uninteresting development of merely formal details devoid of philosophical import.
2. Extensa are atomic not because they are very small but because they are themselves irreducible wholes. In other words, they are not composed of atoms; they *are* "ατομώσ" or "uncut."
3. The theory of extension is irreducible because it provides if not *the*, then certainly *a* fundamental theory of basic spatial experience.
4. The theory of extension is irreducible because it stands at a foundational nexus of an astounding array of seemingly disparate scientific disciplines that are currently dogged by a lack of thorough-going philosophical interpretation.

We submit that on account of the above and more, we in the interpretive community have not been doing Whitehead's work *sufficient* service. Following the taxonomy of the *Function of Reason*, while we have done well, we need to do *better*. We need to give the theory of extension its due.

6   The Principle of Relativity

> The misconception which has haunted philosophic literature throughout the centuries is the notion of 'independent existence'. There is no such mode of existence; every entity is only to be understood in terms of the way in which it is interwoven with the rest of the Universe.
>
> —Whitehead, *Essays in Science and Philosophy* (64)

Introduction

We now turn our attention to Whitehead's arguments regarding "relativity." This examination will serve several purposes. On the one hand, it will allow us to distinguish between Whitehead's metaphysical sense of the term, as opposed to the almost exclusively physical meaning that is now associated with it. On the other hand, it will also serve to commence another aspect of our critical enterprise by examining some of the deeply problematic issues that have come to be entrenched at the heart of contemporary physics. Our primary focus will be on the standard model of gravitational cosmology,[1] the core principles of which stem from Einstein's general theory of relativity. The problems with general relativity we will highlight here serve to draw attention to the difference between Whitehead's approach to nature and science, and what we are calling "model-centrism."

By the term "model-centrism," we mean something similar to the critique of the language of "models" for the sake of "explanation" in science. Old-style scientific realists who were persuaded by Einstein's decision to claim that his theory of General Relativity was not a model of the world, but the way the world *is*, used such language (e.g., such as Hans Reichenbach, Grover Maxwell),[2] This is misplaced concreteness at its worst.

But what *is* a good model? For us, following Paul Ricouer, a "model" is a heuristic fiction which may grow from a root metaphor, be brought to bear by imaginative work on an experience of a discursive world, as a symbol, having both semantic and non-semantic aspects, and being then brought back to the world *as* a model. Ricouer says, "Thanks to this detour through the heuristic fiction, we perceive new connections among things."[3] Ricouer may use the term "perception" as Whitehead does. Ricouer's method is a version of coordinate and genetic analysis, undertaken as a new approach to explanation. This is in keeping with Whitehead's use of models and language. We also see the problem of reference, when employing models in science (and philosophical cosmology), as implying that "the suspension of referential function of the first degree affects ordinary language to the benefit of a second degree of reference, which is attached precisely to the fictive dimension revealed by the theory of models. In the same way that the literal sense [in Ricouer's technical usage] has to be left behind so that the metaphorical sense [also in technical usage] can emerge, so the literal sense must collapse so that the heuristic fiction can work its redescription."[4] With Ricouer we affirm that metaphors, when they ground models, add something new to our knowledge. The "as if" of scientific explanation, understood as "discourse," does not undermine realism, it avoids model-centrism—the mistaking of the model for the thing it models. But models are indispensable to scientific knowledge. This account includes the role imagination plays in *forming* those models.

It is important to realize that the support for general relativity is far weaker than one might suppose from reading the popular or scientific literature. There is no real *experimental* evidence for general relativity, and the *observational* evidence is severely undermined by the large number of untested (or even untestable) assumptions applied in interpreting those observations.[5] There is a significant amount of contemporary scientific work—beyond Whitehead's 1922 developments—challenging general relativity and the framework it imposes when interpreting that observational data.[6] Our approach in criticizing general relativity will be fairly intuitive; a more detailed and technical discussion may be found elsewhere.[7] It is sufficient to understand a little of what motivated Whitehead in building his theory of nature, so that we can then see how that concept integrates with his larger project.

So, having focused in the last three chapters on generically logical issues, we turn now to consider the metaphysical principle of relativity. On the one hand, shifting our attention to this principle will deepen our appreciation for Whitehead's use of the term "atomic" in the sense of "uncut." Whitehead's use of the term "relativity" is also "old school"; it is not based upon the Einsteinian sense of the term that has emerged from physical science. This physicalist usage is radically limited in comparison to what Whitehead means by it. For Whitehead, "relativity" has to do with the densely *relational* character of all that is real. On the other hand, the narrowly Einsteinian sense also has its place here; together with the role of extensive connectedness in presentational immediacy, "relativity," in the Whiteheadian sense, is an essential link connecting Whitehead's metaphysics with his natural philosophy.

On the side of nature, Einstein created an insurmountable problem for physical cosmology by building his theory of space and gravity upon an irreducibly relational basis, yet all the while taking for granted a non-relational epistemological and metaphysical framework for interpreting that theory.

This approach has had the disastrous effect of institutionalizing a fundamental incoherence in the theory of measurement at the very heart of the standard model of gravitational cosmology. The "heart" of that standard model is, of course, Einstein's general theory of relativity, "GR." We will begin with the exemplification of the principle and then turn to the principle exemplified.[8]

The Logical Problem with Cosmological Measurements

There is a fundamental problem at the heart of general relativity that relates to the very possibility of generating meaningful cosmological measurements; measurements that are fully interpretable, empirically adequate, and logically coherent within the framework of general relativity and its presupposed philosophy of nature. The above claim does not relate (directly, at least) to the empirical results that are obtained when numerical calculations based upon GR are compared with various observations. Rather, our claim here relates to the underlying *logical* basis by which it is *assumed* that those numerical calculations and refined observations are, and *can* be, *meaningfully* interpreted—interpreted in a way that is scientifically valid. The critical aspects of this argument were developed by Whitehead in his triptych on natural philosophy between 1919 and 1922. Our interest here is not exegetical. It is with the *logical* problem of measurement persisting to this day in contemporary physical cosmology, and the implications of which reach further than that specific discipline.

This "measurement problem of cosmology" is easy enough to state and understand, and is directly related to what has commonly been presented as one of the principal strengths of GR. This, ironically, is part of our difficulty today: The "measurement problem of cosmology" is so straight-forwardly described and so effortless to understand, that we are inclined to conclude that it must be mistaken. General relativity is far too "successful"—or so we have been assured—for any such difficulty to be real. But these are our *assumptions*. (We remind readers of the Einstein quote about "mother's milk" from the previous chapter.) We must set aside our certainties if we are genuinely to examine what we really know.

Throughout this book we have been using the term "logical" in the broader and more traditional sense one finds in such thinkers as Aristotle, Dewey, and Hintikka, as the general theory of inquiry and not merely as a formal subdivision of mathematics. So in characterizing this problem as a "logical" one, we do not mean to suggest that there is a flaw in the formal, mathematical reasoning involved. The logical issues here are erotetic ones about the kinds of questions we ask, and the presuppositions we make about what qualifies as an answer. These erotetic/inquiriential issues raised by the measurement problem of cosmology go far beyond any specific facts about physics *per se*, and raise questions about the kinds of things that must be presupposed by *any* form of measurement-based inquiry. We will find that a substantive degree of uniformity of subject matter must be presupposed throughout the inquiry itself by any act of measurement, in order for said inquiry to terminate in meaningful operations and logically valid conclusions. This uniformity has a very well developed formal/mathematical structure. We will have more to say on that later.

With regard to GR, there are a few salient facts we should review. GR is a genuinely relational theory of space—it holds that space is not an independent "thing-in-itself," but rather a relational structure that is an emergent property of such realities as bear spatial relations to one another. This is not a new idea, even in physical science: Newton's proxies argued this matter with Leibniz in the early eighteenth century. However, as far as physical science was concerned, Newton "won" that argument for 200 years until Einstein and GR came along. Where GR is unique is in the degree of dependency of space upon other realities (we are making an effort to avoid using such loaded words as "things" or "objects") that the innermost characteristics of space are as contingent as the most arbitrary non-spatial entities, such as might relate themselves spatially. There is nothing necessary in this contingency—space can be relational and yet still *uniformly* spatial in all of its relations, without ever being independent of the relational totality from which it might properly be said to emerge.

General relativity denies not only the independence but the uniformity of space as well. Indeed, GR collapses space and physical properties (mainly in the form of gravity) into a single structure. This structure is the "metric tensor" of GR, usually represented as "$g_{\mu\nu}$," where Greek subscripts are indices along which the character "g" is "summed up" into a single physical quality. But this physical quality varies from one "point" of space to another, because the physical factors that operate on any given "point" of space will differ—at least slightly—from those operating at other "points." What *makes* them all "points" is no easy question. On account of this "mono-metric" structure of Einstein's metric tensor, where both physical and geometrical characters are bound into a single feature, there can be no genuine uniformity of space. Space itself, on this flawed account, is totally dependent upon the physical—and hopelessly contingent and various—factors that impinge upon space at different locations, which in turn lend every local "$g_{\mu\nu}$" exactly the characters it possesses. These characters will include the basic geometrical shape of space at each and every individual "point" of space.

To see how this contingency of geometrical structure undermines the logic of measurement, let us step back from the details of GR and ask "What kinds of logical relations are presupposed by spatial measurements *in general*?" Any measurement, whether it is quantitative or of a merely qualitative character (e.g., an ordinal comparison of more or less) requires the comparison of like to like. Thus, for spatial measurement, the comparison of one spatial segment to another is an essential aspect of this kind of measuring *operation*. In other words, some established spatial segment must be held up as the "standard unit" to which others will be compared. Whether this standard unit is taken to be universal for all such comparisons, or specific to some single act of comparison, it must nevertheless be *functionally* uniform throughout the act of measurement. In other words, the standard unit of measurement must "mean the same thing" throughout the act of measurement. Moreover, this standard must ultimately be manifested in the form of an extended physical object, such as a yardstick, so as fully to ground the *spatial* standard of measurement upon that of a rigid *physical* one. It is this physical standard unit that establishes the baseline of spatial extension that is the necessary basis in any spatial measurement. This standard unit is the possibility of "conjugacy," of a meaningful act of comparison, within the logic of measurement. But as vital as this physical standard unit is, by itself it is not enough.

It is also a logical requirement of measurement that it be *possible* to bring the likeness of the standard *into* comparison with that which is to be measured. There may be no practical way of achieving this direct comparison. But we must be able—in some indirect manner or another—to bring our "yardstick" up to that which we intend to measure. If our unit of measure is locked in a vault in the National Institute of Standards in Washington, DC, while the bit of extended space we need to measure is in the form of a 2×4 piece of wood in a lumber yard in Los Angeles, then either we must directly carry that standard unit from the one location to the other to make our measurements, or we must have a more readily transported surrogate we can use for the job. For the purposes of measuring yards of wood on this planet, such surrogates are readily found in the forms of the vast array of acceptable measuring devices permitting us to project our standard unit to our points of interest. This is the second, absolutely essential, relational factor in the logic of measurement. While we must have a standard unit of spatial comparison for *conjugacy*, we must also have standard(s) of spatial *projection* such as will allow that unit of comparison to be uniformly brought into comparison with—i.e., meaningfully projected onto—those things we mean to measure. The conjugate relation must be formally possible.

It is this second step which GR denies us. The requirement of uniform geometrical relations that make such projectively conjugate comparisons possible are rendered dubious if not impossible by the fundamental assumptions of GR. These assumptions include the idea that the necessary and uniform relations of geometry are collapsed into the contingent and heterogeneous physical relations of matter, energy, and gravity. Thus, the very geometry of space is dependent upon the varying physical effects of matter and energy at each "point" of space. Einstein's move of collapsing the formal and *logical* structures of geometry into the contingent and variable relations of physical interactions, has long been viewed by physicists as one of Einstein's most brilliant postulates. Einstein himself viewed this conceptual collapse as an act of essential insight, and looked with disdain upon such alternatives as that formulated by Nathan Rosen. By eliminating the distinction between geometry as a purely mathematical discipline and physical space (as it is studied in empirical cosmology), Einstein also compromised the relational structures that make a coherent theory of measurement possible.

The question now arises: how are we to bring into comparison our chosen standards of measurement and the objective spatial extensions that are to be measured? We have no *a priori* basis for making a comparison of a unit of measurement to a structure that is to be measured. This is because in GR, the geometrical structures of physical reality are dependent upon the *contingent* distributions of matter and energy throughout the universe. We do not know in advance what the geometrical characteristics of any piece of space happen to be, so we have no direct or indirect way of applying our units of measurement either to that stretch of space, or to any object where our measurements would essentially pass through that mentioned stretch. In the absence of a uniform system of geometrical relations, we cannot know in advance how that stretch of space—with its geometry contingently modified by the influences of matter and energy upon it—will in turn affect our attempts at measurement that are being projected upon or through it.

If we had direct access to every relevant point of physical space affecting the stretch in question, we could determine by those direct means the projective relations that would connect, in logical terms, that spatial region with our standards of measurement. We do not have such direct access. We have only barely transited beyond the reaches of our own solar system with a few primitive probes. Our only access to distant space is by purely formal and "*a priori*" processes of projecting our standards of measurement. Yet, by identifying the *contingent* factors of physical (notably, gravitational) nature with the geometrical relations of space, Einstein eliminated the uniform and logically *necessary* relations that would make it possible *to* thus project our standards.

Once again, according to Einstein the very geometry of space is contingent and variable, its nature at any given point being dependent upon the varying influences of matter and energy throughout the universe upon that point. And here we have the crux of the measurement problem of cosmology. If the structure of space is a contingent aspect of physical influences, then we must first know the nature and distribution of those physical factors before we can know the geometry of any spatial region. But in order to know this distribution of physical factors, we must be able to make accurate and reliable spatial measurements properly to place and relate those contingent, physical influences upon any given point of space. But in order to make accurate and reliable spatial measurements, we must have a robust understanding of the geometry of the spaces in and through which we are measuring. Only with this latter can we understand the effects *on* our standard unit of measurement of the non-uniform and contingent projective relations of those spaces, and thereby establish a logically meaningful system of conjugacy with the things to be measured. Yet such a robust understanding of the geometry of space is precisely what we do not have, and cannot establish, for it is exactly what GR refuses to grant us. We must know the complete distribution of matter and energy in the universe prior to knowing its geometry. But we must have a comprehensive grasp of this geometry in order to discover this distribution. As Whitehead pointed out, with GR as our theory of space and gravity, we are saddled with a situation where we must first know everything before we can know anything.

The very simplicity of the problem can disguise its profound nature. So permit us to restate the situation from a slightly different angle. When we carry our tape measure from our house to the lumber yard, we are confident that the tape measure continues to *mean* the same thing at the lumber yard that it meant at the house. Measuring out x-number of square feet of plywood, for instance, possesses all of the necessary functional characteristics at the store that we used at home when we first determined how much wood was required for our project. How is it possible that the tape measure should achieve such a continuity of meaning from the one locale to the other? Well, the tape measure is itself an extended segment of space, and when we carry it from one place to another so as to exploit its characteristics as a measuring instrument, we are asserting that the space itself—at the two locales and as represented in the tape measure—is appropriately comparable. We are asserting that there is a functional uniformity of projective relations that permits a workable conjugacy to occur.

Even if the respective spaces changed in some manner, this alteration would still not be excessively problematic, provided we could actually go to all the relevant points of space and directly determine what our tape measure means at each of these new locales.[9] This direct determination would enable us to continue to use our measuring tape, because the rules of its application would nevertheless be knowable, although somewhat more complex, than if the spaces were all uniformly of a kind. But on cosmological scales, even on scales only slightly beyond the boundary of our own solar system, we simply do not have this direct determination option. We cannot directly test such spaces to see what the projective relations are; projective relations that general relativity tells us are no longer uniformly knowable from Earth. We must come up with some reason to believe that our Earth-based measurements can be legitimately projected to these distant spaces in order to have even the hope of a meaningful cosmology. But if the very geometry of space is something we cannot know until *after* we can confidently engage in measurement, then cosmology as a science teeters on the brink of nonsense. For while we must first measure before we can know, GR requires that we know before we can measure.[10]

This is our philosophical quandary. GR saddles physics with a general theory of space that renders the possibility of measurement questionable, because the essential requirements for the possibility of spatial measurement are explicitly denied. And yet, GR appears to be successfully employed in formulating and evaluating cosmological measurements all the time.[11]

Indeed, general relativity and quantum mechanics are often held up as the premier examples of the most successful physical theories ever conceived. How are we to reconcile such practical successes with the supposed philosophical issues raised above?

The question almost undoes itself in the asking, for it is a well-known fact that these two theories are mathematically irreconcilable. But it is not just at the small scales of micro-physics that GR runs into problems. At the very large scales of physical cosmology, GR has also proven to be inadequate. In an effort to account for observed phenomena—phenomena of which physicists assume they have meaningful measurements—it has been necessary to reintroduce the idea of the "cosmological constant," as well as to invent such extravagant new ideas as "dark matter" and "dark energy" in an attempt to account for the evident behavior of the cosmos.

Even at the most mundane levels, GR is problematic. For example, the mathematical theory of general relativity is thoroughly non-linear. This means that GR as it stands is almost useless when it comes to producing numbers which can be compared with actual measurements. Rather, relevant sections of GR must first be approximated with a *linear* theory, and then this linear approximation is used to calculate theoretical values. Furthermore, this failure of linearity within GR is a significant contributing factor to the irreconcilability of GR with quantum mechanics. This latter theory *is* linear in important aspects, and as such defies any direct resolution with a macro-physical theory that lacks such linearity. Once again, let us emphasize that there is nothing controversial in the immediately preceding statements regarding general relativity.[12]

The Logical Problem with Model-driven Cosmology

General relativity offers a conceptual framework for scientific cosmology. However, this framework is shot through with multiple, *fundamental* logical difficulties. If an alternative framework were to be proposed that avoided and/or minimized some of those difficulties, then surely that framework would merit consideration. And such framework*s* are in hand. They include Whitehead's theory of relativity, first proposed in detail in 1922. But while Whitehead proposed his own applied, scientific theory, of greater significance is that he described an entire *family* of alternative theories. It is this family of theories which is of interest to us. Let us note that one will search popular discussions of cosmology in vain for any mention of either the logical failures of GR or the well-established and scientifically viable alternatives that can readily be found in the scientific literature on the subject.

Now, Whitehead had no knowledge of the recent troubles with GR that were noted above; his sole reason for criticizing it was the measurement problem of cosmology. It was his observation that if we do not maintain the separation between geometry and physics, then we lose the logical basis of the rules for conjugacy and projection that make spatial measurement possible. Whitehead's solution is reminiscent of the old joke:

Patient: "Doc! Doc! It hurts every time I do this!"
Doctor: "Well then, stop doing that!"

General relativity collapses geometry and physics together by enveloping both within a single metrical tensor. Whitehead's argument was that we should stop doing that.

So instead of Einstein's "mono-metrical" approach, as we have termed it, Whitehead proposed what has now come to be known as a "bimetric" solution. Instead of shoe-horning all of the metrical relations into a single tensor (the "$g_{\mu\nu}$"), collapsing geometry and physics, Whitehead's theory utilized *two* metrical tensors: His "J" tensor representing the contingent physical relations of gravity and other forces, and his "G" tensor or the necessary spatial relations of geometry.[13] As Whitehead claimed, and Eddington demonstrated, Whitehead's theory was equivalent to Einstein's at the limit of special relativity.[14] For almost fifty years after the publication of Whitehead's theory, no one was able to propose an empirical test that could decide matters between Whitehead's theory and Einstein's general relativity. Clifford Will finally suggested an effect known as "earth tides," which gave a hope of a measurable difference between the two theories.[15] But Will's proposal was predicated upon the dubious assumption that all the matter in the galaxy is concentrated at the center, while no account was made for the rest of the matter in the universe. Yet, despite these questionable simplifications, most physicists—or, at least, most physicists who are even aware of Whitehead's theory—assumed that Will's argument was conclusive. And this assumption came after several generations of physicists gave no attention of any sort to Whitehead's theory, despite the fact that there were neither formal nor empirical reasons for its neglect.

All of this is largely moot. Even if Will's purported 'refutation' could be modified so that it was genuinely valid, Whitehead's own theory was never more than an example of an entire class of theories. There are now many other members of this class than just Whitehead's, and many of these theories are known to be viable alternatives to GR. Each of these bimetric theories separates the geometrical relations from the contingent facts of physics, and thus they have the required level of uniformity available to them to permit of the projective relations needed for meaningful measurements.

It is a matter of no great philosophical relevance whether Whitehead's specific scientific proposal still offers a viable alternative to GR. Whitehead himself never viewed this as anything more than an application of his general philosophical principles. What we have instead is much more interesting: a developed system of philosophical ideas that are well vindicated by an entire family of scientific exemplars that are known to be formally and empirically valid. Moreover, these alternative theories are also much more linear in their formulations. This means that testable predictions can be calculated *directly* from these theories, and that they are far more easily reconciled with quantum mechanics.[16]

The uniformity criterion that Whitehead argued for, and which can only be achieved by separating the necessary relations of geometry from the contingent ones of physics, lends itself to a precise formulation. Indeed, this uniformity means that the geometrical relations exhibit a high degree of *symmetry*, such that the formal spatial relations have what is known as "constant curvature." Euclidean space is a paradigm example of a space with constant curvature—namely, zero curvature. But it is the constancy that is the essential structural feature of Whitehead's uniformity criterion. Given such constancy, Whitehead is quite explicit that non-Euclidean geometries are fully capable of meeting the necessary level of uniformity to avoid the measurement problem.[17] At the time that Whitehead was writing, the formal understanding of spaces of constant curvature was not yet fully developed. However, now, the relations between classes of symmetries and spaces of constant curvature—and hence, Whitehead's uniformity criterion—are well known. The basis of this symmetry is to be found in the mathematical theory of groups.[18]

This complex relationship among uniformity criteria, group theory, acts of measurement, and space is an important discovery. As we saw above, the measurement problem is not quarantined within physical cosmology. Rather, Whitehead's uniformity criterion is a generic requirement of *any* spatial measurement. While the kinds of relevant uniformities may vary in different situations, the need for stable relations of comparison and conjugacy between the things measured and the standard of units of measurement will always be present (see chapter 8). Even when the act of measurement is as simple as counting, the need for some uniformity is so manifest that we mark it by cautioning against comparing apples to oranges, even though apples and oranges have many common features. One might reasonably long for such common characters when bringing pulsars and background radiation under a common unit of measurement.

Formally, such uniformities will always be exemplified in those structures most appropriately modeled by the mathematical theory of groups. We thus find ourselves in possession of a fully developed, mathematically rigorous, universal *logical* requirement of all metrical inquiries. Group theoretical structures are astonishingly omnipresent. They are the beating heart of most, if not all, physical laws of nature, and are even of vital importance in understanding perceptual invariances.[19] Just how far a study of the logical requirements of uniformity might take us is an open question. But it is one worthy of pursuit, because these self-same *formal* criteria of measurement are radically empirical facts of experience.

Einstein's commitment to his "clever" idea (space and gravity are the same thing, as represented by the mono-metric structure of GR) came at the expense of both experience and logic. We again emphasize that "experience" always and everywhere is *radical* while "logic" is about *inquiry* and not the merely formal manipulation of symbols. Einstein had no sense of either of these irreducible facts, and as a result he reduced the principle of relativity to a merely formal criterion that disdained experience while making meaningful measurement essentially impossible in principle. These failures have been inherited, largely unquestioned, by the standard model of cosmology.

The standard model of cosmology generally goes by the name "ΛCDM," where the lamda stands for the "cosmological constant" (a number necessitated by GR as a consequence of the interpretation of the observed red-shift of the spectra of various objects in the universe), while "CDM" stands for "Cold Dark Matter." This latter is an entirely hypothetical entity—which has consistently defied all attempts at detection or observation—necessary to account for the structure of galaxies, structure which would otherwise serve as a direct falsification of GR. Notice that both key components of ΛCDM are entirely about saving GR from observational failures. These failures are not a matter of cranks or crackpots who have been left behind by the cosmological mainstream. Rather, these failures are artifacts of a cosmological mainstream that is more in love with its formal model—its *standard model*—than it is with the fundamentals of logic in general and scientific inquiry in particular.[20]

While we will explore it in more detail in chapter 10, we will preview the core of the problem now: namely, gravitational cosmology has outrun the empirical content of the available observational evidence in its rush to justify its adored standard model. Rather than evidence-driven science, gravitational cosmology has become model-driven speculative mathematics. It has ceased to be science at all; it is "old-school" metaphysics. And by "old-school" we do *not* mean anything like Whitehead's return to the pre-Kantian roots of modern thought. We mean the wholesale return to medieval speculation ungrounded by testable consequences and governed only by formal coherence. Gravitational cosmologists have not only lost track of the *genuine* principle of relativity, but also of their roles as scientists.

The Principle of Relativity

Chapter 1 of Whitehead's *The Principle of Relativity* begins with these observations:

> The doctrine of relativity affects every branch of natural science, not excluding the biological sciences. . . . Relativity, in the form of novel formulae relating time and space, first developed in connection with electromagnetism. . . . Einstein then proceeded to show its bearing on the formulae for gravitation. It so happens therefore that *owing to the circumstances of its origin a very general doctrine is linked with two special applications.*[21]

Whitehead uses *Principle of Relativity* to develop the gravitational form of the application in a direction that was (and is) scientifically robust, avoiding Einstein's dogmatic commitment to a mono-metric model in favor of a more logically sound bimetric approach. The electromagnetic application was essentially completed by Clerk Maxwell, and was encoded in Special Relativity. Whitehead's proposal matched Einstein's at the latter, special limit.

However, the mature statement of the *general* principle waited for *Process and Reality*, where it appears as the fourth category of explanation:

> (T)he potentiality for being an element in a real concrescence of many entities into one actuality is the *one general metaphysical principle* attaching to all entities, actual and non-actual; and that every item in its universe is involved in each concrescence. In other words, *it belongs to the nature of a 'being' that it is a potential for every 'becoming.'* This is the 'principle of relativity.'[22]

In other words, nothing is so cut off from anything else that it at least *might* exercise some influence on that other entity's process of coming to be. Being potential is, for Whitehead—and for any radical empiricist—explicitly recognized as a *real* form of relatedness. So even if an actuality manages to play no definite role in the determination of another actuality, the two are still not cut off from one another in their mutual interconnectedness. It is here that the *a-tomic* (undivided) character of reality most fully reveals itself. Whitehead states (and we emphasize) that the principle of relativity is "the *one* general metaphysical principle attaching to *all* entities, actual and non-actual." He makes it clear that this principle stands out from the others. The first three Categories of Explanation—that the actual world is a process, that the becoming of an entity is the achievement of real unity, that there are no novel eternal entities—are not more basic than the principle of relativity. Rather, these are "axioms" needed to contextualize the meaning of relativity itself.

Relational atomicity means something *categoreally* different in a world of substance rather than of process. This categoreal distinction shows itself in three ways, corresponding to each of those categories of explanation that precede relativity. The first category of explanation tells us that the fundamental structure of inquiry has an entirely different logic if one's focus is upon how things stay the same rather than upon how they change relationally. The second category of explanation tells us that one must understand how relational structures are intrinsically focused upon the method or the process of their *achieved* unity, rather than upon a merely given "what" of that unity. The third category of explanation emphasizes the fact of the abstract (as opposed to vacuously *recondite*) "potential" of the relational world as algebraic relational possibilities, which cannot be treated as slipping and sliding about without thereby calling down upon ourselves the revenge of Zeno in a *logically* unanswerable form. In short, eternal objects cannot change, otherwise, Zeno returns. Together these three categories illustrate how and why we must have certain *logical* fundamentals in hand before the real force of the principle of relativity can show itself.

Whitehead's approach is entirely recognizable from the mathematical perspective. The serial order in which the categories are presented is not the same as the logical order of their importance. Central players in a mathematical argument will *not* be presented in an order that aims at "narrative drama"; they will appear in such a way as to construct the most pedagogically effective logical development. So Whitehead's highlighting of the principle of relativity as "the one general metaphysical principle" might seem so understated as to diminish the principle's centrality to the full categoreal scheme. But having highlighted the principle *at all* is the step that must stand out. Indeed, on only one occasion in all of *Process and Reality* does Whitehead so much as modify the attribution of "atomic" with an adverb. (That would be "incurably," of course.) Such cautious rhetoric is standard with mathematicians. The narrative order of presentation, or what we have called the order of exposition, is not to be confused with a hierarchy of logical determinations that characterize the systematic aspects of the philosophy of organism.

Let us respect Whitehead's rhetorical restraint without trying to limit ourselves to it. The principle of relativity means that everything that is real—not merely "actual," but *real* even if only unactualized possibility—exists in an *incurable* web of relatedness to everything else. *Regardless of whether that web immediately expresses itself as the definiteness of actuality or as the determinateness of potentiality, the web is equally real.* Steps need to be taken to internalize the central point of relational connectedness—Whitehead's "relativity." We cannot *cut off* any part of reality as though that part—even if "only" potential in its relatedness—were somehow a mere "*tomatic*" effluvium to the whole. *Relativity* is the *one* metaphysical principle that counts everywhere and at all times.

This is the principle that Einstein and his devotees have abandoned: not the mathematical expression of their physical model; that model is itself only an *application* of what has become standard dogma of orthodox cosmology, with its narrowly defined approach to the interpretation of a truncated representation of experience. Rather, physical cosmology has left behind the *full* principle of relativity and its unqualified commitment to the incurable relatedness of the real. That abandonment comes in the truncation of experience at the root of their largely unexpressed theory of experience. For one cannot have a universal principle of relativity—applicable to *all* that is real—unless one takes experience in its real, relational totality. Experience—both actual and potential—is exactly the kind of reality that falls under the principle of relativity. One cannot take the metaphysical principle of relativity seriously unless one is a *radical empiricist*.

But the approach of Einstein and his followers has been that of traditional empiricism, an approach that takes for granted what Dewey characterized as the "spectator theory of knowledge."[23] This approach was effectively built-into GR with its treatment of time as little more than an adjunct to (and additional dimension of) space. Such an approach is an overt denial of the phenomenological reality of temporal experience. Within GR, time is "already there" in the same way that space is, and the experience of change is an illusion to be "corrected" by our mathematical model. Brian Greene, for example, refers to time as a "frozen river," and confidently insists that reality is essentially a Parmenidean block.[24] "Experience," in the view of Einstein, Greene, and present scientific orthodoxy, is not a relational participant *within* the world, but merely an external observer *of* the world, and a poor one at that! We will explore this topic further in chapter 10.

It is sufficient to notice that when experience in its *radical* fullness and connectedness is disdained in favor of clever mathematical models, then the resultant medievalism of contemporary physical cosmology is an inevitable result. For *what* are we left with to *test* our models, other than the formal and recondite cleverness of those models? What standards might we apply to test our models when our model-centric approach demands that we measure experience *by* those models, rather than those models *by* experience? What yardstick shall we apply? Do we achieve anything other than a preordained "fit" with a "reality" that bears no particular relation to experience? Do we just promiscuously add new parameters as needed and convenient? Under such assumptions, experience can offer no meaningful feedback, since it has been shut out of the process of model-centric science when it dared to challenge the models that experience would be called upon to test.

These are not the rhetorical concerns of philosophers but the active criticisms of working physicists, such as Ratcliffe, Eastman and others.[25] While such critics remain a minority, they are currently challenging orthodox cosmology and ΛCDM on exactly these grounds. It is sufficient to highlight how the logical problem with model-centric inquiry of any kind leads to the abandonment of its own "relativity" for a position of spectator-like authority by leaving behind the relational reality of experience. Meaningful inquiry cannot proceed in this manner, and we see one manifestation of this in the logical problem with cosmological measurements. Having dismissed experience from any constructive participation in its model-building, cosmology has turned a blind eye to the basic assumptions that make spatial measurements possible.

We have now set the stage with relativity, and it is time to push deeper into Whitehead's relationalism. The next two chapters will explore some vital aspects of this relationalism, which in turn enable more detailed criticism of the desiccated metaphysics currently underlying contemporary physical science.

7 Genetic and Coordinate Division and Divisibility

Mickey Mantle: "Hey Yogi, what time is it?"
Yogi Berra: "You mean right now?"

How, then, does the *a-tomic* (the undivided) come to be divided in Whitehead's philosophy, and in keeping with what we have just said about the principle of relativity? The concretely indivisible is abstractly divided by two kinds of analysis: coordinate and genetic. We have been using these terms in quite specific ways prior to this point in the book, but here we seek to make that usage explicit. First let us rehearse a bit of history.

Background

From the time he finished volume one of his *Universal Algebra*, up until about 1913, Whitehead was distracted by the emerging proof-theoretic logics of Dedekind, Peano and Frege, before he realized it was a dead end.[1] He returned at that point to the algebraic logic he had been working on before—not the extensional logic of the *Principia*, but the idea of universal algebra and its associated theory of extension. In short, Whitehead resumed his concern with the problem of space.

Whitehead's reading of Kant, Bergson, and others, had convinced him that spatialization was an indispensable part of any formal explanation, and to have a *philosophical* explanation, one needed a description of the ground of our *acts* of spatialization (more specifically, our "presentation" *as if* we existed spatially in a variable flux). Whitehead's theory of extension, which is a non-metrical, axiomatic systematization at the most general level of the structural characteristics of all actual spaces (perceived, prehended, implied, or potential[2]), is finally offered in Part IV of *Process and Reality*.

For several of his books, Whitehead adapted a form of mathematical inquiry to the task at hand, which he called genetic and coordinate analysis. Since the time of Plato, and even before, philosophers have been borrowing formalization techniques from mathematics to address philosophical problems. The greatest steps forward in philosophical inquiry are often of such a kind. Descartes' search for the *mathesis universalis*, Spinoza's geometrical method, and Leibniz's quest for the universal characteristic are well-known examples from the modern period, while Bergson's qualitative calculus and Cassirer's use of group theory and projective geometry are less well known efforts in the spirit of making philosophical methods mathematical. Jaakko Hintikka belongs to this line of thinking as well, distinguished by creative philosophizing about adapting formal tools of inquiry to varied purposes.[3]

Genetic and Coordinate Division vs. Genetic and Coordinate Analysis

We now explain the how and the why of the two major movements in the structure of Whitehead's method of inquiry in *Process and Reality*, showing the relation of genetic and coordinate division in the phases of inquiry. There is a difference between the way that the actual entity divides the extensive continuum, and the way that prehensions realize the potential of divisibility in the *analysis* of actual occasions. The reason for the difference is that the division of the extensive continuum by an actual entity *is* a realization—it is treated as actually occurring. The division of actual occasions into prehensions is entirely hypothetical or virtual; prehensions are relational entities, relating the actual entity's transition to its conscrescence, and thus, prehensions cannot really be genetically divided from process except in abstraction. To treat them otherwise invites Zeno's paradoxes and commits the fallacy of misplaced concreteness (see the previous chapter).[4] Here we understand why Whitehead uses the term "merely real" to describe eternal objects (possibilities) and actual entities (the quantum of explanation) when we attempt to consider their independence of one another.[5]

Genetic and coordinate *analyses* (as distinct from methods of division), as *methods* of analysis, are indifferent to whether the entities selected and grouped (or specified, generalized, and coordinated) are divisible, real entities, or actually indivisible components *of* entities. Rather, we seek only to learn what description is yielded if we examine varying rules and relations that, if specified one way rather than another, point to a *unit of explanation* (and if that unit be irreducible and necessary, a *quantum* of explanation) that is exemplified throughout the entire *explanandum* (which in the case of metaphysics, is our entire cosmic epoch). Whitehead was not alone in using this kind of dual analysis. Cassirer expressed it clearly in 1925, saying "it has become increasingly obvious that genetic problems can never be solved solely by themselves but only in thoroughgoing correlation with structural problems."[6] This latter issue of structure is handled as a coordinate analysis, not only by Cassirer and Whitehead, but also by Bergson, Royce, Dewey, and other process philosophers.

Genetic *specifications* (or suggested proximal units, relations, or rules) function as what mathematicians call "implicit definitions,"[7] which become more explicit as analysis proceeds. They are "axioms," not formally but in the original sense of that word, i.e., "thoughts worth thinking."[8] Once these genetic specifications are adequately explicit (or "definite" in Whitehead's language), the principles of their determination, as possibilities (i.e., intelligible structures on their own account) are coordinated to create a model of the whole from which the genetic divisions were taken. The coordinate whole precedes the genetic specifications formally, regardless of whether the whole was developed from a hunch about the parts and then empirically adjusted until the parts and the whole fit together tolerably. That is how a categoreal scheme is created. The coordinate version of the phenomena precedes the genetic specification in the way that the whole precedes its parts, since in order to be "parts," a whole (to which they belong) must be presupposed, *logically* (in our peculiarly erotetic sense of that term).[9]

This modeling can be done in many ways, but first, we would like to offer a very simple example. We will use genetic and coordinate analysis to create a model of an existing physical structure. Be warned that this example oversimplifies and renders overly static the actual way that Whitehead carries out genetic and coordinate analysis, but these deficiencies are heuristically bearable. The most important distortion is that our example is essentially metrical, and the whole point of Whitehead's theory of extension is that it is *non*-metrical. However, the metrical nature of our extended analogy below also provides further illustration of how metrical presuppositions can creep into our conceptual schemes and overwhelm our powers of judgment, and so serves as a cautionary tale as well.

We intend to show that Whitehead's concept of the actual entity (the "quantum of explanation") is a kind of mereotopological unit, adjusted to the descriptive purposes of metaphysics, that rises to the standard he sets for any such description: adequacy, applicability and universality of exemplification, logical rigor (as a matter of inquiry), and maximal consistency. The *outcome* of the analysis is the categoreal scheme. Interpreters have taken the categoreal scheme to be a set of conditions for metaphysical method. It is nothing of the kind. The order of exposition is not to be conflated with the order of logical importance, as Whitehead points on numerous occasions. The whole result of *Process and Reality* is *in* the categoreal scheme, but it takes the entire book to provide the reasons justifying just *that* scheme, and offering a comprehensible narrative about it. The categoreal scheme is a model of our cosmic epoch, a non-dogmatic alternative to the received model of General Relativity. A cosmic epoch is a far-flung object of study. Let's do something simpler first.

Building a Model: Basic Lessons in Genetic and Coordinate Analysis[10]

If we wanted to create a scale model of a famous sports stadium, say, Fenway Park in Boston, we would need to specify measurements in three dimensions, at least. We could use any unit of measure, but we should choose something that facilitates the coordination of the parts when projected to a different scale, since the purpose of our inquiry is to recreate the structure on a smaller scale. That scale need not be numerical or geometrical—indeed, it could be historical, or even mythic, and projected according to many horizons of meaning, many ontological regions. Yet, all such models would be algebraic in structure. There is some x (our model) such that F (the real Fenway) is x under some projection α. Narratives (whether they be stories or instructions, or programs, etc.) are hardly exempt from "structure" as we explained earlier. Here we will work with numerical and geometrical structures, but we make that choice for the sake of simplicity. Genetic and coordinate analysis can be done without numbers and without geometrical space. One assumes there would be *some* sort of extensive relatedness and some sort of unit, but that is sufficient for analysis.

Since we are modeling a three-dimensional, physical structure, it is not enough for genetic specification to specify measurements in three spatial dimensions and project them on a different scale. We also need a procedure for classifying the actual measurements into their types—to distinguish x, y, and z axes for height, depth, and width, and group the measurements in ways that will hold constant their relative positions through the course of the analysis, through any series of transformations. With these rules and entities, we have only begun our task. We need not choose an infinitely precise unit of measure, nor do we need an absolutely distinct set of axes. We need enough specification to keep separate what will be separate subsequently, and that is determined by what we plan to *do* with our analysis, our *purpose* in inquiring.

If we intend to reproduce a small model of Fenway Park for a shelf at home, in which much detail will be lost, macro-measurements rounded in feet or meters will be adequate. The function is decorative or symbolic and needs to capture only the sorts of stadium features that enable a person to recognize it as a *model* of Fenway Park. The purpose of the inquiry in this case is determined at a high level of generality because the intended function is easily achieved. That is, Fenway Park is easy to recognize even when quite distorted. The basic geometrical relations may be inexactly reproduced, inexactly specified, and, with the addition of one or two identifying features (e.g., the left field wall and the famous "angle"), the model can be created, and in various materials, for reasons quite independent of any analogy they bear to the original. We can make a model of gold to impress people, or of some material that takes paint for the sake of representational correspondence, and so on. Our coordination and genetic specification can be very sloppy in this instance. It is the aim or purpose that determines the level of precision in the analysis itself. In this case, the *purpose* allows for great inexactitude.

But if our aim is to build a model of Fenway Park that will be usable by baseball-playing youth, we need much greater specificity in measuring. How can one insure the discovery of the proper scale for *that* purpose, and choose the right unit of measure and the right *level* of specificity? One needs to go to *experience*. We will have to experiment with units and adjust them for functionality before we could fairly characterize the coordinate whole. This can be done in many different ways. We can, for example, compare the height of the average 12-year old youth with that of the average major league baseball player, create a scale of ratios, according to *some* method of averaging, and treat the resulting ratio as the base unit, call it a "Feneter," and subdivide or multiply those Feneters as needed for greater specificity. We can tweak the parameters as needed. By this method we would get a stadium that sort of "looks right" when we see the smaller players running around in it. For greater precision (not necessarily greater applicability or adequacy), we can modify the Feneter to include height-weight proportions of 12-year olds and major leaguers to get a more refined unit. But it seems to us that such refinement is unlikely to make the model more functional, and it wouldn't help us "see" the stadium as being any more *like* the original—if anything, it might introduce distortions in the visual aspect ratios of the model because youngsters weigh less in proportion to their heights than do adults, especially professional athletes. So we should reject this proposed refinement of our Feneter.

The point is to recognize: (1) refinement of any unit is available along many lines; and (2) not all refinements will be relevant to the given aims and purposes of an individual inquiry. To overlook a refinement that would be of clear value in achieving one's given aim will lead to a less than optimal model (such as Einstein's General Relativity, as we explained). In this case we have suggested that a relative scale be developed based on average height, and we need to define what we mean by "12 years old," what we mean by "height" (e.g., with or without shoes, in centimeters or inches), and most importantly, what we mean by "average."

But is there any reason to suppose that the simple size (whether average height, or combined height and weight, etc.) will enable us to create a unit (of measure) that allows us to construct a model that *performs* for small players in just the way the full sized-stadium *performs* for major league players? Actually, there *should* be some relation between the two situations, but the method of using height might not work well because we need greater precision than visual congruence to achieve functional modeling. In fact, we may need to *sacrifice* some visual congruence for the sake of function. The Feneter just isn't enough. A purely model-centric view could easily disguise this flaw. Finding a single unit, such as the Feneter, and insisting that *whatever* happens in the park thus constructed *must be* the true and accurate game itself is the sort of reasoning we find in model-centrism. Einstein's cosmological model may picture a cosmic Fenway, but no one can play ball there, not even God, and we ought not adjust our ideas about what really happens to fit this block universe of curvilinear space-time. Experience is the measure, not the model. Now we are beginning to grasp where a full genetic and coordinate analysis will be helpful.

The meta-methodology of genetic and coordinate analysis might not even be the best way to treat the problem, but working under the "theory" (in Whitehead's sense, see chapter 1) that *it is possible to create something that functions like Fenway Park, but on a smaller scale for youngsters*, we have not yet *begun* to explore the riches available to us from genetic and coordinate analysis. So far, in the Feneter, we have only an arbitrary set of suggestions, drawn haphazardly from an abduction that the height and/or weight of players may have *some relation* to the way the game unfolds dynamically in the original stadium. Surely it does. But alone, height, or any purely visual unit, is too crude to guide genetic specification or express coordination (the existence and intelligibility of the real Fenway Park, functioning as the "whole" for this inquiry). We must ask ourselves what we *want* from our completed model. Only then will we know what the functional "parts" might be, and how they might be genetically specified.

When we said it *should* be possible to create Fenway Park on a smaller scale, we meant that there probably exists a complex set of relations such that, *if* we get the right determinations, *then* we can put these determinations through a series of transformations that preserves the functional features that serve *our current* purposes. We must pay attention to the adequacy, logical rigor, and applicability of all our reasoning about the problem. We must seek out just those determinations and relations that allow our reasoning to follow the relevant trends of transformation that will, finally, provide us with parts that make up the coordinated model we sought, and also to discover our errors. The model cannot be allowed to trump our experience, both as the experience is *had* and *aimed at*. Whitehead is, after all, unapologetically teleological, even if he rejects fixed ends as resolutely as any pragmatist.

We now suggest that the example at hand requires a more complex process of genetic specification to achieve our coordinated model. We think we need to take account not only of the

1. height (and maybe weight) ratio of the players, but also
2. of what materials the youngsters' bats and balls will be made, and how their materials, sizes, and weights relate to the materials, sizes, and weights of the major league analogues;
3. the exact materials of which the stadium is built (including the type of grass and its average height);
4. the measurements of the stadium in height, depth, and width according to a degree of absolute measurement in millimeters. By "absolute measurement," we only mean that the units (millimeters, etc.), have been imported from other problems, not developed for this problem alone, with conventions not adapted in advance to this inquiry. We suppose that the tiny distortions that results from using a metric not developed from the special character of our problem will not have any

important effect on the creation of the model—a fallible supposition, but we choose the millimeter instead of the inch because we think the metric system is functional in a way that the older English system is cumbersome; we choose the millimeter rather than the centimeter because, regarding this problem, specifying an external or conventional metric only down to the centimeter might actually create functional problems with the model.

These units are not "eternal objects" in Whitehead's sense because our problem is a concrete problem, and hence the units are fixed not metaphysically but in relation to a particular inquiry. This is not metaphysics, but a physical analogy. Metaphysics is harder. Thus, let us call this genetic group of four categories the "static" characters. But to these, for a more complete genetic specification, we must also add the "dynamic" characters:

1. how fast each group of athletes (professionals and youths) runs, on average (this affects distance between the bases);
2. how far they hit the ball, and how often (watching closely for trends of variation—for example, major league players may hit the ball a great distance more *often* than youngsters, so it is not enough to set up a scale of relative distances; the proper procedure involves mixing a static with a dynamic measurement, and major leaguers may, on average, get more or fewer hits per times at bat than youngsters. In fact there really is a discrepancy here, so we must adjust the units).

We are very far from the Feneter now. These two genetic groups, the static and dynamic, provide the basic relations we need to know in order to create a model of Fenway that functions roughly as we wished. Some inquirers might want to call this the search for the "true Feneter," and Grand Unified Feneter Theory, but such people need to become cosmic Cubs fans. *There is no true Feneter*, any more than there is a single mythic or cosmic reason the Cubs can't win. Cosmic Cubs fans keep hoping for a String Theory narrative about their problem, but some problems simply cannot be solved by a finite inquiry.[11] Good inquirers know this (as indeed do real Cubs fans). We do not need to become Fenway Fundamentalists or Wrigley Writhers when all that is needed is a place to play ball—which is as fair an analogy to honest inquiry as we can imagine.

In the final phase of inquiry, attempting to *create* the model (in a completed concept) of the coordinate whole with which we began, will integrate these static and dynamic characters; this integration requires some pretty fancy algebra, and some differential geometry, and some calculus to turn the dynamic measures into some units that can be geometrized, since we have assumed that every dynamic variation implies some *geometric* adjustment, possibly even large adjustments, because ultimately every variable dictates that we adjust the *physical* structure of the stadium (even where physical adjustments are only aimed at achieving dynamic ends, such as preventing too many or allowing too few homeruns by the youngsters). We must remember how many empirical measurements are required to determine the "average" player, and average the speeds, frequencies of hitting relative distances, and so forth. A serious mistake anywhere could spoil our results. Also, if there is some empirical factor of crucial importance we have missed, it may spoil our applicability, regardless of our logical rigor. Or, perhaps we may omit something crucial but the omission ends up cancelling on either side of the equation in just this one instance, and we succeed in creating our model only by fiat, and the results are unrepeatable. This is just how inquiry goes. Ask any Cubs fan. After all, they *could* win.

We need not measure everything we can think of. We can *reason* about our inquiry *as* we inquire. For example, we do *not* think we need to take into account how hard the young pitchers throw the ball in comparison to the pitching velocities of their major league counter-parts. Why? Because the hitting distance calculations will already contain the most useful information that might be gleaned from measuring the pitching speeds. We also do *not* think we need to measure climatic variations between the average air temperature of Boston and the temperature *wherever* we build the model—unless the model is to be built in a place with wildly different weather. Temperature affects play, certainly, and the ball flies further on cold days. But the differences here would actually be minimized since youngsters play baseball on a smaller scale. The smaller scale model reduces the effect of air temperature, which is barely enough to affect the major league game (although this is sometimes debated).

We might be ready to build our model. But there is crippling news:

> The orientation of the model relative to (1) the prevailing wind patterns and (2) the position of the sun are extremely important in our creation of the model.

And here we begin to understand the limitations of genetic specification, even when it takes its direction from a prior coordinate whole. Let us examine these two factors. We can measure the orientation of Fenway Park to the sun very precisely, but we can only recreate that pattern with fair exactness along the same parallel of latitude as Boston when we build the model. That is a very great limitation. Apart from that parallel, we will have to settle for an approximation. The further away from the Boston latitude we place the model, the more we will have to accept a variance in orientation that will certainly affect the way day-time games unfold. After all, what times and which fielders battle the sun matters greatly. Many baseball leagues (e.g., American Legion, Babe Ruth, etc.) even regulate the orientation of the field so that no unfair advantage accrues to one team or another in a home stadium.

We can probably counter these effects to a limited extent by playing the games at different times of day than they would be played in Boston, or better yet, by playing all games at night, with lights situated in precisely analogous positions with the ones at Fenway, but at the *same* degree of brightness (not an analogous degree, scaled down). Here we discover something that must be designated as *invariant* in our genetic specifications: the amount of light needed for night games. Young eyes may see more keenly than old eyes, but not among baseball players, all of whom must have excellent vision.[12] Always in inquiry, empirical factors arise that need to be treated as invariant, and sometimes those factors are surprising, as in this instance—the light poles could be shorter in the model, but the candlepower needs to be close to the same. (There could be some increased illumination from closer grouping of the poles, but the lighting *effect*, to remain functional, must be equivalent, not scaled.) We do not say the candlepower factor actually *is* invariant, only that it must be treated that way in this inquiry. We do not think that Whitehead would allow that we can make sense of the idea of invariance *an sich*. Invariance is relative to a coordinate whole (see chapter 5).[13]

We can probably overcome the challenge posed by orientation to the sun, doing so within the acceptable bounds of our original purpose in inquiry. But we are up against factors that no unit of measure can quite capture. It is easy to know very precisely the variation between the original and the intended model, but the variation cannot be scaled and converted into a unit that dictates an exact structural adjustment in the model. We simply turn the entire model to the best approximation of the sun's position and adjust more by starting the game earlier if we are south of Boston, or later if we are north of it.

These two sorts of calculations, amount of light required for night games and orientation to the sun, do not then affect the development of our "master unit" of measure that governs all transformations (all those that can be captured in a physical alteration of the stadium itself). Instead, genetic specification and reasoning have shown us some limits of our modeling process, and adjusting the unit of measure will not assuage the difficulty. These are non-metrical factors in the model, and analogously, creating a metaphysical model of our cosmic epoch, in words (i.e., a system of philosophy), in a finite set of concepts, is not a wholly metrical, i.e., scaled, representational process, of achieving correspondence. Some relevant factors will be irreducibly metaphorical, but there are better and worse metaphors, and metaphors can be improved upon:

> Progress in truth—truth of science and truth of religion—is mainly a progress in the framing of concepts, in discarding artificial abstractions and partial metaphors, and in evolving notions which strike more deeply into the root of reality.[14]

We learn here that genetic specification encounters obstacles, and some of them can be addressed with straightforward reasoning (e.g., pitching velocity is accounted for in the metric of various hitting measurements), and some characters have to be treated as constants (light needed for night games), and some factors mutually determine others (frequency of night games and game times need to vary with distance from the latitude of Boston), while some may seem to work at cross-purposes to others.

An example of the latter might be the very thing we thought was "straightforward." In fact, little league baseball fields are *not* proportional to major league fields, especially regarding the pitcher's "mound" (it's really just a rectangular piece of rubber flat on the ground, in Little League ball). The distance and height of the pitcher's mound to home plate really *cannot* be adjusted so that the velocity at which youngsters throw the ball results in a proportionate frequency and distance of hits by youngsters.[15] Kids hit the ball less frequently than major league players, since, hitting the ball is more difficult to learn than is pitching. Thus, young pitchers with good fastballs dominate among 12-year-olds.

You cannot build a model of Fenway Park with an elevated pitching mound at a proportionate location and expect the functional analogy to hold. Varying the pitcher's mound by setting it further back and not elevating it will not wholly solve the problem. Pitchers will still dominate Little League baseball. But treating the mound as an independent variable (meaning that this genetic specification is neither dynamic nor static, having its own independent metric developed from actual use of the model and not from the original Fenway) can minimize the problem—we have to vary the pitching rubber and elevation experimentally and not dogmatically (and model-centrically) to keep its geometrical proportions constant with the rest of the model. This is how Little League regulators found a distance of 46 feet to be right for 60-foot bases. Perhaps if they moved the bases to 66 feet, they could get a closer set of ratios. It's an empirical problem, but it varies with the coordinate model in relation to its genetic specifications. If we move the bases to 66 feet, we must adjust all variables (invariant factors would remain the same). There is probably some improvement available to the Little Leagues. It might bring a revolution in the youth game analogous to that created by Bill James's method of keeping statistics.[16]

By the time the players reach 18, the pitchers lose their advantage, but they will be playing at some park by then. Thus, we have learned some creative ways to deal with genetic specification in light of the prior coordination (the real Fenway) and the model we aim to create for our stated purposes. Metaphysics works almost the same way. You might think of Descartes' and Leibniz's use of algebraic relations in philosophical method as early attempts at Whitehead's genetic and coordinate analysis of our cosmic epoch. However, the "whole" with which the Modern philosophers began would not be designated as a "cosmic epoch." That adjustment is pretty telling, as are the other distinctions introduced in Whitehead's cosmology that strike "more deeply into the root of reality," due in no small degree to the progress in physical science and mathematical methods since the Modern period.

But perhaps some problems in recreating Fenway cannot be solved, even with all the tools currently available. An example, in our analogy, is the prevailing wind. If we set our scale model just anywhere, we risk having the actual games proceed nothing like the games in the original scale—because the wind has a huge effect on baseball games at any scale, turning foul balls fair, and home runs into put-outs, and vice-versa. We must know the wind patterns at Fenway, and we have to have a comparable average wind, in velocity and direction, in building our model. So we must add that factor to our list of required genetic specifications above—and this is a dynamic specification.

But what if (1) there is no place with an acceptably close prevailing wind pattern within the region we want to create our model? Or what if (2) our capacity to measure average winds is too crude? It may cease to be a model of Fenway at all in many respects. We may need simply to understand that the wind bloweth where it listeth, and leave the matter at that. Genetic specification here meets a fundamental limitation in our analogy. The coordination of all our wonderful specifications has proceeded without a solution to the problem of the wind. Thus, having begun with the coordinate whole we chose, we also acquired limitations, perhaps unconsciously, and the discovery of the implicit limitations have to do with its relation both to what is actual and what is possible. This is an objective limitation to which our whole inquiry must conform. Everything that is actual is certainly possible, and we chose an actual whole, but no coordinate whole exhausts what is genuinely possible, let alone what is abstractly possible for an inquiry. Failing to note this characteristic limitation of inquiry is the principal failure of model-centric thinking (see chs. 6 and 10). To ignore this characteristic of models (i.e., coordinate wholes and their structural limitations) is to risk failing in our purpose, and perhaps without knowing it.

In this analogy to Whitehead, trying to do metaphysics *in natural language* is like dealing with the wind at Fenway, unfortunately, which is to say that we are always coping with at least one very important aspect of our analysis that, as far as we have yet learned, cannot be expected to conform suitably to the demands that our goals have placed upon our chosen task—that of modeling the order of the universe (whether moral, or natural, or aesthetic order, or some other); more precisely, modeling the order of *experience*, that Fenway-of-the-living, in language. But in genetic analysis, the difficulty is creating a set of categories (e.g., entities, obligations, explanations) which can be *used* to specify along multiple axes (some variant, some static, some dynamic, some invariant, some independently variable), exactly the entities, processes, and relations that will enable us to say in language the presupposed (and coordinate) whole of what is experienced.

The analogy also shows how genetic specification is undertaken for the sake of clarifying a model and making it serve a purpose, although the coordination is presupposed in the process of specification. If we keep firmly in our minds the requirements we will later need in order to *model* all of these genetic specifications (of the coordinate whole), and if we also manage to keep in our awareness *how* we will carry out our modeling, we then know something about how to specify them genetically. Failure might leave us with a bunch of genetic specifications that are incommensurable—perhaps even with no clue about our mistake(s). We avoid this result by appealing to a common ground for each inquiry. In *Process and Reality* this is the idea of how time gives rise to spaces of various kinds.

Do our categories above, Feneters, independent variables, invariants, etc., actually contain the needed information to recreate Fenway Park on a smaller scale? We don't know, but we probably got some of it right. We could carry out a lot of the measurements and decide intelligently how to average them. Our principle that every significant dynamic and static ratio should be concretized as an adjustment in the structural (in the sense of spatial) characteristics of the park provides us with a clear idea about the common metric that will be used to coordinate all axes. The Feneter will *have to use* a three-dimensional geometry such as architects commonly use. After all, it is an architectural analogy. The math to carry out this kind of coordination already exists, as do the computer programs that can adjust a huge number of variables in relation to each other. Still, this problem of recreating for 12-year-olds the dynamics of the major league game has surely never been programmed precisely enough to build what we imagined. We will need a good mathematician to reduce these numbers to geometrical proportions, converting Feneters into meters and millimeters. We need a structural engineer to apply the resulting geometry to the problems of constructing the stadium, and a builder who can serialize the engineer's plans. And we need a clever accountant and some investors who love the Boston Red Sox enough to pay for it all.[17] And we suppose children who want to play baseball are easy enough to find, even some willing to wear a Yankees uniform, and to be paid better (for one needs both good *and evil* to generate the drama, if the drama is to look like our experience of either baseball or the universe). Creating a metaphysical system requires all the analogous components, but it is not nearly so easy.

Of Hasenpfeffer and Toenails

But consider this: There just *might* be a single unit, for all we know, very simply turned into a metric unit, that would do *all that work*. For example, it is *possible* that starting with the ratio of the weight of the bats used by each group (kids and professionals), we could calculate everything we need about how to vary the model we are creating. Maybe there is, then, a "master principle," if only we could discover it. This is the dream of so many philosophers and scientists. But it is fetishism, we think. The twentieth-century philosophers fetishized language as though we might solve all philosophical problems by analyzing the way we talk about them. But it is superstitious to suppose this approach could work. Our language about the world has only the loosest hold on the world, which is *why* we value it—its versatility and adaptability. The cost of the promiscuity of language is that it makes a poor tool for understanding our concrete experience with precision. The idea that refining our understanding of language will necessarily make us better at philosophy is akin to believing that paring our toenails will help us run faster. If you aren't careful, you can spend your life trying to find the perfect length for your toenails and never join the race at all. There surely is a relation between our language and how well we philosophize, just as there is a relation between toenail length and foot-speed, but the relation is loose.

If such a trick worked, if we apparently got faster by trimming our nails just so, or if we got better at philosophy simply by analyzing language, it would be magic and we would not learn anything from our luck. The discovery of the perfect unit would be an accident: luck is not a method and hope is not a plan. We don't know how to get lucky in metaphysics. Whiteheadian philosophers and radical empiricists do not fetishize language and do not search for a magic master principle that will save us from all that work of genetic and coordinate analysis. We trim our toenails with modesty, study our logic, read our poetry, and proceed with the real business of philosophizing, which is slow and difficult. It requires disciplined thought, not language tricks and duck-rabbits.

Someone (probably with an expensive pedicure) may say that Fenway Park cannot be recreated using the method of genetic and coordinate analysis— in short, our original "theory" may be disputed. We disagree. We think it can be done, adequately, rigorously, and applicably. But in building scale models, the proof would be in the execution. In metaphysics, we don't get off so easily. That someone (i.e., Whitehead) has created a fantastic genetic specification and coordination of a problem as tough as *the order of the universe-as-experienced* is difficult to demonstrate to a resolute trimmer of toenails. And how good is this system of Whitehead's? It does contain distortions. It is not the perfect philosophy, as he well knew. Remaining radical in our empiricism is our best hope of discovering our mistakes, and of avoiding mistakes to begin with. But we certainly will err. Error is the engine of progress according Whitehead, and there is error in every symbolic reference. We don't know how to avoid all error, but we do know how Whitehead uses genetic and coordinate analysis in his inquiries. Let us err then, and move forward.

We mentioned that the method of genetic and coordinate analysis is indifferent to whether the entities, processes, and relations under discussion can be *actually* divided in the ways required for the analysis. Genetic and coordinate analysis provides us with ways of *thinking* about our problem, without telling us anything final about what we can or cannot really *do*. In the analogy, the truth is that Fenway Park is a fact of the universe and nothing else can be just like it.[18] That does not mean that *modeling* reality in genetic and coordinate analysis is anti-realist. It is emphatically, radically realistic.

This extended analogy of Fenway Park is based on the idea of scalar modeling. Scaling is only one application of modeling and we want to make it very clear that genetic and coordinate analysis is not limited to such applications, and indeed, Whitehead's cosmology is not an instance of scaling at all. The kind of modeling Whitehead does in *Process and Reality* is mereotopological, and that is Whitehead's approach whenever he is dealing with the problem of space. But the other problem that fascinated Whitehead throughout his career, the problem of "the accretion of value," as we call it, is a dynamic factor in the theory of space that cannot be wholly harnessed.

Thus, Whitehead's books that take the accretion of value as the coordinate whole, books such as *Science and the Modern World*, *Religion in the Making*, and *Adventures of Ideas*, do not attempt to *solve* the problem of space. Rather, they *assume* that presentational spaces capture the growth and decline of forms of valuing, while leaving aside the general problem of space. Genetic and coordinate analysis is just as useful in one sort of inquiry as the other, and *Process and Reality* comes as close as any of Whitehead's works to accommodating the dynamics of the accretion of value within the coordinate whole presupposed by the problem of space (and answered with the theory of extension). In the idea of the actual entity, as employed in *Process and Reality*, Whitehead has, at the same time, a temporal divider of the extensive continuum *and* an accretion of value. The question is whether a coherent whole can be specified to which an actual entity as a divider and an accretion of value, can belong and whether this whole realistically *models* both our experience and the physical universe as our science reveals it. Both are requirements of any serious realism. Philosophies mis-labeled "scientific realism" ignore the accretion of value are nowhere close to being genuine realisms. They are atemporal pipedreams. Hilary Putnam did much in his final years to unseat the pretension of such realisms, but even his message is not realistic enough to satisfy us.

Long May You Run

We have not yet learned all we can learn from the activity of constructing models. Scalar modeling is an exercise in projective geometry[19] and linear algebra, combined, and with the algebra limited to the greatest extent possible by the aim of converting dynamic problems into geometrical relations in three dimensions. Similarly, we may say that Whitehead's axiomatizing of his theory of extension is an algebra developed under the limitation of the general theory of value is an effort to *model* experience.

Another example of scalar modeling, having almost the same features as the Fenway analogy, but with revealing differences, comes from Neil Young, the Canadian singer/songwriter. Young is a serious model train enthusiast and he became a co-owner of the Lionel model train company when it fell into financial trouble. Being a sound engineer (as many rock musicians are),

Young began to develop a sound system that would make the trains sound like the trains they were modeling, and this became part of what was called the LEGACY Command and Control system, the development of which Young financed and helped to design. Having taken his project that far, Young writes:

> The next thing model trains need to do is abandon modeling sounds by user input [apparently, the controller presses a button to provide the sound] *and become real*. The effort involved in pulling a load needs to be measured, the algorithms that used to be based on user input need to be newly based on the locomotive's effort measured to pull the load or perform the task. Then there is little to do but drive the train down the track, allowing it to measure its own efforts and trigger sound and smoke effects and speed changes to reflect the laboring that is being measured. That is the next step, the future of modeling, or at least part of the future. Everything is there now in the Lionel system to make this happen on a basic level, except a good measurement of the effort being put out by the locomotive model to pull its load. Not just some gross measurement like measuring the electric motor effort, but an electromechanical high-resolution capturing of each nuance of the laboring. That will be nirvana and I will be celebrating my ass off when it happens![20]

Note how Young presupposes that the geometric projections have all been done in such a way that the "laboring" of the model already captures the laboring of actual engines (which is almost certainly false), and that he has not considered whether, say, the materials from which model tracks are made offer the same resistance to the wheels of the models as steel on steel, gauged as it is, operates in the original. It is not clear that making the new model from the *same* materials as the original trains would provide the same resistance ratio—gravity doesn't work that way, and Neil *needs* the cooperation of gravity here.

Furthermore, it really may not be possible to recreate that resistance ratio without some serious experimentation. Young apparently hasn't realized that the *unit* of measure he is seeking is a resistance ratio that varies proportionately with the variations in the full-sized train. For his purposes, it makes no difference at all about the load except as it affects resistance between track and wheel. A little engine pulling 20 pounds on a toy track does not stand to gravity in a relation analogous to a 200-ton locomotive pulling hundreds of thousands of tons of freight. The ratio of loads, effort, and grades cannot be reproduced (due to the gravitational difference in mass) at the smaller scale, but what *can* be varied to capture those dynamics is the resistance offered by the tracks in various places. A tiny adjustment in the width of the rails themselves increases the amount of contact between wheels and rails and would slow down the train proportional to its load. Using the broader rails where there is a grade to climb can easily become a bit of information a sensor on the train interprets to adjust sound effects. Neil, your nirvana is within your reach, but you haven't found your unit of measure, your quantum of explanation.

So Young is on the right track, so to speak. But for our purposes, he is struggling with adapting dynamic (effort), relative invariants (gravity), and independent variables (grade, relative load) to a projective geometry that was established experimentally over many decades, as modelers all tried to make the trains more "real" in appearance, without sacrificing too much function. Now Neil wants to bring the function, the algebraic relations, up to the standards set by those fine-tuned geometric projections. He wants it to be "real." Later, he continues:

> There are so many ways to model the actions and sounds of a machine like a locomotive, it is endless—and the complexity involved is like a drug [this is something Neil Young knows about]. For instance, every action has a sound and every sound has variables. Every sound variable needs an algorithm based on an action, and every action needs a variable control mechanism and a sensor to monitor its position or at least predict its position, possibly based on the positions of other related moving parts of the machine's systems. To me, this is a stimulant [this is what Whitehead called "reason," which enlivens the body]. I am fascinated by it, by all the possibilities. Every sound needs to be recorded in such a way that it is variable by an algorithm based on the mechanical action or by the controller. You can see how I get hung up.[21]

Neil Young has discovered, all on his own or with enhanced consciousness, the problem of coordinating dynamic space, and he has seen how a solution has to be organized. But note that he is taking his point of departure from *mechanical* movement. The fact that all of this behavior in the train could easily just be programmed wouldn't appeal to Neil because he really wants to *drive* the train. It would be very, very easy to develop a software application that recreated on a given model track the dynamics of a train negotiating that set of obstacles. Making that software interactive would be a more complicated task, but it could be done. Yet, Neil wouldn't want it—any more than playing a video game of baseball with variables modeled precisely on Fenway Park would satisfy our analogy above. It is clear that Neil, along with the whole human race, wants something *real*, gritty and physical, something electromechanical, not a virtual recreation of the *appearance* of those effects. Neil wants the effects *themselves*. Otherwise, it isn't "real."

In many ways, Neil Young's brand of realism is a lot like Whitehead's. The cosmological model created by the narrativizing of a genetic and coordinate analysis must capture, in its proportions, the cosmic epoch, its structure and its story, and it needs to create in us a model of an experience of that epoch. That is what cosmology does. As Neil says "It's cosmic, dude."[22]

The analyses aim at a macro level description of our situation, a whole, and at developing a unit, analogous to a metric, and a procedure for generalizing and parsing the unit, all of which results in a model *of* the whole. The whole is presupposed as a whole with a purpose for us, and for our inquiry, and thus limits the modeling. In *Process and Reality*, the whole is the extensive continuum and the unit is the actual entity. The actual procedure is division, and the extensive continuum's patience before the activity of division (called "decision" by Whitehead when it is being realized).

The concept of the actual entity is not the stipulation that makes cosmological analysis possible, and it is not a mere tool of analysis. The concept of the actual entity is the *unit* Whitehead suggests for cosmology. This relation was illustrated above with the Feneter, or with the Youngian loco-metric Nirvana, but in metaphysics it is the quantum of explanation. Its definition and limitations are what make it function analogously to a unit of measure (it doesn't measure anything numerically, of course, it is more of a logical unit; but to divide with a unit is to differentiate part from whole). Some of these aspects of the concept of the actual entity, e.g., that it is actually only one place actually *and* potentially must be everywhere, provide structural limits upon how we *use it* for modeling. Some aspects, such as its phases of concrescence show us how it may be used in genetic specification.

From Extension to Space (Space . . . Space . . . Space)

We argued above that Whitehead's concept of "extension" is an irreducible aspect of his metaphysical system, and this concept is the very source of organicity in the philosophy of organism. Whitehead says:

> Extension, apart from its spatialization and temporalization [in the present cosmic epoch], is that general scheme of relationships providing the capacity that many objects can be welded into the real unity of one experience. Thus, an act of experience has an objective scheme of extensive order by reason of the double fact that its own perspective standpoint has extensive content, and that the other actual entities are objectified with the retention of their extensive relationships. *These extensive relationships are more fundamental than their more special spatial and temporal relationships. Extension is the most general scheme of real potentiality, providing the background for all other organic relations*. The potential scheme does not determine its own atomization by actual entities. It is divisible; but its real division by actual entities *depends upon* more particular characteristics of the actual entities constituting the antecedent environment.[23]

The "more particular characteristics" to which Whitehead refers are, most importantly, the temporal and spatial standpoint of a given actual entity.[24] It is easy to lose sight of the repeated claim that time and spaces (as experienced) are *particular* to the present cosmic epoch (by hypothesis), and that they supply only *one* way in which a divisible but undivided extensive continuum *can be* divided. There is no necessity—either logical or metaphysical—that the extensive continuum be divided according to the determinate orders of time and space, but once this type of order *is actual*, further conditions are imposed upon all other actual divisions within their scope (i.e., the present cosmic epoch). If the first division of the extensive continuum is temporal, then all spaces will be under the limits imposed by temporalization.

This point establishes the priority and extreme generality of extension for this inquiry, and almost by itself refutes the claim that Whitehead's mature metaphysics gives priority to what Ford calls Whitehead's "discovery" of "temporal atomicity," as we have discussed earlier in this book.[25] But an actual entity simply *is* a contingent temporal division of the extensive continuum in our own cosmic epoch. The actual entity is a *unit* (a quantum) that Whitehead has devised for bringing genetic and coordinate analysis together in categoreal and narrative language. All temporality, so far as it concerns the metaphysics of the present cosmic epoch, is a contingent specification of a maximally general extensive continuum. This is a position Whitehead articulated as early as 1898, restated in explicit terms in the triptych, and never abandoned. The idea of an extensive continuum as a coordinate concept of the whole, while variously described in different inquiries, is a constant feature of Whitehead's philosophy.

In *Process and Reality*, time is a contingent specification of extension, and extension is not "space"; extension is undivided divisibility (this is a functional definition, not a localizable existential assertion)—which is to say that extension is logical, not ontological. What other modes of division, apart from time, and *then* space, *might* be brought to bear on the extensive continuum? This is a question we are not in a position to answer, at least in the domain of cosmology—we have no experience of any non-temporal mode of genetic division beyond our cosmic epoch. But that lack of experience on our part is precisely the *reason* that time and space should be treated as contingent rather than necessary modes of division. For a radical empiricist such as Whitehead, where evidence is lacking, the very *last* thing we should do is declare the evidence we *have* to be universal, utterly exceptionless, or "ontologically necessary." That would indulge the ultimate *argumentum ad ignorantium* (see chapter 10). However, the limits of experience are not identical with the limits of imagination. Hence, we find Whitehead carefully offering conjectures about what *might* hold of other cosmic epochs. This process of offering conjectures is familiar to mathematicians, and it does carry a certain weight of expectation, but its principal function is to remind us that the present inquiry exists in a context wider than we can encompass with our hypotheses. Are there other cosmic epochs than our own? What right have we to suppose there are *not*?

Knowledge comes by contrast, and lack of contrast implies lack of knowledge. To insist upon the necessity of whatever we have or have not yet learned to see in a contrast indulges the dogmatism Whitehead criticizes in the Introduction to *Process and Reality*. The *absence* from our experience of other equally primary modes of dividing the extensive continuum is not evidence for the *necessity* of time and then space as its dividers (that is, ultimately, actual entities); it is rather the best *reason* to treat them as contingent, and to hold open the possibility of other (as yet unknown) modalities of division. It is not an accident that Whitehead moves directly from the above passage to a long citation from William James, his favorite radical empiricist.

The issue to be treated presently is not the irreducible generality of the undivided-but-divisible extensive continuum, but the sense in which time and then space, as the modes of division in *our* cosmic epoch, impose *further* conditions upon the metaphysical description of the present as experienced. The extensive continuum does not determine the structure of time and then space; rather, time and space, that is, actual occasions as described in the theory of transition and concrescence, divide the continuum in two successive modalities:[26] first, temporal determinations, yielding causal efficacy in perception, and then the spatial division of past from future, which is the space of presentation (or presentational space), and the only conditions under which "contemporaries" can be perceived from any single perspective (there are no true contemporaries; rather, contemporaneity is the illusion created by presentational space). Hence time and then space, i.e., actual entities in transition and concrescence, divide the extensive continuum and thereby constitute all the relations among the entities they condition in our cosmic epoch.[27] That is *why* the actual entity functions as a quantum of explanation.

Yet, time and space divide the extensive continuum in two different ways, genetically and coordinately. It is easy to become confused about what divisibility (that which is susceptible to division) and actual division are. Consider again this passage from *Process and Reality*:

> (P)hysical time or physical space . . . are notions which presuppose the more general relationship of extension. . . . The extensiveness of space is really the spatialization of extension; and the extensiveness of time is really the temporalization of extension. Physical time expresses the reflection of genetic divisibility into coordinate divisibility.[28]

In what follows, we will treat in some detail the senses in which genetic divisibility provides a basis for coordinate divisibility. Divisibility is not division itself, but a susceptibility of *possible* division. Note that in our long analogy of Fenway Park, no stadiums were actually divided. It wasn't necessary for our purposes to divide anything, although it was very important that the stadium be divisible in various ways, in multiple modalities, and under discoverable limitations. Similarly, Neil Young's loco-metric Nirvana requires divisibility in thought, not actual division of existing trains, but these distinctions are in no way a sacrifice of cosmological realism.

A Prehensile Tale

Divisibility is a mode of *determinate* order (the order of constellations of possibilities in their mutual determinations[29]) which can also be made *definite* in innumerable ways. Genetic divisibility in the present cosmic epoch is the "theory of prehension," for Whitehead; when we consider prehension we have arrived at the sense in which all things in the cosmic epoch are concretely selected and related, whether negatively or positively, totally and without remainder. It is the *relation of relations*, crossing all actual and possible divisions. Nothing is left out of an account of prehension, no actuality, no possibility. That kind of completeness cannot be expected where *actual* division is the basis of the whole analysis, as would be operative in any "atomistic" philosophy.

But prehensions are not the unit of measure for the universe, and the reason they aren't is because they are not actual dividers of the extensive continuum. A unit of measure has to correspond to a concrete act of measuring, not just relating (which is too embedded in experience to correspond to a single act). To recreate a model of Fenway Park, we have to be able to get at Fenway Park actually. A set of blueprints might be substituted, but then we would be making a model of a model, not of Fenway. In the same way, actual division is the task of the actual entity, which is a prehending subject-superject, a contextual actor. It leaves un-done (not enacted) what it does not actually divide, and what is un-done is eliminated from its actual world. Prehensions, on the other hand, as a collection (which Whitehead infelicitously calls a "group"[30]), include all the possible modes of division and the genetic structure of division itself. There are infinitely many ways of prehending, although most are not productive of greater intensity. A theory of prehensions models the whole and makes possible the coordination of every actual division.

Prehension itself, as a totality of relations, is a substitute for the common sense notion of "space," as we would normally understand it in thinking about cosmology: a total scheme of possible relations of everything that exists to everything else that exists simultaneously, in every modality, with nothing left out. The totality of prehension then becomes a *concrete* model of the entire extensive continuum. It has the full relational character of the original, as ordered according to the purpose for which we created the model (in this case to use the basic ideas of Modern philosophy to build an adequate, logically necessary, and applicable model of our experience *of* the universe). The model retains the appropriate functional structure of the original and its dynamism: it is the little Fenway Park, executed mereotopologically. So, applying the analogy, the original Fenway for *Process and Reality* was the extensive continuum, a characterization of our cosmic epoch for the original problem we posed in the analogy.

Therefore, the theory of prehension can be used to coordinate all of the more abstract levels of actual division. But prehensions do not *actually* divide the extensive continuum; rather, the actual entity does that. It is the master unit of the cosmology, the Feneter, the Youngian loco-metric, the quantum of explanation. In the Fenway analogy, we proposed a master principle, corresponding to Whitehead's ontological principle. We said that every factor, in every modality, should be converted into a geometrical adjustment in the physical model. We then discovered a few things that had to be investigated empirically, and at least one thing (light for night games) that was invariant, and one feature we couldn't make into a geometrical adjustment (the wind), but rather would simply have to seek favorable conditions and hope for the best. Metaphysics has the same sorts of slippage. These limits apply as surely in metaphysics as in baseball or physical science (see chapter 10).

Thus, prehensions are the relational entities we discover when we analyze the actual entity in its actual world, and the theory of prehensions then describes how prehensions, which are relations, "concrete facts of relatedness"[31] demonstrate that each actual entity is a complex unity,[32] even though no actual entity can actually be divided; prehensions virtually divide *and* coordinate the *actual entity in its actual world* (not the extensive continuum). In Whitehead's words (bracketed interpolations are our clarifications):

> The selection of a subordinate prehension from the satisfaction [actual occasion] . . . involves a hypothetical, propositional point of view. The fact is the satisfaction as one. There is some arbitrariness in taking a component from the datum [the actual world for the actual entity] with a component from the subjective form [which is indivisible], and in considering them, on the ground of congruity, as forming a subordinate prehension. The justification is that the genetic process can thereby be analyzed.[33]

The sentence immediately preceding this passage is one of the places where Whitehead uses the phrase "incurable atomicity," and it is undoubtedly one of the places that misled Ford. The passage in context reads:

> The prehensions in disjunction [such as genetic division creates] are abstractions: each of them is a subject viewed in that abstract objectification. . . . There are an indefinite number of prehensions, overlapping, subdividing, and supplementary to each other. The principle, according to which a prehension can be discovered, is to take any component in the objective datum of the satisfaction; in the complex pattern of the subjective form of the satisfaction there will be a component with direct relevance to this element in the datum. Then in the satisfaction, there is a prehension of this component of the objective datum with that component of the total subjective form as its subjective form.
>
> The genetic growth of this prehension can then be traced by considering the transmission of the various elements of the datum from the actual world, and—in the case of eternal objects—their origination in the conceptual prehensions [this shows the dependence of prehensions

on functionally coordinated transition]. There is then a growth of prehensions, with integrations, eliminations, and determination of subjective forms. But the determination of successive phases of subjective forms, whereby the integrations have the characters that they do have, depends on the unity of the subject imposing a mutual sensitivity upon the prehensions [i.e., they are not actually disjoined because the subjective form determines them, which shows the functionally coordinated dependence of prehensions on concrescence]. Thus a prehension, *considered genetically*, can never free itself from the incurable atomicity of the actual entity to which it belongs.[34]

Far from saying that the theory of prehensions has somehow replaced the theory of transition, this passage says that individual prehensions are found or discovered by following a principle: we take a bit from the macroscopic transition of the actual entity (some part of its datum) and a bit from concrescence (some portion of the subjective form), where they seem congruent, that is, we *consider them genetically*, and then, importantly, we repeat the passage from the main text:

> The selection of a subordinate prehension from the satisfaction [indivisible actual entity]—as described above—involves a hypothetical, propositional point of view [because an actual entity cannot really be genetically divided]. The fact is the satisfaction as one [because actual entities actually divide the extensive continuum, but nothing actually divides actual entities; this is the finality or incurable atomicity of the actual entity.]. There is some arbitrariness in taking a component from the datum [the actual world for the actual entity, a part of its transition] with a component from the subjective form [which is what is actually indivisible in the actual entity], and in considering them, on the ground of congruity, as forming a subordinate prehension. The justification is that the genetic process can thereby be analyzed.[35]

Hence, we see that the theory of prehension obliges us to treat as divisible what is actually indivisible. The theory of prehension aims to get at what is most concrete in our cosmic epoch, but discovering a prehension requires us to postulate, for the sake of analysis, as separate, on the basis of "congruence," that which is indivisible. This postulation creates some arbitrariness, but without it we can carry out no analysis.

The analyses made possible by this way of considering prehensions are of two kinds (see the end of the passage). These are genetic and coordinate analyses, which Whitehead calls the two "descriptions" in the Categories of Explanation:

(viii) That two descriptions are required for an actual entity: (a) one which is analytical of its potentiality for 'objectification' in the becoming of other actual entities; and (b) another which is analytical of the process which constitutes its own becoming.

The term 'objectification' refers to the particular mode in which the potentiality of one actual entity is realized in another actual entity.

(ix) That *how* an actual entity *becomes* constitutes *what* that actual entity *is*; so that the two descriptions of an actual entity are not independent. Its 'being' is constituted by its 'becoming.' This is the 'principle of process.'

Genetic and coordinate analyses are explanations in complementary but different senses. The genetic explanation presupposes a coordinated *whatness* or constitution, while coordinate analysis presupposes a procedure of division even if that procedure is virtual rather than actual. Explanation depends equally upon both descriptions.

The actual division of the extensive continuum in our cosmic epoch (time and then space) *conditions* the further division of the actual entity, the entity that is itself a basic spatio-temporal division of that extensive continuum. To be clear, we are asserting that prehensions do not *do* anything, they are entities of possible doings, while actual entities, each in its own actual world, which *do* divide the extensive continuum, are exemplifications of those possibilities of doing. Thus, a *theory* of prehensions is both a genetic specification of the types of things that *can* be done by an actual entity in our own cosmic epoch, and a coordination of what prehensions actually do in relation to what they did not do, but might have done.

The theory of prehensions is, in the first place, a theory of *divisibility*, not of actual division, and only in coordination does the theory of prehensions move from the possible to the actual. Feelings, which are the content of prehensions, are concrete relational beings, constitutive of actual entities, and not the sorts of things that float around in empty space. But the *theory* of prehensions takes aspects of feeling genetically and coordinates them in both inclusion and elimination, as a *model* of the actual entity that can be generalized, without committing fallacies, back to the fullness of the extensive continuum. In coordinating a transitional aspect of feeling with a concrescing aspect of feeling, according to certain criteria (relevance, intensity, width, etc.), the theory of prehensions abstracts them, but the distortions are minimal.

The limits upon effected division in the extensive continuum, as found in the prehending actual entity, further determine and condition coordinate divisibility. This is to say, in simpler language, that the space of the actual world in our own cosmic epoch just *is* the highly mediated space of a prehending actual entity. This idea of space is lively compared to the dead geometries of the Euclideans and relativity dogmatists like Einstein. Whitehead's space is organic. It is the space of lived experience, not of recondite measurement. Thus, the most concrete entity, the prehension, is the appropriate entity for the coordination of the cosmic epoch itself—whatever order is imposed upon a prehension should be reliably discoverable in all intermediate entities throughout the cosmic epoch (which is another way of saying that an actual entity must be actually somewhere and potentially everywhere in a cosmic epoch). Such mediated space is highly ordered, at numerous levels, and thoroughly determinate. Of all the things it might possibly have been, the actual entity realizes just one collection of those possibilities.[36] The limits encountered by prehending actual entities in achieving their satisfactions (and there are many such limits) are conditioning factors upon the organic space of the actual world of any given actual entity. Each actual entity is, of course, hypothesized to be one standpoint (not perspective) on the cosmic epoch, although this cannot be fully demonstrated; it is assumed. Thus, "extension" is rich in potentiality and "space" (for all its liveliness) is impoverished *by comparison*. In Whitehead's terms:

> The satisfaction of each actual entity is an element in the givenness of the universe: it limits boundless abstract possibility into the particular real potentiality from which each novel concrescence originates. The 'boundless abstract possibility' means creativity considered solely in reference to the possibilities of the intervention of eternal objects, and in abstraction from the objective intervention of actual entities [actual division] belonging to any definite actual world, including God among the actualities abstracted from.[37]

On the other hand, space is rich in determination and order, while extension is "chaotic" by comparison, in Whitehead's sense of the term "chaos."[38] The *possible* intervention of eternal objects upon creativity, considered in abstraction from the objective intervention of actual entities, has a more familiar name: it is called "logic." We will take this up in the next chapter.

The key transition between the extensive continuum and space is genetic division (and note that genetic *division* is not genetic divisibility—we will explain the difference shortly), and the most concrete analysis of genetic division is the theory of prehension. The theory of prehension follows a logic that is common to all levels of generality in *Process and Reality*, but that logic is given its exposition in the theory of prehension. Our strategy is to assess the ways in which genetic division comes to be reflected in coordinate division by examining the limits exerted by "contrast" and "comparison" (in Whitehead's sense), which is the final stage in the theory of judgment, or propositions, with which the theory of prehensions is completed (*Process and Reality*, Part III), and the way opened for the theory of extension.

The transition (from attained value to value in attainment) between the extensive continuum and its eventual spatialization is genetic division of the continuum by an actual entity, and the *analysis* of genetic division *is* the *theory* of prehension.[39] A significant part of the secondary literature has confused the tool (the theory of prehension) with *what* is to be analyzed (experience as related to transition). We want to examine the conditioning limits exerted by temporality (or, the temporal determination of the extensive continuum) on the prehending actual entity. If we could come up with a form of initial division apart from time-determinations, then surely the tool of analysis would also be different. But we have no experience beyond our own cosmic epoch with which to contrast the temporal mode of determination. We can feel relatively confident that the universality of time in experience points to its importance in our cosmology, but we cannot conclude its exceptionless necessity. Thus, other modes of determination of the extensive continuum may exist, and if they do, the tools of analysis adopted by those who analyze such determinations will need to conform to *their* purposes in inquiry.

The apparently universal conditioning role of temporality in our cosmic epoch is a matter that can operate only contingently in our cosmology precisely because we lack contrast. Here we work with the instruments of analysis that Whitehead developed, but those tools must be seen in their contingency, as selected instruments, as well as in the relational necessity of their modes of effectiveness. The limits have to do with the ways in which eternal objects ingress, and how the subjective form of a prehending actual entity constitutes an indivisible unity conditioning subsequent spatial division. In short, by grasping what, in principle, the actual entity *can take in* macroscopically, from "attained actuality into actuality in attainment"[40] (its transition), and what it *must eliminate*, we learn both how any actual entity hangs together, and about the determinate structure of the space of our cosmic epoch.

A Pretty Tough Page

Actually, there are two pages in the corrected edition of *Process and Reality* that have to be visited, pages 283–284. It is Section I of chapter I of Part IV, entitled "Coordinate Division." This passage says that an explanation of an actual entity requires two descriptions, one of its transition and one of its concrescence. That distinction has been partly explained earlier in *Process and Reality*.[41] The theory of prehensions enables us to consider together the results of both of these descriptions, by using the tools of genetic and coordinate *analysis* (not to be confused with genetic and coordinate *division*). We may understand the actual entity genetically in both descriptions, wherever we find a congruence of subjective form and objective datum, and then we may coordinate those hypothetical entities—the prehensions are real, but their virtual division, which we will discuss the coming pages, for the *sake* of analysis introduces arbitrariness.

Whitehead has spoken of genetic and coordinate division prior to this point, but he has not indicated their full relation. Few students of Whiteheadian thought have ventured into these waters, most of them skipping from page 280 to page 337.[42] We must grasp where in the course of the inquiry the logic of prehension lies, and why it is situated just there and not elsewhere. At the end of Part III, Whitehead has achieved an outcome, which is a functionally complete description of *how* an actual entity, in achieving satisfaction, *divides* the extensive continuum. Thus, the satisfaction is the entire or completed "part" or particular, and it exemplifies a *full* set of relations (which is the same as to say that it is a unity) to the whole of our cosmic epoch, exhibited at every level of generality, as Part III ends. Thus, as Part IV begins, Whitehead says (please forgive our interpolations in brackets):

> There are two distinct ways of 'dividing' the satisfaction of an actual entity into component feelings, genetically and coordinately. Genetic division is division of the concrescence; coordinate division is division of the concrete. In the 'genetic' mode, the prehensions are exhibited in their genetic relationship to each other. The actual entity is seen as a process; there is growth from phase to phase; there are processes of integration and reintegration [at various levels of generality, such as perception and judgment]. At length a complex unity of objective datum is obtained [that is, the satisfaction], in the guise of a [generic] contrast of actual entities, eternal objects and propositions, felt with corresponding complex unity of subjective form [that is, the actual entity is a unified togetherness of all these other entities].[43]

Note that the satisfied actual entity, genetically described, comes as an appearance of contrasts between what it actually *did* in unifying and contributing itself to the world, and the world *to which* it has contributed itself. The contrast is not ultimate precisely because it is and can *only* be *generic*. There is no ultimate or final atomization of the whole by the parts, which is to say that we are never in any way raising the question whether the whole is equal to or greater than the sum of its parts. None of us exists concretely in a completely atomized condition (in the sense of self-sufficient, independent existence). Our contrast with our actual world is concrete, but generic. We are genuine parts of a genuine whole. We are simply observing some of the ways in which parts entail, in appearance, the wholes to which they belong, while wholes really do contain their parts. For now the point is only that the *contrast* of whole and part we achieve in the theory of prehensions is a generic rather than an ultimate contrast.[44] Whitehead continues (and again, our interpolations are in brackets):

> The genetic passage from phase to phase is not in physical time: the exactly converse point of view [that physical time is *in* the genetic passage from phase to phase] expresses the relationship of concrescence to physical time. It can be put shortly by saying, that physical time expresses some features of the growth, but *not* the growth of the features. The final complete feeling is the 'satisfaction.'[45]

This passage further elucidates the cryptic statement, so often quoted by the defenders of the "temporal atomism" that there is a becoming of continuity, but no continuity of becoming.[46] Note that physical time is in genetic passage, not the other way around. It is in the genetic passage that one finds the "atomism" or "undividedness." Whitehead goes on:

> Physical time makes its appearance in the 'coordinate' analysis of the 'satisfaction.' The actual entity is the enjoyment of a certain quantum of physical time. But the genetic process is not the temporal succession: such a view is exactly what is denied by the epochal theory of time.[47]

Interpreters have taken these lines as confirmation of the primacy of temporal atomism, but they have done so without working through the methodological relation between genetic and coordinate division.[48] The very next line provides the needed clue for working through that difficult relationship. The key term is "subjective unity":

> Each phase of the genetic process presupposes *the entire quantum*, and so does each feeling in each phase. The *subjective unity* dominating the process forbids the division of that *extensive* quantum which originates with the primary phase of the subjective aim. The problem dominating the concrescence is the actualization of the quantum *in solido*.[49]

Why does "subjective unity" forbid the division of the quantum, and why is the quantum here referred to as "extensive"? Recalling that extensiveness can be divided either temporally or, subsequently and conditionally, spatially, we have arrived again at the hypothesized irreducible extension of our cosmic epoch (see chapter 5). Lest there be any doubt, Whitehead immediately reminds us that the quantum in question is not only a temporal quantum but also a spatial quantum: "There is a spatial element in the quantum as well as a temporal element. Thus the quantum is an extensive region."[50]

The spatio-temporal quantum, which is another name for the satisfied actual entity, will now be treated as a "region" in the final coordinate *division*. Whitehead further says that the "region is the determinate basis which the concrescence presupposes."[51] The *forbidding* of the concrete division of this quantum by its own subjective unity (which would be a kind of *negation* or total elimination of possibilities), for the sake of *maintaining* the solidarity of the actual entity with its actual world, is obligatory in genetic division, in conformity with subjective unity;[52] that forbidden division would render hopelessly abstract what it seeks to discover concretely. The prohibition is neither temporal nor spatial in character, it is categoreal (i.e., hypothetically presupposed by the terms of this inquiry, and confirmed in the results of the inquiry). This is another way of saying that the satisfied actual entity is in every sense a sufficient "part" of the whole to which it belongs, regardless of how we treat it in our *analysis* (not division). To imagine the actual genetic division of a satisfaction would be the same as to say that what is finished is not finished, that it is only an abstract, but never a *wholly* concrete, part of the whole to which it belongs, its actual world. This violates Whitehead's solidarity thesis. For now, it is enough to note that something is being forbidden by the requirement of subjective unity, which is a categoreal obligation of genetic division.

But coordinate division (which depends upon the character of coordinate *divisibility*) proceeds without regard for the categoreal obligations of genetic division:

> The coordinate divisibility of the satisfaction is the 'satisfaction' considered in its relationship to the divisibility of this region. The concrescence presupposes its basic region, and not the region its concrescence. *Thus the subjective unity of the concrescence is irrelevant to the divisibility of the region.* In dividing the region we are *ignoring the subjective unity* which is inconsistent with such division.[53]

This is why, for instance, we can discuss, intelligently and realistically, frog parts without harming any frogs. The region, in such a case, coordinates a frog-process that is a genuine feature of the region, a concrescence. There is no deeper division of that concrescence which does not presuppose the coordinate region. Is the frog just a temporal event, or string of events? No to both, because there is a subjective unity that is the frog's experience which not only presupposes the temporal process (and coordinate region), but also admits of a genetic division, and together these divide the extensive continuum, to the extent it *is* divided. In short, we do not deal with satisfactions *qua* satisfied when we engage in coordinate division, and coordinate *divisibility* (not division) was the presupposition of our earlier genetic *divisions* and their exposition (from perception to judgment). If it were not possible to *treat* an actual entity *as if* it were potentially divisible in ways it *actually is not divisible*, there could be no genetic division.

In plainer English, if we could not suppose hypothetically what is actually impossible to do, we could not get an understanding of any concrete part-whole relationship. This is not like trying to create a model of Fenway Park. We are now dealing with organically related space. Everything is now not just a mix of static and dynamic characters, etc., but functionally *alive*, as indeed our perception and judgment *are*, and we are not permitted to kill the organicity to study it. We want access to the character of this living space, and it is too complex to model with statics and dynamics.

Coordinate division is more like dissecting our friend the frog, except without harming it—one takes the frog whole, but *supposes* a dissect-able living frog. Then one provides a description of the main phases of its life, at multiple levels of abstraction, from its evolution, environmental requirements, physiology, major systems, anatomy, cell structure, etc., and longitudinally, one describes its linear and cyclical process from conception (origination) to death (satisfaction), and then one says "now I have a complete frog." But one actually does *not* have a complete frog because one has supposed a divisible frog whose subjective unity forbids such division as our genetic account has carried out—in words rather than with sharp knives.

Thus, *the* coordinate hypothesis behind the entire inquiry in *Process and Reality*, the philosophy of organism, is a regionalization into extensa of the entities we will exhibit genetically as satisfied actual entities in solidarity with their corresponding actual world. We then further specify those extensa until we have an adequate, applicable, and logically rigorous account of their relations, from whole to part (transition) and from part to whole (concrescence). The result is a pre-metrical topologization of experience.

The ground of this initial move in inquiry, the hypothesis or "as if" character of inquiry, is the willingness to take the possible as prior to the actual—not in fact, but in our thinking. Coordinate divisibility is a presupposition of our thinking, not of our concrete existence, and genetic division is what provides us with the materials for coordinate division (logically, not ontologically), imaginative dissection along multiple planes of experience. We ignore the concrete unity of whatever is actual when we coordinately divide, but we already treated the actual entity as divisible when we entered upon the inquiry. As Whitehead says, "the region is, after all, divisible, although in the genetic growth it is undivided."[54] You can dissect the frog and you can actually divide organic spaces, but not while maintaining the organicity. *It is often overlooked that perception and judgment have the same sort of organicity at a different level of generality*. In short, meaning genuinely exists, but its type of presentational space doesn't survive actual division into subjective and objective sides. Value accrues organically as well, although consideration of that topic would take us away from the problem of space (and extension).

One of the main points of interest in this complex relationship of genetic and coordinate analysis is the role played by possibility. Whitehead continues: "So this divisible [possibility] character of the [actually] undivided region is reflected into the character of the satisfaction."[55] In short, the fully realized part with which we ended our genetic account, the satisfied actual entity, is shot through with possibility already—not metaphysically, but because it was treated as a coordinately divis*ible* extensum from the start. We don't as yet *know* whether the possibility we find in the satisfied actual entity is a function of the *inquiry* or of the character of the actual entity in solidarity with its actual world. In other language, we don't yet know whether it *must be* what it is or whether it simply appears *to us* as it does as a function of the conditions of inquiry. Thus, as we proceed into coordinate division itself, we must be careful how we understand the results of our efforts. As Whitehead says:

> When we divide the satisfaction coordinately, we do not find feelings which *are* separate, but feelings which *might be* separate. In the same

way, the divisions of the region [we presupposed at the outset of inquiry] are not divisions which *are*; they are divisions which *might be*. Each such mode of division of the extensive region yields 'extensive quanta': also an 'extensive quantum' can be termed a standpoint.[56]

What follows this passage is a rehearsal of a set of logical relations. For our purposes, enough has been said if we now grasp the priority of coordinate divisibility, the intervening task of genetic specification, and then our carrying-out-in-analysis of coordinate division itself. We wish to learn something about the relation of possibility to actuality, and we wish to find it not only presupposed but also exemplified in the most general way in our metaphysical descriptions. Only if the exemplification comes up to the level of what we presupposed—in adequacy, applicability, and logical rigor—will we have met the demands of our own inquiry. There must be a genetic description of possibility, as it enters into the actual entity's transition and concrescence, providing a complete picture of the actual entity in solidarity with its actual world, the world it unifies and enriches.

What of this possibility? How can we get at it? Consider now the following from earlier in *Process and Reality*, while Whitehead is working his way toward a completed description of a satisfied actual entity: "The novel actual entity, which is the effect [of conformal feelings of the present to the past], is the reproduction of the many actual entities of the past. But in this reproduction there is abstraction from their various totalities of feeling."[57] That is, we take actual entities out of their own actual worlds and coordinate them as elements in the actual world of the actual entity we are analyzing into prehensions. He continues:

> This abstraction is required by the categoreal conditions for compatible synthesis in the novel unity [in other words, the solidarity thesis]. This abstractive 'objectification' is rendered possible by reason of the 'divisible' character of the satisfactions of actual entities. By reason of this 'divisible' character causation is the transfer of a feeling, and not of a total satisfaction. The other feelings are dismissed by negative prehensions owing to their lack of compliance with categoreal demands.[58]

Negative prehension (the principal topic of the next chapter) in the sense mentioned here is the process by which an actual entity eliminates or dismisses (Whitehead uses both terms) those aspects of its actual world which it cannot possibly unify in its satisfaction. The actual entity is under a kind of compulsion here, but the limitations are logical, not ontological. These limits have to do not with what an actual entity is in its actual world, but rather with the presuppositions we have made in order to *analyze* an actual entity in its actual world, and to bring a genetic description to completion.

Furthermore, if the use of the terms "cause" and "effect" is confusing here, bear in mind that just one page earlier Whitehead said: "The reason why the cause is objectively in the effect is that the cause's feeling cannot, as a feeling, be abstracted from its subject which is the cause. This passage of cause into effect is the cumulative character of time. The irreversibility of time depends on this character."[59] In other words, concretely speaking, we cannot prioritize actual relations in which we are interested in over the merely possible relations we are studying without denying time's arrow.

More can be (and should be) said, but if we have adequately stressed the complexity of the relations between coordinate and genetic division—how each falsifies experience in an effort to model it, and then offers its resultant model as a clarified version of experience, adapted to purposes we choose—then enough will have been done for the present. Yet, we have come upon the issue of the elimination of those possibilities that cannot be brought into conformity with the subjective-objective categoreal demands of the actual entity. While it is true that these eliminations are logical rather than ontological, it is also crucial that we understand the logical function of the elimination of possibilities in order to grasp the full logic of the quantum of explanation. Hence, we move to a consideration of possibility in its determinate forms, both negative and constructive.

8   The Problem of Possibility

> The subjective form lies in the twilight zone between pure physical feeling and the clear consciousness which apprehends the contrast between physical and imagined possibility.
>
> —Whitehead, *Process and Reality* (263)[1]

By "eternal objects," Whitehead means possibilities. We will use the terms "possibility" and "eternal object" interchangeably (as indeed we have been doing up to this point), but we favor the term "possibility." There are those among Whitehead's interpreters who are misled by the name Whitehead chose for this category of entities. Some seem to think that Whitehead must be "naturalized," and that eternal objects are somehow non-natural or even unnatural. The term "possibility" is much less confusing on this point. Anyone who wants to rid Whitehead's philosophy of eternal objects is actually attempting to cut out the category of possibility. It is neither necessary nor possible to do that while preserving any semblance of the philosophy of organism. Taking possibility seriously is a requirement of all process philosophy.

We begin with Whitehead's statement in Part III of *Process and Reality* that "all the actual entities [in the actual world of a prehending actual entity] are positively prehended, but only a selection of the eternal objects."[2] This statement may make it *appear* that in excluding some eternal objects, some *possibilities*, the actual entity renders them irrelevant to its subjective form, but that does not follow, and is emphatically wrong. Negative determination is a variety of relevance, especially when we are speaking of prehension. Just prior to the sentence quoted above, Whitehead has limited the entire theory of prehension to the genetic account of what he terms the "cell-theory of actuality," while promising that the "morphological theory [of these cells] is considered in Part IV, under the title of the 'extensive analysis' of an actual entity."[3] One can see that prehension divides in analysis what, for Whitehead, cannot be divided *in experience*, as we have emphasized in the previous chapter.

144   *The Problem of Possibility*

Many Whitehead scholars have not given appropriate weight to the idea that Parts III and IV of *Process and Reality* are complementaries in the philosophy of organism. The theory of prehension genetically divides and specifies what will be treated morphologically in Part IV. This morphology is what we have been calling, and indeed, what Whitehead presupposed as his "coordinate" account. We emphasize that the genetic account *presupposes* the coordinate account, but the order of *exposition* may be different, as noted earlier. In fleshing out the idea of a morphology of the actual entity in its actual world, or what we call the quantum of explanation, Whitehead argues that the *objective* character of the actual entity is pragmatic, since this objective perspective allows the world of the actual entity to be plural and diverse just to the extent that this world remains a real potential for the concrescence of that actual entity—its real consequences. On the other side is a series of transformations that must maintain the formal (i.e., subjective) unity of that same actual entity. This is the aspect of temporal passage he calls concrescence. When we consider the objective and subjective perspectives of concrescence in its solidarity with its actual world, we have the actual entity as transition. That is a "standpoint." It is all one process. Only when these objective and subjective requirements are considered *together* do we have a "standpoint." A standpoint is *definite* only in light both of what it has incorporated and what it is has *eliminated* in its morphology so as to maintain solidarity with its actual world. Whitehead says:

> The satisfaction of each actual entity is an element in the givenness of the universe: it *limits* boundless, abstract possibility into the particular real potentiality from which each novel concrescence originates. The open 'boundless, abstract possibility,' means the creativity considered solely in reference to the possibilities of the intervention of eternal objects and in abstraction from the objective intervention of actual entities belonging to any *definite* actual world, including God among the actualities abstracted from.[4]

We have italicized the term "definite" in this passage. The distinction between definite and determinate order has not been understood by Whitehead's interpreters. We will explain it in detail below. For now simply note that the term used here is "definite." We are *abstracting* from the actual world of any given actual entity in Part IV in order to consider boundless, abstract possibility. This is coordinate division. In other words, we really are dividing what is a solidarity in actuality so as to consider the possible in its relation to the actual, and not vice versa. But that is not what Whitehead is doing in Part III.

Why is the morphology of the actual entity, coordinately divided, in Part IV, not simply an example of misplaced concreteness? After all, we will be considering actual entities *as if* they could occupy a common space (not space in the physical sense, but in the logical sense, at least). Yet, Whitehead says this account of extension is about *limits*, in the logical sense, placed upon relations of inclusion and exclusion, and proceeding from a regionalization of standpoint, not from an ontological consideration of perspective, whether objective or subjective (i.e., together as concrescence). Limitation of possibilities *as such* (i.e., logical), then, requires us to consider the relation of the possible modes of order, and as mutually *determining* (not *defining*), without losing sight of their function as real potentials for *some* actual entity. We are not here required to say *which* actual entity we are considering because the theory of prehension has been completed at this point. We have the genetic account of the actual entity as concrescence in Part III. We are now in a position to consider actual entities as relational and not as atoms in Part IV, without risking fallacies of simple location or misplaced concreteness. We already know what sort of relational entity a prehension is, and we know how to *find a prehension*, in any actual world.[5]

In short, we sum the functions of Parts I-III and carry on, with our results confined to their proper places. We will not violate the principle of relativity by over-stressing atomicity, and we won't violate the ontological principle by asserting ontological simultaneity of actual entities in Part IV, but that is because those categories have already done their work. What we are now interested in is how the structural features that have been demonstrated in Parts I-III can be used for purposes other than maintaining their own integrity and solidarity in genetic analysis. We now use the results to get at some features of extension that are pre-spatial. It was for just this purpose that the account in Parts I-III was set out, and it is not easy to explain *why* the genetic account reads as it does *unless* one realizes that the theory of extension was conceived and worked out ahead of the genetic account.[6] The genetic account can be used for other purposes, but it will be distorted if pieces are removed from the coordinate analysis and thus, coordinate division for which they were constructed. There are many ways to detach the actual from the possible, but most lead down the wrong roads; this coordinate division of possible and actual is limited in ways that prevent our getting lost in our abstractions. The ideas in Parts I-III simply are not independent of that coordinate division. To treat the ideas as having a sufficiency independent of their coordinate analysis, to treat them as if they could be lifted from the philosophy of organism and employed in other service, is like thinking you could pull a heart out of an organism and all systems would go on functioning.

We do not, with the theory of extension, attempt to divide the extensive continuum; we now consider its parts in relation, both as determinate and as definite. All definite relations are determinate, but not all determinate relations are definite. Thus, we are not considering in Part IV the *whole* of Whitehead's account of the actual entity. We are considering only the ways in which the actual entity "is exhibited as appropriating for the foundation of its own existence, the various elements of the universe out of which it arises."[7] Those elements are God, eternal objects, and the act of valuation that constitutes the actual entity as a *definite* existence. Therefore, "each process of appropriation of a particular element is termed a prehension."[8]

A prehension is a genetic version of the universe from the standpoint of a given actual entity (not actual entities in general). We do not, when thinking about prehensions, consider anything like "actuality as such," or "possibility as such." The genetic account proceeds under the limitation that it shall be complemented by a morphological analysis of the actual entity in Part IV. The latter is of a very different character, but its function is to integrate feeling *as felt* (prehension, traced all the way through to final generic contrasts) along with and as conformable to the extensiveness of the determinate orders of nature, their law-like character, as required by our experience *of* nature. The aim in *Process and Reality* is to provide a full description of *experience*, not of nature, but where *all* finite experience, so far as we know, is experience *of* nature, clearly some account of nature *in* experience should accompany the description of the nature *of* experience, as felt. (See chapters 10 and 11.)

Here we see the importance of the idea that we feel *all* actual occasions in our actual world, but only a *selection of the possibilities* is felt, as we quoted above. It does not follow that the unfelt possibilities do not exist, or exist only in a Platonic realm of forms, as the caricature of Whitehead maintains. Such views trade in the obligation to account for the reality of determinate modes of order in our cosmic epoch for a crypto-determinism in which possibility has no role in limiting real potentiality. To dump the eternal objects is the same as reductionism, and many of the so-called naturalists are committed to the materialism Whitehead criticized in *Science and the Modern World*. That was not their intent, but eternal objects are indispensable to Whitehead's cosmology, and they are *not* "Platonic forms" or "universals," if by that one means that they are non-natural.

Regarding our experience of nature, we cannot say with any great assurance what definitely *will* happen, but we have a better hold on what is *not* going to happen, i.e., is excluded from possibility, by comparing the ways in which competing selections or constellations of determinate order exclude one another from the standpoint of a prehending actual entity. Our account must explain why some possibilities are felt and are not. How is the "felt selection" accounted for while the unselected possibilities are, genetically speaking, *absent*? And how can we situate unselected possibilities relative to the extensive continuum, as a part of extensive order? *This* is the "problem of possibility." How can we understand the exclusion of some possibilities from feeling such that this exclusion may have a bearing on the order exhibited in our cosmic epoch?

We must venture a terminological innovation to supplement Whitehead, although we would rather not. We are contrasting the word "constellation" with "collection" in order to capture the distinction Whitehead articulates as follows, with our modifications in brackets:

> In "presentational objectification" the relational eternal objects fall into two sets [we will not use the term "set" for obvious reasons; we use

"clusters"], one set [cluster] contributed by the "extensive" perspective of the perceived from the position of the perceiver [which we call a constellation], and the other set [cluster] by the antecedent concrescent phases of the perceiver [which we call a collection].[9]

Considered as collections, possibilities will be just those included (ingressing); but an eliminated *constellation* is as absent as any reality can be in the actual world of an actual entity. It is unfelt. An eliminated constellation has no real potentiality ever after. (see chapter 9) The act of *eliminating* a constellation *is* experienced, however. Whitehead continues:

> What is ordinarily termed "perception" is consciousness of presentational objectification. But according to the philosophy of organism there can be consciousness of both types of objectification [causal and presentational]. There can be such consciousness of both types because, according to this philosophy, the knowable is the complete nature of the knower, at least such phases of it as are antecedent to that operation of knowing.[10]

Whitehead has defined causal objectification as "what is felt *subjectively* by the objectified actual entity transmitted objectively to the concrescent actualities which supersede it. . . . In this type of objectification, the eternal objects, relational between object and subject, express the formal constitution of the objectified actual entity."[11] In short, the collections of possibilities belonging to the actual world of an actual entity, as a result of their contributions to the forms of all that is actual in that world, are subjectively felt causally, regardless of whether they are objectified presentationally. Whitehead's sense of the term "perception" is unusual because he includes unpresented experience under the term. This is part of what distinguishes him from Kant and post-Kantian philosophy (making this an exercise in *modern* cosmology).

Thus, *collections* of possibilities can be experienced both causally and presentationally, while "mere" *constellations* (determinate modes of order, i.e., clusters, that are experienced in their extensive connectedness only) can be presented, and objectified thus, but have no causal efficacy. But to deny that these constellations are indeed knowable is to deny that "the knowable is the complete nature of the knower." Thus cluster of presentational objectifications called "constellation(s)" is distinguished from the cluster that has both presentational and causal objectification, i.e., "collection(s)."

Possibility in Modern Cosmology

Leibniz's struggles with possibility are well known, but the problem of possibility and its relation to necessity and eternity was general in the seventeenth and eighteenth centuries. In re-connecting cosmology to these Modern roots (so as to avoid Kant's transformation of the problem of space into the problem of the transcendental conditions governing the space of *presentation* [*Vorstellung*]), Whitehead is obliged to revisit this problem and to bring possibility under a new kind of scrutiny with new tools.[12] Readers should not conclude prematurely that the idea of possibility was a monolithic problem for the Moderns. In fact, the problem was rich with nuance in the discussions of the best Modern thinkers.[13] One thing that has crippled Whitehead scholarship is the erroneous notion that somehow *Process and Reality* belongs to the post-Kantian discussion in philosophy. But both cosmology and finally logic went off the rails after Kant. The Moderns were still working on the problem of the togetherness of the world, and still seeking a logic that would tie extension and intension into a single calculus or algebra.[14] The moderns had something *right*, in Whitehead's estimation.

Whitehead delved back into the foundational texts that treated the problems of space and time, both from the point of view of "natural philosophy" and from that of the most influential theories of perception (Descartes, Locke, Leibniz, Hume, Kant). He teased out of these texts the portions of those theories that would withstand scrutiny those who had learned the math, logic, and empirical studies of perception during the time elapsed since Hume's *Treatise*. Especially important to these considerations were Leibniz's recently published ideas. Whitehead's readings of these figures are at first baffling to those of us who learned their thought as part of a narrative about rationalism and empiricism in the Enlightenment. He notices all sorts of passages that our interpretive tradition ignores He is a good reader, but Whitehead frequently also confesses to "reading in" the meanings he finds. His purpose is not to provide a pristine, historical account of the Modern figures. He wants them to join with him in tackling the same problems anew, with better tools.

The question we now confront is what must we do with the issue of possibility in light of how we might now (in 1929 and beyond) handle a reformulated cosmology. The major developments affecting our efforts to naturalize possibility have to do with advances in mathematical thinking, especially the implications of universal algebra as an *idea*, as a way of handling the ordering of possibilities as series of transformations, in short a morphology of the coordinate relations of parts and wholes (now called mereotopology, also recognizable as the general algebraic theory of Categories).

Leibniz planted within possibilities a striving or *dynamis*, a sort of longing to *be actual*, with final realization made actual under the sufficiency of reason and final sorting carried out according to a principle of the "best possible." Whitehead made possibilities (eternal objects) relatively independent of actuality in the philosophy of organism, although he retained a teleological structure in his theory of concrescence and transition. He replaced the striving possibilities with an ultimate category of creativity, and a trinity of entities that are together exhaustive of actuality and possibility. The category of the Ultimate *presupposes* this triad of formative elements in the universe, actual entities, God, and possibilities. These formative elements achieve the purpose of the universe by mutual limitation, but the "striving" is God's exclusive contribution, a purpose for the world, imparted to everything actual, but not to the possibilities, *qua possible*, which have no actuality but *do* exist. This latter is the distinction between existence and actuality of which Hartshorne later made so much (see chapters 12 and 13).

Leaving aside medieval universals here, clearly we do come hard up against the same problems when the metaphysical chores are differently parsed by Whitehead. The Moderns had several strategies for working their way out of that old discussion, but it wouldn't be accurate to say that they *solved* the problems surrounding universals, real relation, individuals, and names. Much of what we will have to say in the final two chapters will depend upon how the medieval theories of universals are variously treated by the original thinkers of the twentieth century, and such discussions *assume* the appropriateness of Modern cosmology tacitly.

Here it will be sufficient to say that Whitehead is a *conceptualist* with respect to these issues, but of an unusual sort—he holds that universals exist as the *conceptual feeling* in everything actual. That is an original way to approach the issue, since it insists upon the ubiquity of the concept, but sees concepts as part of primary relations (the special meaning of the term "feeling" for Whitehead).[15]

Situating Whitehead in the Modern framework as a conceptualist focuses for us the issue of possibility. We might ask: what would Leibniz's "striving possibles" look like if they didn't strive? Would there still be order among them? Whitehead's answer to that question is an emphatic affirmative. The order among possibilities *qua possible* allows for infinitely many modes of formal relatedness, and our imaginations fail when we press the question very far. But there is reason to think that the mathematical modes of order we are able to create have a positive relation to those infinitely diverse modes of mutually determining connection.[16] The geometrical and the algebraic modes of order are the two most important, and they are not wholly commensurable. There is no point in imagining that there is only one master mathematics, the *mathesis universalis* Descartes sought, or the elusive universal characteristic of Leibniz, but make no mistake: the quantum of explanation, i.e., the *idea* of the actual entity, plays an analogous role in Whitehead's philosophy.

This is a mathematical philosophy, but mathematical order never has been and never will be reduced to a single principle. The problem of the One and the Many is solved only by learning to live with the sense of both the Parmenidean and the Heraclitean insights into being as process. The many become one and are increased by one. The question is not *how* this category of the Ultimate constitutes order, but rather what *modes* of order does it presuppose and how are those modes best described?

In hitting upon the idea of extension, Descartes separated the problem of *internal dynamis*, or concrescence, from the problem of the order of possibility. Descartes did not *know* he had accomplished this.[17] There was still an effort in his thinking to make mathematical order the sole explanation of the order of nature. In the twentieth century it became clear that there was no such thing as a univocal order in mathematics, let alone in nature. There were hints earlier—Hamilton's quaternions, the development of non-Euclidean geometries, and so on. But the real lessons of these developments were not driven home until the final failure of univocity, as manifested in the *Principia Mathematica*, became indisputable. Subsequent interpreters have focused on our failure to reconcile the mathematical order, which seems shot through with genuine possibility, with the mechanical order to which mathematics must be subordinated in the act of measuring. This problem is, if anything, worse for Newton.[18] Whitehead says:

> [T]he classical theory also assumed . . . that any actual occasion only lies in one duration; so that if N lies in the duration including M's immediate present, then M lies in the duration including N's immediate present. The philosophy of organism, in agreement with recent physics, rejects this . . . though it holds that such rejection is based on scientific examination of our cosmic epoch and not on any more general metaphysical principle.[19]

Here we see why the measurement problem in twentieth-century cosmology was so crippling. The growing significance of the idea of a "cosmic epoch" is becoming clear in the passage above, more of which shortly. But the Modern insight that extension is the essence of body, if it is *not* conflated with the meaning and existence of *physical space*, becomes the central insight in Whitehead's cosmology. In a very real sense both Spinoza and Leibniz recognized, in different ways, how this move to extension liberated mathematics from being nothing more than a tool for measuring physical objects in relation. Mathematical *thinking* allows human imagination to approach the question of order in a more general, nay, *universal* way (in the sense of "universal" in which it is used in "universal algebra," having to do with the coordinate whole presupposed by a genetic description).

But what, then, are we studying when we are considering mathematical modes of order? The answer is simple. We are considering formal possibilities as modes of *determinate* order in abstraction from *merely* physical laws (in the sense that seventeenth-century philosophers thought of this—limitations from which twenty-first-century physicists and metaphysicians are scarcely immune; see chapter 10). But how can that study be carried out by creatures (this word is chosen deliberately) who cannot genuinely remove from their thinking the assumption of a physical world the reality of which conditions every *act* of thinking about the possible?[20] The answer Whitehead gives is that in the *idea* of extension, *undivided divisibility*, we have a whole to which all the possibilities, including unrealized possibilities, belong, under the condition that we not speak of cosmic epochs beyond our own, except conjecturally.

Whatever possibility *as such* may be, possibility *as it belongs to our own cosmic epoch*, even though it is wholly non-actual, is describable according to principles of inclusion, exclusion, and selection so long as *something* is actual. We will take up the relation between actuality such as it belongs to the divine (as distinct from the finite creature) in the final chapters. For now it is enough to consider the idea of the cosmic epoch as a way of designating the whole to which eternal objects, that is, the possibilities that *belong* to our cosmic epoch, can be described using the mathematical tools we currently possess (especially genetic and coordinate analysis, or mereotopology).

Cosmic Epochs and Natural Philosophy

The idea of the cosmic epoch as an empirical limit is one of the most important and least discussed ideas in Whitehead's philosophy. The modifier "cosmic," apart from its etymological nod to the idea of order, is a significant marker regarding the limits of the inquiry Whitehead is pursuing in *Process and Reality*, an essay in *cosmology*. The subtitle of the book intentionally limits this book to what was called "natural philosophy" during the modern period. That form of study, in which the theories of the natural universe and the theory of the *measurement* of the natural universe were still in reciprocal conversation, has passed away. Pursuant to Newton's discoveries, physical science gradually eschewed all ideas about causality except efficient cause and reduced its sphere of concern to the quantifiable modes of measure. This is to say that "natural philosophy" no longer existed by the time Whitehead wrote *Process and Reality*.[21]

Today we distinguish cosmology as it is pursued by astrophysicists from metaphysics as a subdivision of philosophy, an *a prioristic* pastime of philosophers with no serious effort to hold its assumptions within the boundaries of nature as humans *experience it*.[22] Philosophical naturalism in the present day tends to be an uncritical and uncriticized starting place for philosophizing,[23] where the main thing we know is that there certainly *isn't* a God, and the investigation of *possibility* belongs exclusively to wackiest philosophers.[24] This futile exercise includes a realism/anti-realism debate, which is a cartoon re-enactment of the Medieval problem of universals, now carried out by the intellectual peers of Elmer Fudd (realist) and Bugs Bunny (anti-realist) instead of true giants of human thought like Aquinas and Duns Scotus. If this cartoon is philosophical naturalism, then it needs a better concept of nature, one that cannot be so easily and completely discredited by the likes Richard Rorty (Daffy Duck?), who did little more than to make some hasenpfeffer.

We take up philosophical naturalism in detail later. Presently, we say again that Whitehead's naturalism is radically empirical. He understands very well that anyone's *current* concept of nature is surely inadequate to comprise nature in any determinate way, and is in fact (whether we are conscious of it or not) a kind of metaphor we use in the present, and which we replace in the future with an improved one, to express the outer bounds of our experience as it is *had*. And Whitehead is very clear that there is "no reason to identify it [Nature] with the boundless totality of actual things"—a mistake made by many interpreters.²⁵ If we are to have a genuine cosmology, that is, if we are to return to natural philosophy, there must be some *determinate* idea of the whole to which both the actual and the possible belong, and yet, the act of specifying *any* determinate whole will occlude at least some aspects of experience *unless* its boundary is intellectually permeable. But "Nature suggests for our observation gaps, and then as it were withdraws them upon challenge."²⁶ Thus, we cannot make unqualified assertions about that "boundless totality of actual things." There must be a whole, but it is not to be identified with actuality as such. This idea of the whole to which our experience belongs *is* the cosmic epoch. It includes a concept of nature of precisely the sort Whitehead sets out in his book of that name, but the concept of nature presupposes that we aim to have a scientific understanding of the physical universe *qua physical* (see chapter 9).

Thus, in the cosmic epoch we have an *idea* of the whole to which the actual *and* the possible belong, but we are not entitled to assume *either* that we will not have to revise the description of the "idea" we may offer at any given time, *or* that our imagined determinate whole is to be identified with the universe *itself* (whatever that may be). Hence, when introducing the idea of a cosmic epoch and providing an initial sense of it, Whitehead is careful to make us aware that the cosmic epoch in which our percipient actual entity temporally divides the extensive continuum, that is, *our* cosmic epoch, is *not the only cosmic epoch*. We have interpolated in the famous passage below reminders of the topic sentence of the paragraph, which considers *only* the physical universe. If this seems adventitious now, please withhold judgment until the end of this section, when we will have justified our emphasis upon the *physical* limitation presupposed by the entire paragraph. Whitehead says:

> [T]he extensive continuity of the *physical universe* has usually been construed to mean that there is a continuity of becoming. But if we admit that 'something becomes,' [in the physical sense,] it is easy, by employing Zeno's method, to prove that there can be no continuity of becoming. [Where we are speaking exclusively of the way that actual entities, considered physically rather than *logically*, divide the extensive continuum t]here is a becoming of continuity, but no continuity of becoming. [Under those assumptions, t]he actual occasions are the creatures which become, and [to that same extent, in the physical sense *only*] they constitute a continuously extensive world. In other words, [physically speaking,] extensiveness becomes, but 'becoming' is not itself extensive.²⁷

Whitehead never asserts that the physical universe *is the whole universe*. He explicitly denies that, and those who read the paragraph above in forgetfulness of the topic sentence are misled as to the character of Whitehead's naturalism, *which includes a mental pole* (and that means possibilities as well as actualities and potentialities). The paragraph above considers the physical pole in abstraction from the mental pole, and hence, refers to the causal and mechanistic order *only*. The "possible" (eternal objects) and God, the other categories of existence that are irreducible in the philosophy of organism, are not here under consideration. That helps us understand what follows: "Thus, the ultimate metaphysical truth is atomism." Recall, "a-tomos" means undivided. This is the assertion of the quantum of explanation, not just in the physical but also in the *metaphysical* sense. The temporalization of the cosmic epoch is the act of the actual entity; it divides, and it does so *physically*. The *logical* status of that act is another question. The physical aspect of the act cannot be reduced to a logical or mathematical relation, and, incidentally, it also cannot be measured *exhaustively*, since all acts of measurement presuppose it. No act is exclusively physical, but every act is irreducibly *at least* physical. Whitehead is a realist, not an idealist.[28] He continues:

> The creatures are atomic [i.e., undivided]. In the present cosmic epoch there is a creation of continuity. *Perhaps* such creation is an ultimate metaphysical truth holding of all cosmic epochs, but this does not seem to be a necessary conclusion. The more likely opinion is that extensive continuity is a special condition arising from the society of creatures which constitute our immediate epoch.[29]

Note, Whitehead is speaking of *extensive continuity* as a characteristic of *a-tomism*, i.e., undividedness. If the "temporal atomism" asserted by Ford and his followers were an ultimate metaphysical principle for Whitehead, it is very hard to understand why he would say "perhaps" here. Even more mysterious is the question of why he would follow it with the *conjecture* that the view of extensive continuity (not to be confused with extensive connection) he has just mentioned is *probably* a function of the way the universe is presenting itself to us *now*. In other words, there *is* society of actual entities that constitutes our *immediate epoch*—i.e., not the whole cosmic epoch, but the *way* in which the causal order is operating *now*, in the form of natural laws which condition our *experience* of the cosmic epoch, and oblige us to adopt a quantum principle to explain it. The order of nature can change even within our cosmic epoch, but it would be difficult to learn about it from the standpoint of the *immediate* epoch. As Sherburne rightly says, "one interesting implication of this [cosmic epochs] doctrine is that Whitehead holds that the laws of nature evolve."[30]

Hence, the quantum of explanation, that is, the *idea* of the actual entity, is the universal characteristic for *our* cosmology, as a type of study, but we are not entitled to assume it would be of any use in other cosmic epoch(s), although we are also in no position to deny that it might be. *Perhaps* in every cosmic epoch worthy of the name, there will be a divider or extensive

154 *The Problem of Possibility*

continuity that we might call "the act," or the "quantum of explanation," but what is far more likely is that our tendency to *think* that way is a function of the *immediate* order of creatures within which we presently find ourselves. We therefore *need* this "quantum," understood as irreducibly physical (even if its mode of existing is *more* than just physical).

Determinate Wholes and Local Quanta

Why? The very next thing Whitehead says is: "The proper balance between atomism and continuity is of importance to physical science. For example, the doctrine, here explained, conciliates Newton's corpuscular theory of light with the wave theory."[31] This is an astonishing claim: Wave-particle duality overcome in the quantum of explanation? Yes. Also overcome is the measurement problem of cosmology, along with the paradoxes associated with General Relativity and the need for any kind of universal constant to find a grand unified theory of the physical universe.[32] Whitehead's theory of cosmic epochs, taken in conjunction with the quantum of explanation, or the theory of the actual entity, as it were, dissolves these issues. It really does.[33] Therefore, one cannot exaggerate the importance of the theory of cosmic epochs, as the determinate whole to which cosmology, i.e., natural philosophy, belongs.

What, then, *is* this determinate whole with which natural philosophy is concerned? In attempting to interpret the passages we have just examined, we think Sherburne understates it. He says that "the notion of our cosmic epoch is the notion of a vast society establishing relations that are admittedly not universal, not perfectly general."[34] The cosmic epoch is not a mere "notion" for Whitehead, it is a limit idea that plays a tremendously important role both logically and scientifically (empirically) in his cosmology. The cosmic epoch is the fact of the possibility of a frame in which intelligibility genuinely occurs. Sherburne is also confusing Whitehead's conjecture of the way the *immediate* epoch is probably conditioning *our sense* of the whole (and leading us to suppose that the act of the actual entity is metaphysically *necessary*), with the fact it merely *seems* so from the vantage of our immediate epoch. Sherburne says that the cosmic epoch "establishes" relations, which is a confusion of the extensive continuum with the cosmic epoch; it is the extensive relations which establish the cosmic epoch, not vice-versa.[35]

Returning to the claim Whitehead makes for reviving natural philosophy, that "for example" it solves the wave-particle paradox, we should follow that example through to the end. Then we may appreciate the kind of determinate whole that this idea (not notion) of the cosmic epoch portends. Whitehead continues:

> A corpuscle is in fact an 'enduring object.' The notion of an enduring object [and *this* is a "notion"] is, however, capable of more or less completeness of realization. Thus, in different stages of its career, a wave of light may be more or less corpuscular.[36]

This example would be utter nonsense unless there were a difference between the *immediate* epoch, including the ways it *appears to be* law-like as experienced *by us*, on one side, and the determinate order of our cosmic epoch *as a whole*, on the other side. Apart from comprising a larger temporal span than any single enduring object, our immediate epoch also includes the order of what is possible and non-actual (eternal objects), and what is actual and non-temporal (God). In short, the cosmic epoch has a mental pole, and natural philosophy must *consider it* in order to have any complete contrast of what is actual in experience with what is possible and non-actual. *Thus, the elimination of possibilities in prehension (that is, negative prehension) does not render them merely subsistent, or merely ideal, or in any other way less than fully real.*[37]

We can now make sense of the following:

> The arbitrary, as it were 'given,' elements in the laws of nature warn us that we are in a special cosmic epoch. Here the phrase 'cosmic epoch' is used to mean that widest society of actual entities whose immediate relevance to ourselves *is traceable*. This epoch is characterized by electronic and protonic actual entities, and by yet more ultimate actual entities which can be dimly discerned in the quanta of energy.[38]

Here we find the same contrast of the immediate epoch with the cosmic epoch, but the former is traceable in the latter, by way of the quantum of explanation, those yet "more ultimate actual entities" which need not be electronic and protonic exclusively, although we expect them to be physical in some sense, since they *do* belong to the determinate order of our cosmic epoch. Whitehead calls the laws of nature "arbitrary" givens because our access to the ultimate actual entities that divide the extensive continuum in our cosmic epoch is *mediated* by these laws, as well as by the categories that limit us to the consideration of one actual entity in its actual world, except and until we abstract from that limit (with the associated costs of rendering our account partial and to the same extent, literally false). The laws are real, but they can change, entirely within the determinate order of our cosmic epoch. *Thus, to seek the ultimate physical laws of the universe in physics is a patent waste of time*, since *whatever* we say will reflect some arbitrariness drawn from the ways in which those patterns of order are manifest to us in our *immediate* epoch, which is only *a part* of our cosmic epoch. The proper study of natural philosophy presupposes a determinate order that contains our immediate epoch, and every other *immediate* epoch (those beyond our experiential reach), in our *cosmic* epoch.

Whitehead could not have known that we would find evidence of the ground-level development of the electronic and protonic order in our subsequent study of the development of the physical world, but surely he would have objected to the claim that the Big Bang was the beginning of our cosmic epoch; at most it was the physical beginning of our immediate epoch, and if its laws appear necessary, given the protonic and electronic physical character of actual occasions, then we must be careful not to assume that there is no arbitrariness implied in the way we are situated within the cosmic epoch as a whole. To deny the arbitrariness in our standpoint is to see contingency as necessity. Exaggerating the Big Bang is a model-centric habit of the literal minded among contemporary physicists (see chapter 10 and our Introduction). We are confident that the *physical* character of the actual occasions in our immediate epoch is at least partly arbitrary, as a bit of cosmology, even if not as a bit of physics, because it is reasonable to think that physical laws *evolve* within our cosmic epoch. Cosmology presupposes a whole to which the immediate epoch belongs, and relative to which the immediate epoch exemplifies a stage in the career of that cosmic epoch. We are not entitled to suppose that the career path the cosmic epoch has followed up to our standpoint is the *only one* possible, given the way that the more ultimate actual entities, traceable by us when we aren't too literal minded, divide it.

This idea of the cosmic epoch, apart from nicely restoring the study of natural philosophy to our discipline, and holding physicists within the honest bounds of their special science, *is* the coordinate whole that is analyzed into, but not actually divided by, prehensions. The theory of prehensions gives us access to the determinate order of our cosmic epoch beyond the limits (and ameliorating the arbitrariness) of the physical laws which dominate the *immediate* epoch. The theory of prehension is the tracing of what is presented in a standpoint through to the "more ultimate actual entities" suggested by our limited experience. *Understanding, therefore, what it means for a constellation of possibilities to be negatively prehended requires that we grasp how prehension, or fundamental feeling, relation, etc., is comprised by the cosmic epoch.* Thus, we see what is, as it were, *interior* to our cosmic epoch. But what of its limits, or boundaries, or what is beyond our cosmic epoch?

Whitehead is quite tentative whenever this question arises, but he does say some things that help us understand the outer limits of the whole to which our determinate order belongs. In his discussion of genetic and coordinate divisibility in Part IV (and hence, the context is the theory of extension, which *aspires to be* universal for *all* cosmic epochs, even if only conjecturally), he says:

> [W]hen we examine the characteristics considered in the next chapter [(Part IV ch. II], it is difficult to draw the line distinguishing characteristics so general that we cannot conceive of any alternatives, from characteristics so special that we imagine them to belong *merely to our cosmic epoch*. Such an epoch may be, relatively to our powers, of immeasurable extent, temporally and spatially. But in reference to the ultimate nature of things, it is a limited nexus. Beyond that nexus, entities with new relationships, unrealized in our experiences and unforeseen by our imaginations, will make their appearance, introducing into the universe new types of order.[39]

The "universe," here, includes *all* cosmic epochs, not "merely" ours. Within our cosmic epoch, it is easy (for Whitehead) to imagine that the photon is a coagulation in the career of a wave, seeming necessary to us in its collapse only due to conditions that are transient characteristics of our immediate epoch. The immediate epoch will be called our "environment" in Whitehead's terms, when we consider it in light of the idea of "relevance," which we will take up shortly. But there is something beyond the nexus which *is* our cosmic epoch. Whitehead continues:

> But, for our epoch, extensive connection with its various characteristics is the fundamental organic relationship whereby the physical world is properly described as a community. There are no important physical relationships outside the extensive scheme. To be an actual occasion in the physical world means that the entity in question is a relatum in this scheme of extensive connection. In this epoch, the scheme defines what is physically actual.[40]

This passage renders "temporal atomicity" a nonsensical idea. To exist physically in our cosmic epoch is to exist within the extensive scheme. There are *no* important physical relationships beyond it. In fact, the scheme *defines* (not just determines) what is *physically* actual. Our emphasis on the term "physical" at the beginning of this section may have seemed forced, but please consider those interpolations in light of what Whitehead says here. He is speaking of the physical world in relation to the extensive scheme of *this* cosmic epoch, but as a piece of natural philosophy and *not* of "physics" as we understand it today. The prehensive analysis of the actual entity is designed to exhibit characters of the actual entity's relation to the extensive scheme in such a way as to leave intact its full physical reality, but without insisting that the physical character *is* the existential ground for all order. The reason for this qualification is that we cannot be certain that the most general characters of the physical order *as we experience it* are even remotely necessitated by the order of the universe, or even by the determinate order of *our own* cosmic epoch. This is a strategy for preserving a full-bodied realism while avoiding adventitious reductionism. It is also radical empiricism, bent upon preserving relations that are experienced along with relations that *might be* experienced.

The extensive scheme relative to the determinate physical order of our cosmic epoch yields a total collection of (algebraic, i.e., coordinate and genetic) relations, and that collection is axiomatized as a kind of geometry in Part IV of *Process and Reality*. Such an approach is motivated, for Whitehead, by an interpretation of Maxwell's electromagnetism theory:

> The whole theory of the physical field [i.e., Maxwell] is the interweaving of the individual peculiarities of actual occasions upon the background of systematic geometry. This systematic geometry expresses the

most 'substantial form' inherited throughout the vast cosmic society that constitutes the primary real potentiality *conditioning concrescence*. In this doctrine, the organic philosophy is very near to the philosophy of Descartes.[41]

The last sentence seems shocking unless one has properly understood the sense of "Modern cosmology" involved in this inquiry. What if Descartes had known about Maxwell's field equations? What would he have said about "substantial forms" in relation to his theory of extension? The question of substantial form is generally ignored by readers of Descartes, assumed to be a part of his medieval hangover, while the theory of extension is too quickly conflated with the theory of "bodies in space." But Whitehead simply read Descartes on his own and noticed what is so easy to overlook. Whitehead wouldn't have been aware of the standard lines on Descartes, and such a person, simply *reading* Descartes, and looking at Maxwell's field equations could say "hmmm, Descartes *almost had this*." Whitehead says:

> For Descartes the primary *attribute* of physical bodies is *extension*; for the philosophy of organism, the primary *relationship* of physical occasions is *extensive connection*. This ultimate relationship is *sui generis*, and cannot be defined or explained. But its formal properties can be stated. [Maxwell is part of that.] Also, in view of these formal properties, there are definable derivative notions which are of importance in expressing the morphological structure. [These are given in Part IV.] Some general character of coordinate divisibility is probably an ultimate metaphysical character, persistent in every cosmic epoch of physical occasions. Thus some of the simpler characteristics of extensive connection, as here stated, are probably such ultimate metaphysical necessities.[42]

Here the theory of extension does strive to be universal, i.e., to apply to all cosmic epochs. Substantial form, as Descartes understood it, now means for Whitehead an ultimate metaphysical character (probably) necessary in every cosmic epoch, which is exactly what both Descartes and the medievals were seeking. The physical aspect would be a part of that, but not as an attribute, which could hardly be the explanation of the whole in any case. Now the physical is a relation primary to extensive connection. Descartes and his peers would not have imagined that this physical aspect of reality could be described in the equations of the electromagnetic field, nor axiomatized in a non-metrical theory of extensive connection, but that does not mean Descartes and his successors (e.g., Leibniz) had no notion of the physical character of what they were doing. Many examples could be given, but one should suffice. Wolfgang Smith says:

> As is well known, it was René Descartes who provided the philosophical basis of "classical" or pre-quantum physics by enunciating the

distinction between *res cogitans* and *res extensa*. One generally perceives this Cartesian dichotomy as nothing more than the mind/body duality, forgetting that Descartes has not only distinguished between matter and mind, but has, at the same time, imposed a very peculiar and problematic conception on the former element. He supposes, namely, that *res extensa* is bereft of all sensible qualities, which obviously implies that it is imperceptible. The red apple which we do perceive must consequently be relegated to *res cogitans*; it has become a private phantasm, a mental as distinguished from a real entity. This postulate, moreover, demands another: one is now forced—on pain of radical subjectivism—to assume that the red apple is causally related to the real apple, which, however, is not perceptible. What from a pre-Cartesian perspective was one object has become two; as Whitehead puts it: "One is the conjecture, and the other is the dream."[43]

Smith calls this the "bifurcation postulate," and then says:

> What vitiates the customary interpretation of physics and prevents that science from being 'integrated with higher orders of knowledge' is none other than the bifurcation postulate. This is the one hidden premise one unfailingly assumes in the explication of a scientific discovery. It is true that this postulate has been uncovered and attacked by some of the leading philosophers of our [20th] century—from Edmund Husserl to Alfred North Whitehead, Nicolai Hartmann, and Karl Jaspers, to mention a few names—and yet that problematic tenet remains to this day unexamined and unopposed by men of science even in the sophisticated arena of the quantum debate, where just about everything has been "put on the table." However . . . the premise can indeed be jettisoned, which is to say that nothing prevents us from interpreting physics on a non-bifurcationist basis.[44]

Smith then argues convincingly that everything relevant to the idea of a physical quantum relation, of the sort currently studied and theorized in quantum mechanics and electrodynamics, was already clearly set out in the physical philosophy of Aquinas, in his idea of the *signata quantitate*, that mode of existence through which every perceptible entity passes, leaving its trace or signature at just the point where it becomes *measurable*. If this sounds vaguely like the collapse of the wave function, it is, at the level of *ideas about physical reality*, about the same thing. Smith says:

> I have conceived of the *signata quantitate*, in light of contemporary physics as a mathematical structure; but is this exactly what St. Thomas Aquinas had in mind? Whatever the Angelic Doctor may have been thinking, it could hardly have been Hilbert spaces and Lie groups of Hermitian operators with which contemporary physics is concerned. I must not, however, judge too hastily. . . . The crucial question is

160 *The Problem of Possibility*

whether these mathematical structures are yet "quantitative" in an appropriate sense; and if that be the case, then the *signata quantitate* admits—by transposition if need be—the interpretation I have given.[45]

Smith is not a religious Thomist, he is physicist and mathematician who happens to know more about the history of his discipline than is common. Expressing such views is more than enough to get a person blackballed from the elite clubs of the "respectable physicists," the ones wedded to the Cartesian paradox, their beloved bifurcation, which they are inclined to whip out whenever the public wants them to *philosophize*—something they are untrained to do, but fortunately for them, the public is unable to tell the difference between snake oil and an actual snake. To use a bifurcation as the foundation of a pure dogma about the progress of science is a shameful way for scientists to behave, as Whitehead made pretty clear in *Science and the Modern World*.

It makes no difference to us whether Smith has Aquinas right in this matter, only that he rightly identifies a trajectory and momentum to classical physics based on a problem with Descartes which, when corrected, provides a Modern cosmology of our cosmic epoch based on demystified extensive relations. It is amazing how much of ancient and medieval cosmology comes back into consideration when Descartes' bifurcation has been set aside and the problem of cosmology restated. It inspires one to see the human effort to know the natural world characterized as a common project, not a series of revolutions in which the history of science is one of silly errors, caught and corrected by later inquirers who simply make different, but equally crippling errors.

Whitehead saw the problem very clearly, his interpreters seem not to grasp that solving it is what *Process and Reality* is actually *about*. His other books are about other problems. The reason that *Process and Reality* is his most important book is because this problem of bifurcation is the most crippling problem we currently have (unfortunately we still have it) in our quest to understand our universe. It is also important because his proposed solution, the quantum of explanation in the orientation of radical empiricism, is the best solution anyone has yet imagined and expounded. In light of Whitehead's solution, we are in a position now to assess the progress made in physical science as a result of our improvements in mathematics, technology, and logic since Kant. The improvements in tools for study have been stunning, but our philosophical command of our own advances remains stunted, given that the medievals and the ancients could better articulate than we can how the world, in the largest sense of the word, hangs together, in the largest sense of the word. We have more pieces, but they had a better grasp of the whole. They were not distracted by a concept of the physical that pointed them to philosophical dead ends.

The fact that it has not usually occurred to Whitehead's interpreters to ask themselves *why* he speaks of cosmic epochs is partly his fault, for over-estimating his readers, and for taking it as obvious that anyone who studies these problems seriously would at least learn the basic history of the debate before pronouncing Whitehead right or wrong. He also naïvely believed that people would take him at his word when he stated plainly what he was and was not doing.

But the responsibility lies only partly with Whitehead. Students of the sociology of knowledge have well documented the processes by which succeeding generations of inquirers come to be habituated into taking as beyond questioning matters that were merely suggestions when first made. Today's practice of cosmology is the normal science of yesterday's conjecture. The configurations of today's problems have been built upon the presuppositions and expectations of an ossifying tradition. Whitehead stepped outside of that stream. Many people recognized he was saying something important, even if they couldn't say what it was.

The theory of cosmic epochs in *Process and Reality* provides the backdrop for a new interpretation of the relation of the physical world with all the real modes of existence that cannot be reduced to (or accounted for within) any current concept of the physical, given the limitations of our immediate epoch. The failure of interpreters to grasp the theory of cosmic epochs also accounts for the inability to understand Whitehead's use of the term "quantum" in *Process and Reality*. Whitehead says:

> This section [Part III, ch. 2, Section 1] on simple *physical* feelings lays the foundation of the treatment of cosmology in the philosophy of organism. It contains the discussion of the ultimate elements from which a more complete *philosophical* discussion of the *physical* world—that is to say, of nature—must be derived. In the first place an endeavor has been made to do justice alike to the aspect of the world emphasized by Descartes and to the atomism of the modern quantum theory. Descartes saw the natural world as an extensive spatial plenum, enduring through time. Modern physicists see energy transferred in *definite* quanta.[46]

Whitehead wouldn't give equal weight to the contemporary quantum physics and to the Cartesian plenum if he were a "temporal atomist" (see chapter 4). But note the way that the philosophy of the physical is at the root of the analysis.[47] This is about the quantum *principle*, not about quantum physics. The same distinction applies to the difference between the *principle* of relativity and its *application* in Einstein's general relativity theory, wherein the principle was empirically discovered (although it had been theorized for a long time). Whether the quantum principle would apply to other cosmic epochs than our own is uncertain. Whitehead said, cited earlier, that "we may assume there is a lower limit within [a finite] region [of the extensive continuum]. . . . it follows that the relevant objectifications, forming the relevant data for any one occasion, refer to a finite sample of actual occasions in the environment."[48] That "lower limit" *may* apply to other cosmic epochs, but its role in *physical* occasions is the only thing Whitehead feels secure about.

It should be clear now why Whitehead sees his view in the Modern context, and what he intends to do in providing it with a systematic geometry in Part IV. By systematic geometry, Whitehead *means* universal algebra—his "geometry" is an algebraized geometry, and it is a kind of solution to the problem of physical space as conditioned by the theory of cosmic epochs.

Definite Order and Determinate Order

The forgoing account of cosmic epochs casts our categoreal "entities" in a clearer light. The entities that stand out with an "extreme finality," i.e., actual entities and eternal objects, become the indispensable tools for interpreting our cosmic epoch, and any cosmic epoch we can conceive. The other categoreal entities, prehensions, nexūs, subjective forms, propositions, multiplicities, generic contrasts, "have a certain intermediate character."[49] They assist us in the interpretation of our cosmic epoch, but only as particular ("certain") paths from the center (a percipient actual occasion) to the periphery (all that is *possible* in a cosmic epoch from that standpoint), called *concrescence*, and from the periphery to the center, all that becomes actual in a cosmic epoch, called *transition*. The actual entity, considered as concrescence and transition, is concrete and provides us with what we know of the "definite order" of the actual world. This definite order does not exhaust order as such, since it belongs to the topic of cosmology, but it must serve as the clue to all other modes of order. (When one speaks of order as such, cosmology is but one of the topics one needs to address.) This insight of Whitehead's is a great advance in philosophy.

As Whitehead says, "the group of seventeenth- and eighteenth-century philosophers practically made a discovery, which, although it lies on the surface of their writings, they only half-realized." He continues:

> The discovery is that there are two kinds of fluency. One kind is the *concrescence* which, in Locke's language, is 'the real internal constitution of a particular existent.' The other kind is the *transition* from particular existent to particular existent. This transition, again in Locke's language, is the 'perpetually perishing' which is one aspect of the notion of time; and in another aspect the transition is the origination of the present in conformity with the 'power' of the past.[50]

What the Moderns discovered but did not quite grasp was that the relationship between possibility and actuality was describable in two fluencies— we prefer the terms *internal dynamis* (concrescence in the microcosmic order) and *external dynamis* (macroscopic transition), but the names we may choose matter very little. What matters is that concrescence and transition provide us with two different orders of actual-possible relations. The difference is that the concrescing actual entity, while it bears a real relation to everything in its actual world, objectifies only a finite sample of the occasions (in its region) in its satisfaction, and hence, not everything possible *for it* becomes a part of its achievement, only one collection of possibilities is included. The entire *physical* world has been positively prehended but only a sample of that world is objectified. The outcome of concrescence is a *definite* actuality. In a very real sense, all that was *physically* possible for that concrescence has been actualized, which is to say that no *ingressing* eternal objects that played a part in the constitution of the actual world of that actual entity have failed to find objectification from the combined subjective and objective perspectives in the standpoint of the actual occasion. Elimination (or dismissal) has done its *physical* work, but this kind of elimination functions on a scale of relevance. We will discuss this in the next section (on concrescence).

Alternatively, the transition that *is* the origination of a given actual entity has a quite different relationship to possibility (eternal objects).[51] From this aspect of the process, the actual entity is *not* all it could have been because infinitely many determinate constellations of eternal objects *might have* ingressed, but only a *selection* of eternal objects, one collection, is exemplified in the achievement of the actual occasion. Those unselected possibilities are as *real* as the selected collection, and both the selected collection and the unselected constellations are utterly indifferent to actuality. Here is a non-causal, non-local, and indeed non-physical theory of possibility, characterized by what Whitehead calls "determinate order." To get a handle on determinate order, we must look at definite order. Whitehead says:

> The 'objective lure' [as distinguished from the 'lure for feeling'] is that discrimination among eternal objects introduced into the universe by the real internal constitutions of the actual occasions forming the datum of the concrescence under review [i.e., this is its originating principal actual entity in its actual world as transition, but as dependent on prior concrescence]. This discrimination also involves eternal objects excluded from value in the temporal occasions of that datum, in addition to involving eternal objects included for such occasions.[52]

The relationship between dismissed physical occasions and unselected eternal objects is complicated.[53] In a sense the unselected possibilities that contribute negatively to the achievements of the actual occasions in the actual world of the actual entity "under review" will *also* be unselected by the latter. They are not possibilities for the entity under review, not because they aren't real, but because their unselected status in the origination of the entity under review, as a characteristic of its transition, is a fact of its actual world. Whitehead follows the passage quoted above with a telling example, in which Napoleon's defeat at Waterloo has become a factor in the constitution of our actual world. "But the abstract notions, expressing the possibilities of another course of history which would have followed upon his victory, are relevant to facts which actually happened." This is the structure of determinate order and a statement of its *relevance* to definite actualities. Whitehead continues:

> We may not think it of practical importance that imaginative historians should dwell upon hypothetical alternatives. But we confess their relevance in thinking about them at all, even to the extent of dismissing them. . . . Thus, in our actual world of today, there is the penumbra of eternal objects, constituted by relevance to the Battle of Waterloo. . . . The elements of this penumbra are propositional prehensions, and not pure conceptual prehension. . . . Thus an element in the penumbral complex is what is termed a 'proposition.' A proposition is a new kind of entity. It is a hybrid between pure potentialities [i.e., possibilities *qua* possible] and actualities. A 'singular' proposition is the potentiality of an actual world including a *definite set* of actual entities in a nexus of reactions involving the hypothetical ingression of a *definite set* of eternal objects.[54]

The reason these eternal objects are described as "definite" (as opposed to "determinate") in our actual world is due to their status as being conditioned by their relation to actuality. The might-have-been, insofar as we can think about it or have emotions relating to it, is conditioned by actuality. We experience the lure of possibility in the actual world in that modality, as propositions. This is the mark of definite order, and the logic of this kind of order is our best clue to learning about the kinds of order that characterize eternal objects *as such* (possibility *qua* possible, in abstraction from actuality). The logic of definite order is a theory of inclusion, but without "membership," and hence a mereology rather than a set theory with primitive membership. Whitehead says:

> A 'general' proposition differs from a 'singular' proposition by the generalization of 'one definite set of actual entities' into 'any set belonging to a certain sort of sets.' If the sort of sets includes all sets with potentiality for that nexus of reactions, the proposition is called 'universal.' . . . The definite set of actual entities involved [in a singular proposition, and, with slight elaboration, general and universal propositions also] are called the 'logical subjects' of the proposition; the definite set of eternal objects involved are called the 'predicates of the proposition.' The predicates *define* a potentiality of relatedness for the subjects. The predicates form one complex eternal object: this is the complex predicate.[55]

Note, the complex predicate of the actual entity as concrescence and transition is one eternal object, but as such, is a "set" (collection) of predicates that have been unified by the activity of the actual entity under review. This activity is a unification relative to which the eternal objects are patient. Definite order, then, is conditioned not by the structure of possibility as such, but by the logic governing the activity of the actual entity as transition *and* concrescence. Yet, what we can learn from hybrid entities, such as propositions and nexūs is our only clue to the structure of *determinate* order.

The analysis of the unification of eternal objects, i.e., possibilities, as a collection that preserves both their unifiability under the condition of real potentiality, along with their plurality as a constellation, provides us with an abstraction which shows, in one direction, their existence *as* a multiplicity, and, in the other direction, their unifiability under some condition of agency. Whitehead calls this complex order, taken in its relative independence, "merely real," but even determinate order cannot be actually divided from the actual.[56] Thus, if we abstract from their given unity as a predicate for the logical subject that is the actual entity, we find ourselves able to consider possibilities individually as candidates for association as constellations. If we consider the complex unity of the predicate in abstraction from the elements that constitute it as a collection, we get the notion of an agent who, in principle, could unify some constellations of eternal objects that are beyond the agency of actual entities as our cosmic epoch characterizes them (see chapter 12). This topic of determinate order is raised wherever definite order is treated in abstraction, and indeed, extensive abstraction is the relevant act.

Determinate order, therefore, is a characteristic of the cosmic epoch considered as transition in abstraction from concrescence, while definiteness is a characteristic of the cosmic epoch considered from the standpoint of a concrescing actual entity.[57] Real potentiality is the idea in the philosophy of organism where these meet.[58] There is no perch from which to view possibility *as such*, or *qua* possible, or as "pure potentiality," in Whitehead's terms, but this limitation does not imply that we can provide no *account* of determinate order in abstraction from any given actual entity. The study of determinate order can be undertaken in many ways.[59] But logic, because it exhibits algebraic characteristics, and because it traverses the spatialization process (by which causal efficacy becomes a presentational space, or "perception" for Whitehead), while retaining some of the structural characteristics of temporal passage (the theory of symbolic reference), and because logic captures at least *some* of what we experience in the intentional structure of spatialized experience, is a likely tool for the analysis of determinate order. Determinate order includes, at a minimum, the *way* in which possibilities ingress in "groups,"[60] and the *way* in which the *structural* features of those "groupings" give rise to generalizations regarding the character of possibility *qua* possible for *our* cosmic epoch.

The principal logical function/operator in this analysis is a form of negation. We here distinguish between "function," which is negation considered as concrescence, and "operation." We introduce this terminology to clarify a real distinction in Whitehead's account of concrescence and transition.[61]

166 *The Problem of Possibility*

"Operation," which we use exclusively with reference to transition, is negation considered abstractly, as the fully developed negative prehension that excludes wholly from the subjective form of the actual entity *all* constellations of possibilities except those consistent with the ingressing constellation, i.e., the collection relevant to concrescence. This operational negation is indistinguishable from that subjective form considered as an *example* of what is *possible*. This latter is what survives of the idea of "substantial form" in Whitehead's cosmology, and it is intended to offer a solution to the problem of universals. We will take up that issue a few pages hence.

The type of negation that is "functional," rather than merely "an operator," is more basic to our analysis here. This kind of negation is a perpetual perishing that we can infer from the comparing the variable relevance of prehended actual entities in the subjective form of our principal actual entity.[62] This function is a simultaneous includer-and-excluder. It is the evaluation of its actual world by A (see below), but within the constraints of subjective form generally, and its own subjective unity as a singular existent.[63] When we consider things functionally, *inclusion is prior to and defines exclusion*, and the conjunction of inclusion-exclusion is not commutative. The inclusion process (grading for relevance) is best understood in the mode of concrescence, while the exclusion process (by which we can approach determinate order) is best described in the mode of transition. These are, of course, *the same process*. We may represent the physical process as follows, showing how *inclusion* functions (and inferring *exclusion* from the result).

"Elimination" (or sometimes "dismissal") is the term Whitehead uses when considering negative prehension in the mode of concrescence, where the mediation of overlapping actual worlds in the physical background of the various prehended actual entities (B, C, and D above) belongs to the actual world of a concrescing actual entity (A above). The point is that

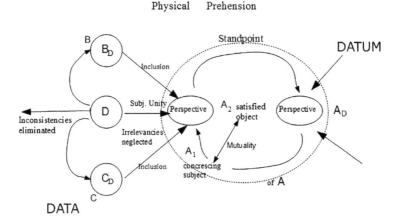

Figure 8.1 The structure of physical prehension

wherever D is part of the actual worlds of B and C, respectively, D will be a part of A's subjectivity *both* as directly prehended by A *and* as mediated (and perhaps amplified) by its presence in B and C.[64]

There is no one-to-one mapping of other actual entities in the subjective form of A because there is no *full* independence of D (or any other actual entity) in A's actual world. D is only an actual *occasion* in A's actual world, even if it is also an actual *entity* in its own actual world. (There is no formula for demonstrating its identity from its presence in A to its singular status in its own world, but there is no empirical hypothesis advanced by denying that identity.) So the presence of D in the subjective form of A is not "causing," in part or in whole, A's *form* to be as it is. (In fact, *causation* is irrelevant to prehension altogether.) Rather, it is the requirement that the subjective form of A *be* a unity that accounts for the *way* in which D is both relevant to A's concrescence and to the limits of that concrescence, wherein neither the determinate whole of D (nor of any other actual entity) is present in A's subjective form. We refer to the "whole of D" as a determinate rather than definite order because there is no concrete standpoint from which D as a whole would be available to any concrescing actual entity, least of all to D itself. To speak of any actual entity in the actual world of A as "a whole" requires abstraction. Hence, so far as we may speak of the order that characterizes D in *its* actual world, from the standpoint of A, we must encounter the possible *qua* possible, or, in short, determinate order.

Whatever D is, as a determinate whole, its definite role in A's actual world involves at least elimination and a great deal of mediation. The elimination is not merely the result of A's act of grading its world for relevance to its aim, but is also a function of how *other* actual occasions in A's actual world have objectified D. No complete criterion will be available in the analysis of A's subjective form to identify D *in itself*, but we can be assured that D's relative insistence in other actual occasions provides reason, if not wholly *sufficient* reason, for the way that D comes to be objectified in A's achievement. The most insistent actual occasion in A's actual world, the one most widely *distributed in* and *mediated by* the other actual occasions in A's actual world is A's immediate predecessor, and is, apart from A, the most *definite* fact of A's actual world. However, there is no basis for claiming that even A's immediate predecessor has not suffered elimination. As Whitehead memorably writes, "a feeling bears on itself the scars of its birth; it recollects as a subjective emotion its struggle for existence; it retains the impress of what it might have been but is not."[65] The scars are evident upon analysis in *every* actual occasion felt by A, but if its immediate predecessor is the least battered, it is nevertheless bowed by the impossibility of finding its way into its successor without suffering elimination. D is the favorite stepchild of A, but a stepchild all the same. The immediate predecessor of A is not, therefore, *wholly definite*, but is the closest thing to definite *for A*.[66]

It is important to remember here that D does not *explain* A; rather, A *explains* D.[67] The question, for Whitehead, is not what accounts for A, but

168   *The Problem of Possibility*

how A can be analyzed to provide an explanation of the way in which D is *in* A. No other kind of explanation is available to finite creatures working within the limits of a cosmic epoch like our own. Thus, the proper way to understand D's influence on A,[68] as A *feels* D, is to use a kind of algebra to get at what of D is absent in A. This is extensive abstraction. Even though we cannot know D in its *definiteness* by analyzing A, we can understand D as an example of *determinate* order in which the *way* D is in A is an instance of how it *might be in A*, and avoiding any assertions that would commit us to how D *must be in A* (which is beyond our ken).

The study of determinate order as it affects concrescence gives us half of the story of negative prehension, which is elimination by perpetual perishing, discoverable in the analysis of the actual entity by comparing the relevance of the various actual occasions in A's actual world, and extrapolating from the immediate predecessor (the most pervasive actual occasion, which we have supposed to be D in this case), to the actual occasion just below D in relevance (let us imagine it is C in this example), and so on until we have a sense of the physical route that *leads* to A. We will not learn much about A's relation to eternal objects by examining concrescence alone, but we will never learn *anything* about determinate order without carrying out this painstaking examination of the physical world.

We do not learn about possibility in our cosmology by beginning with, for example, mathematical or logical relations and then hoping they apply to whatever we are trying to understand. There are ways to study determinate order (as in a given instance of mathematical structure) in abstraction from cosmology, but the only way to learn the scope of the application of such modes of order to physical reality is to study relevance in a single actual entity in its actual world. This limitation, otherwise known as the Ontological Principle, is required by the category of subjective unity—a complex unity that cannot be explained but is presupposed in the quantum of explanation, i.e., the *idea* of the actual entity.

The *full* study of the workings of relevance in the philosophy of organism would take us too far from our present subject, but some attention must be given to it.[69] In cosmology we learn about determinate order very slowly and carefully, by examining the relationship between the actual entity in concrescence and in transition, and recalling that the actual entity as data will not violate the requirements of subjective unity and will also provide a datum for all other actual entities, a datum that is both identical to itself in their objectifications of the datum, and different from every other datum in the actual worlds of every subsequent actual entity (Whitehead calls these the categories of "objective identity" and "objective diversity").[70] The demand that the data become a datum provides the *definition*, in the most literal sense, of the actual entity A. It is the most complete account of any percipient actual entity that a finite mind can achieve empirically. A more *determinate* account can be given of the prehending actual entity, but it is abstract.

Eternal Objects in Concrescence: Relevance
(Part IV, ch. 1, sect. VI)

In the previous section, we laid great stress upon the idea of relevance in moving toward a clear understanding of what is eliminated (dismissed) in the actual entity considered as transition *and* concrescence. However, there is no way to avoid the relationship between the actual entity as percipient (which is one aspect of transition and concrescence) and the actual entity as prehending its actual world.[71] For Whitehead, the latter, a prehending actual entity, is indivisible *in concreto*, but analyzable *in abstracto*. We must tread carefully here, avoiding any discussion that blurs this important distinction between perception (which presupposes an actual dividing of the extensive continuum into transition and concrescence) and prehension (the concrete and actually indivisible modes of relatedness, or "feeling" that are characters of everything actual). But the relation of transition and concrescence entails both the act of dividing and the indivisible actual entity. Later Whitehead says that genetic division is the division of the *concrescence* while coordinate division is the division of the *concrete*.[72] We will explain what this means in subsequent discussion. Our question here is what we may learn from the idea of "relevance" when we see the actual entity *as* concrescence. Relevance crosses the border of perception and prehension.[73]

That which is physically absent (negatively prehended in the sense provided by the diagram above) in the achievement of a concrescence *has been* felt, but is of negligible importance to the concrescing actual entity. The eliminated characters contribute so little influence as to have greater relevance in their absence than in their presence. That which is wholly *unfelt* by an actual entity in *transition* belongs to the cosmic epoch as among the unselected eternal objects, but that is a feature of the coordinate division and must be inferred from the *functional* definition of the concrescing actual entity as genetically divided, *as if* it were a temporal unfolding.[74] This is the correct relationship even though the genetic division presupposed the coordinate division and is a derivative. But in exposition, the genetic derivatives are offered as pieces of a puzzle that has been cut beforehand. Proximately, in the immediate epoch (and with the limitations discussed), there is a sense of negative prehension that belongs to the genetic specification of the definite order exemplified by the prehending actual entity, most generally contained in the theory of concrescence, but exhaustively analyzed in the theory of prehension, taken as a whole.

That which fails to appear in the objectifications of the prehending actual entity lacks relevance to that entity, but is still a part of the physical universe (and the physical universe is an all-pervading character of the actual world of that actual entity). That "failing to appear" has this meaning in spite of Whitehead's statement later, in discussing the coordinate whole, that *all* physical actualities are positively prehended. This is certainly true in coordinate division as well, but in considering the genetic account of a *percipient* actual entity (as distinct from the *prehending* actual entity), we cannot *find* the whole universe in its objectifications, notwithstanding the principle of relativity. In short, irrelevancies are functionally eliminated in the physical pole, and more completely than they can be eliminated in the mental pole, i.e., incurable unity of the mental pole[75] in the subjective form of a *percipient* actual entity, and as distinct from the subjective unity of the *prehending* actual entity as concrescence. The prehending actual entity, *qua* prehending subject, is neither an actual division of the extensive continuum, nor is it actually divided by its prehensions. The prehending actual entity is a concrescence (genetically) and a concrete relation (coordinately), not a divider. When itself virtually divided (whether genetically or coordinately), the *prehending* actual entity suffers a distortion for the sake of analysis. This restriction—that dividing the actual entity for the sake of analysis, although it cannot be divided *actually*—is the final sense in which the idea of the actual entity is *a-tomos*, an undivided unity, and the quantum of explanation.

Whitehead did not possess all of the formal tools needed to accomplish the task he set for himself (see chapter 9). Yet, his discipline as a mathematician and his imagination as a philosopher were closely attuned to each other, and he would not allow his imagination to be strangled by the comparatively crude proof-theoretic tools that were current in his time but which are demonstrably inadequate to the task of reconciling calculus and physics.[76] The model theoretic approach developed by logicians from Tarski to Hintikka provides vastly superior tools. We will discuss this in subsequent chapters, but here we beg anachronistic license to sort out Whitehead's insights about the relations between definite and determinate order.

The key idea for Whitehead is that the *definite* order does not exhaust the conceptual feelings available for concrescence in an actual entity. More simply, the order of nature does not exactly determine physical reality. Whitehead rejects physical determinism even though he is a scientific realist. The association of scientific realism with mechanistic physics is an unfortunate episode in the history of science and the philosophy of science. Many think this view was overcome in Einstein's General Relativity, but that theory is thoroughly necessitarian, even Spinozistic; it metaphysicalizes necessity. We say, more accurately, that the quantum revolution problematized the necessitarianism of General Relativity, and that controversy continues into the present.

Whitehead criticizes "old school" scientific realism, and that realism survives and pervades current scientific journalism, popular science, and even some serious science. We discuss this problem later, but for now it suffices to point out that, like Peirce, Whitehead rejected the types of necessity presupposed by mechanistic and model-centric scientific realism. Both Peirce and Whitehead held that physical laws change. This view means that the determinate order of the immediate epoch, if we can learn anything about it, cannot be identified either with the definite order exemplified in the satisfaction of an actual entity *or* with the order of eternal objects in the cosmic epoch. And in terms of prehension, it means that the determinate order of possibility as such also cannot be *deduced* from the definite order of a satisfaction. A different kind of inquiry is needed.

Henceforth we use the term "constellation" instead of "group" (Whitehead's word) for relations among possibilities as exhibiting determinate order. A constellation of possibilities ingressing as the mental pole of a prehending actual entity *is* a selection of the eternal objects (possibilities) available to that actual entity. The ingressing (i.e. selected) eternal objects is a "collection." Thus, every collection is also a constellation, but not every constellation is a collection. The *un*selected eternal objects are negatively prehended, eliminated altogether from the definite achievement. But we may *infer* from the logical order pervading the ingressing constellation of the collection that there is also intelligible order in the constellations of *un*selected eternal objects.

That order of *un*selected eternal objects is important for our understanding of science, since it is our only clue to the relationship between the immediate epoch and the cosmic epoch, and more generally, to what is possible *for our cosmic epoch*. What is possible *for us* is certainly possible *as such*, and the determinate order (if we can learn what it is) in the constellation of eternal objects possible *for us* is modeled *as* an emergent structure we *infer*. The difference between the immediate epoch and the cosmic epoch is also a matter of relevance, but our grasp of it is very abstract guesswork and conjecture. In moving from a discussion of the prehending actual entity in its definiteness to a discussion of its implications for determinate order, we move from our designations above as "A" and "D in A" to an abstract version of the entity "A." That entity, insofar as it exemplifies a determinate order of eternal objects (and we emphasize that this requires an abstraction) will be designated in lower case Greek letters. Hence, A above becomes α. The lower case Greek characters are abstractions.

From the standpoint of transition, if the natural laws that characterize the cosmic epoch change, then their changes *should* follow a pattern that leaves a trace of its history in what remains functionally recognizable (in analysis) through a series of transformations. This is a conjecture and we will discuss it further, but the order we are supposing here is an *evolution*, a growth of some kind, even if that growth is sometimes, from the perspective of physics, what would be called an entropic decay. If this is a warranted conjecture, then it should be possible in principle to trace the outlines of that order by looking for variations of least degree in the definite order of a satisfaction and abstracting from it a constellation of eternal objects. We may then look at the route by which the actual entity came to be what it is by contrasting it with its immediate predecessor. In other words, if we can learn *which* actual entity *is* the immediate physical predecessor of the prehending actual entity we are studying, and learn also the immediate predecessor of that predecessor, and so forth, then it should be possible to contrast the order evident in the mental poles of those predecessors with one another to discover instances of elimination in the successive constellations of eternal objects we have abstracted from this sequence of actual entities. This is "the egress of possibility," to be treated in a later work. This contrast of constellations of eternal objects is part of "coordinate division."[77]

Genetic and coordinate division are delicate intellectual operations. The first requires us to separate a constellation of eternal objects from a concrescence for the sake of analysis, treating the constellation of eternal objects in our principal actual entity as analyzable in abstraction from the physical route. That move has to be kept in mind in evaluating the veracity of any result we may get. We are modeling the route for the sake of creating a contrast that will be quite abstract but highly informative. We can only *infer* the immediate predecessor of our prehending actual entity, but the inference is very clear on the basis of *relevance*; it becomes less certain as we move backward in history.[78]

Coordinate division, on the other hand, treats the world in its solidarity, but as exemplifying many contrasting determinate orders of eternal objects. The contrast of the constellations of eternal objects from the prehending actual entity to its immediate predecessor will usually be slight, and indeed, *that* contrast will be the closest we can ever come to saying "x is a possibility" in any *univocal* sense. The eliminated eternal object in the contrast of the earlier and the later *is* the limit of that possibility "x," insofar as it *has* a limit. To *define* something requires the specification of its limits, and insofar as possibilities have boundaries, they are found in the sorts of contrasts we have described.

Considering only the eternal objects involved in the world as a solidarity, or as concrete, is the operation Whitehead calls "coordinate division." We will discuss below what sorts of limitations apply to this analysis, but for now it is enough to note that certain further distortions of the concrete actual entity are levied upon it by coordinate division. Whitehead says, "Coordinate division of the actual entity produces feelings whose subjective forms are partially eliminated and partially inexplicable. But this mode of division preserves undistorted the elements of *definiteness* introduced by eternal objects of the objective species."[79] We will cover the three species of eternal objects below, but for now, this way of analyzing the actual entity has a name, coordinate division, and the distortions brought on by our abstractions do preserve the *structure* of the possibilities involved in the concrescence as genetically divided.

Unfortunately, the "possibility" or "possibilities" we discover through these kinds of analysis can be predicated only as "might-have-beens," something *un*selected by the prehending actual entity that is a characteristic of its immediate predecessor. We should not be surprised, however, that the barebones account of a possibility is essentially negative. The bane of contemporary modal logics is confusion about the status of counterfactual predicates. The limitations introduced by Whitehead's distinction between genetic and coordinate division prevent us from assuming that the necessity (types of negation involved in abstraction) invoked in comparing determinate with definite possibilities is the same sort of necessity. There can be a difference between a possibility seen as *unselected* as opposed to being *negated through neglect*. In both instances, the possibility is taken account of, but in the former case the possibility belonged to the route being analyzed but not to the actual entity as a concrescence. After all, we start from what we *do* have, actually, not from what we *don't* have when we are analyzing the actual entity as concrescence. Other possibilities are real but do not belong to the route being analyzed, and those will be handled when we consider the actual entity as transition.

The *reality* of non-identical constellations of possibilities in a prehending actual entity (considered as concrescence) and its immediate predecessor, whether considered as genetically or coordinately divided, suggests all by itself the inadequacy of mechanistic philosophies of nature and science, including mechanical interpretations of quantized energy. We can push this operation of contrast-by-relevance back further in the history of the prehending actual entity, but we do so with increasingly tentative acts of abstraction. The *differences* among the constellations of eternal objects in a given route can then be generalized into an abstract model of transformation and functional stability. That series of transformations will show that there is a real sense to the idea of evolution.[80] Keeping our contrasts in contact with the physical route of predecessors holds us within the limits of genetic division. We continue asking only what the eternal objects have contributed to the *definiteness* of the prehending actual entity as a concrescence. It is a story about the disappearance of certain possibilities based on their prior inclusion in the same physical route. Evolution, to the extent that it can be studied scientifically, is about this kind of "extinction."

We do not have a good basis for assessing the positive inclusions of eternal objects that seem not to have been *included* before (i.e., by other actual occasions in the actual world of the prehending actual entity)—it is impossible, apparently, to tell whether the presence of an eternal object in a collection is novel or was latent in an earlier collection, and it is impossible to separate the real potentiality of that character from its possibility *as such*. The telling difference among collections resides in what has been eliminated, what we *don't* find later *as characteristic* that we *did* find earlier. Things that were potential for an actual entity earlier in its route come to be no longer possible. This is another way of saying that physical time and change describe an arrow that, so far as we know, is irreversible. But we must not make physical time the ground of genetic division. As Whitehead says, "The genetic passage from phase to phase is not in physical time: the exactly converse point of view expresses the relationship of concrescence to physical time. It can be put shortly by saying, that physical time expresses *some* features of the growth, but *not* the growth of the features."[81]

Our tendency of thinking about time as something that contains all physical routes is the bad habit of thought we must break. It is not accurate to say that all becoming is a reduction of potentiality to actuality; this old-fashioned characterization is insensitive to the distinction between definite and determinate order, which is to say it is necessitarian. Physical time isn't the medium, the container, or the cause of growth, evolution, or the transition of physical forms. Physical time is an *expression* of *some* of the features of that larger process we are modeling with genetic division.

The study of modeling by the elimination of possibilities is the study of the *determinate* order of the cosmic epoch insofar as it contributes to the *definite* order of the actual entity as a concrescence. It is formalizable. In a paper on Peirce's very similar view of nature, David R. Crawford formalized essentially the same process, which we will draw on and modify here. But there is a crucial difference between Crawford's Peirce and our Whitehead. The process we have explained above, by which we learn about the immediate predecessor of an actual entity, is not like Peirce's account, and neither does Peirce have the theory of extension, nor clear distinctions among the levels of generality (cosmic epochs, transition-concrescence, perception, and prehension) that make Whitehead's theory so much more applicable and detailed. Peirce also does not have quite the same distinction between determinate and definite order, and nothing like the relation between genetic and coordinate division, but he does reject necessitarianism for the same reasons as Whitehead, and he does have a formalization of relevance that operates (in our sense of that word) well for Whitehead.

The common threads that hold Peirce's and Whitehead's accounts together is that both are realists, both reject physical necessity, and both philosophers were aware that the challenge of defending such a realism is to find a way to model change such that natural laws themselves change according to an *evolutionary* principle.[82] Whitehead thinks we won't ever know that principle with much clarity, whereas Peirce is optimistic, but they agree that we have to try to get at it. Genetic division provides a basis for that study without confusing the determinate order of possibility with the ways in which possibilities contribute to the definiteness of the actual world we are studying.

The idea of concrescence, then, enables us to study of the immanent order of the actual entity, and to use insistent relevance to infer immediate predecessors. When we consider the contrasts among collections of eternal objects in the prehending actual entity as concrescence, and those collections contained in its immediate predecessors, and when we carry that contrast back into its history, we study relevance *itself*. The study of relevance is the study of *growth by elimination*. This work of modeling tells us all we will ever really know about the conservation and transformations of energy in our cosmic epoch. We use the word "evolution" almost with a wry smile, since it seems to express our hope that entropy does not reign supreme. But perhaps the ending of a cosmic epoch is no tragedy. Who knows?

9   The Algebra of Negative Prehension

> Necessity requires accident and accident requires necessity. Thus the algebraic method is our best approach to the expression of necessity, by reason of its reduction of accident to the ghost-like character of the real variable.
>
> —Whitehead, *Essays in Science and Philosophy* (97)

Looking Forward by Looking Back: Leibniz and Modality

Before proceeding with our discussion of David Crawford's advances on Peirce's logic of possibility, as well as our own extended proposals for approaching such a formal analysis, we need to take a moment and cast an eye back at the beginnings of contemporary logic in the modern period, and to suggest some of the ways this return to the modern—that Whitehead insisted upon—might aid in our look forward at potential advances upon the current state of the philosophical logic of modality. Merely naming Leibniz as the inventor of the idea of a "universal characteristic" is not enough. We must make a true return to the moderns, not to worship or damn them, but to tease out a few ideas that they may have entertained but rejected or left aside. While Descartes introduced the idea of extension, and Locke reminded us of the inexpressible *fundamentum* of experience, it is Leibniz who wrestled most explicitly and interestingly with the issue of intensionality in logic, and who gave rise to the further treatment of it in Holland and Lambert, and in Castillon (who succeeded to a significant degree in formalizing reasoning in intension without sacrificing extensive applicability).[1]

Over a period of years, Leibniz devised a way of looking at what we would now call class membership that depended on the idea that one could assign unique numbers to irreducible concepts and gather those irreducibles together into a fundamental class of simple concepts. We might then build more complex concepts from the simples by some sort of recursive operation, captured in a calculus or an algebra. This was his effort to discover the discipline of Universal Characteristic. He made some progress on this mapping system, but, as C.I. Lewis observes, he hit a number of recalcitrant problems, especially with combinations of concepts that have no existential instantiations in the world.[2] One could use his numeric maps to find and calculate both combinations of concepts that *are* instantiated in the world, and ones that don't exist at all, such as "invertebrate humans." Part of the problem was that the numeric mapping applied only to substances and not to relations, and the single relation Leibniz was considering at that time was genus-species—that species were complexes of logically simpler and irreducible genuses, not simply regarding animal, but a generic concept should take in (in both intension and extension) more complex specific concepts that have been built from the simpler.

We can see here the principles of Cartesian analysis, along with the conviction that all genuine knowledge is necessary, grounded in a universal and an analytic concept of truth and meaning. These assumptions would take a beating at the hands of Hume and Kant, but Whitehead looked to find a way to reconnect with a kind of thinking that, without the reductionism, seeks an intelligible relation between intension and extension that, if not ontologically necessary, still preserves logical necessity. The important logical writings of Leibniz most relevant to this quest first became available to scholars in the 1850s through the 1870s, so there was no history of scholarship in Whitehead's time.[3]

Of particular importance in this analysis of possibility and intensionality is Whitehead's idea of division: There is something about division as an operation that corresponds to irreducibility. If one divides to get *to* the irreducibles, then one has not only metaphysical simples that yield to actual division, but also an operation that has some flexibility. So Whitehead's sense of division is interesting. But there is *division*, and then there is *divisibility in analysis* in abstraction of real dividing. In the actual world, the actual entity is a real cut, but it can be analyzed *in abstracto* along with entities that cannot be cut, such as prehensions and generic contrasts. So the actual world has both divisible and non-divisible actualities, and the principle of divisi*bility* may be separated into actual and *in abstracto* based on the relation to possibility. Possibilities ingress in collections and are not divisible in actuality—i.e., not separable from the actualities whose mental pole they constitute. Perhaps possibilities taken as constellations would yield to some kind of abstract division, if we mapped them. Our discussion of Crawford's scheme below is such an attempt. Crawford's model will open up into what we are calling the "continuum of possibility." However, the concept of "continuum" here entails more subtlety than just brute-force set theory with the dogmatic assertion of the (set theoretic) continuum hypothesis.

Leibniz saw that he could carry out such mapping by using primes because one creates irreducible combinations that have unique quotients (and enable one to follow a thread to any given constellation of possibilities) by assigning a prime to collections. This approach failed for Leibniz because unlike the situation with Gödel and arithmetic, the analysis of possibilities does not proceed in a fixed and narrowly limited universe of discourse composed of purely extensional entities, which in Gödel's case were simple numbers. ("Extensional" here means the traditional, logical form of extension, not

Whiteheadian extensa.) When dealing with possibilities, one needs tools with which to navigate an intensional universe, and as such cannot take meanings as analytically fixed in advance.

A Formalization of Concrescence under Relevance: From Genetic to Coordinate Division

Rather than operating as Leibniz did, starting with fixed meanings and attempting to map these onto the world, the Crawford method starts with the world and analyzes possibilities until meanings are discovered in division. In this way, Crawford's analysis of Peirce helps us understand what Whitehead is doing with constellations of possibilities as they exhibit determinate order. Crawford says:

> In "On the Number of Forms of Sets"[4] Peirce defines the diversity of a set (or 'collection') of objects as the number of object types it contains. This is distinct (in quality, but perhaps not in quantity) from the plurality of a set, which is equal to the number of object tokens it contains. An object type is uniquely identified by some set of properties which tokens of that type possess (here, β and *not*-β are equally good properties). An ordered set (or sequence) is one in which the place for each object is numerically individuated.[5]

Here the distinction between the "diversity" of a collection and the "plurality" of a "collection" (we prefer this to "set" since this is not set theory) corresponds closely to what we would call the determinate as opposed to definite order.[6] We do not need the type-token distinction, however, since we have replaced it with a procedure for finding the immediate predecessor of a prehending actual entity as a concrescence; we do not need an overarching type-token abstraction for sorting properties, qualitative or quantitative.[7] Thus, for us, *not*-β (and this use of lower case Greek letters we already adopted above and will use below) is an operational contrast of eternal objects in two collections, between a prehending actual entity and its immediate predecessor, and likewise back into the past.

Crawford points out that carrying out an analysis of these "diverse" collections or, as we call them, constellations, includes "positional properties," which are postulates, or, as we would say, possibilities/eternal objects: "Peirce's inclusion of positional properties of objects in his analysis . . . is important not only because it captures a more complete account of diversity, but also because it is integral to Peirce's treatment of causal independence."[8]

We would distinguish the "causal" independence in Crawford's summary from our account, since the independence we are investigating is logical, or logico-mathematical. We believe that Peirce thought the right "model," if we can get it, would be internally related to the truth *itself* in the infinitely distant future. Whitehead does not see the creation of mathematical models

178 *The Algebra of Negative Prehension*

that way. He sees them as metaphors we replace with better ones, and logical rigor, adequacy, and applicability are the standards by which we assess the model.

Peirce handles the successive transformations of constellations of growing entities as analyzable in the *distribution* of characters, including positional properties. Crawford formalizes it this way:

Given the set of objects $\Gamma = (\gamma_1, \gamma_2, \gamma_3, \ldots)$ and the property α where A is the distribution of α in Γ, the distribution of α in Γ is uniform if there is some property β, where B is the distribution of β in Γ, such that A is dependent on B.

$$\alpha_n = 1 \longleftrightarrow \beta_n = 1$$

$$\frac{A = (1010111001010101010)}{B = (1010111001010101010)}$$

Crawford defines dependence (for Peirce) differently from our earlier account of dependence as successor to the prehending actual entity, but for our purposes, the dependence relation that is shown in the structure of the immediate predecessor of the prehending actual entity, as concrescence, will suffice.

With Crawford's algorithm, one follows a path until there is a variation. When there is a variation, one assigns a prime for the constellation "completed" and then "forces" the variation back into conformity with the sequence that presents actual divisions (in analysis, *in abstracto*). So the actual world is being used as the model and the constellations of possibilities are being discovered by the deviations from the actual and mapped *as* constellations with primes, to be rediscovered by means of this kind of division. Each deviation is to be taken as marking a limit and all the pairs entailed are to be treated as an actually indivisible whole. This is the definition, then, of a constellation of possibilities.

The last act which broke the pair defines both the limit of the whole and its simple constitution, because the Constellation minus that act is a map of the physical world with its actual conceptual feelings, i.e., its mental pole—so these are possibilities genuinely exemplified in the actual world of some actual entity. The collection becomes only a "mere" constellation by the agency of that relational "failure" or "deviation," and the inclusion of it as a part of the whole. Without that deviation, the collection has *some* actually divisible elements, but with the inclusion, it has *none*. Thus, there is an asymmetry of division/divisibility between the possibilities and the actualities. This may be stated in the negative by saying that no possibilities are actually divisible; they exist (intelligibly) only in indivisible constellations. One can state the law positively as constellations of possibilities are indivisible wholes (of varying complexity) and derive their unity from their limit—i.e., the last member of the sequence. This provides a way of formalizing minimal deviation from the actual world without presuming that possibilities and actualities can be genuinely separated, apart from analysis. It is a realistic model that treats the difference relation of the last element (and elements only come in pairs, i.e., a physical feeling with its conceptual feeling or its mental pole), and it privileges the physical. But the deviation is considered a real asymmetry between the actual and the possible that exists *as* a possibility, but is actual *as* a limit to the physical description of the world.

The constellation, so defined, may also be seen, from the standpoint of the eternal objects as a "thought." From the standpoint of the collection defined by "constellation minus that last pair" this is what Whitehead calls a "concept." From the standpoint of perception, the concept is a presentational space, and the difference between that space and the Fact of the World (i.e., the failed pairing that limits it), is the error inherent in all symbolic reference. Meanwhile, the Crawford technique restores our attention to the causal efficacy of the physical world. It isn't an arbitrary imposition of meanings; rather, it is a rejoining of the possible to the actual, "spacelessly." This rejoining has no effect on extension (in the Whiteheadian sense). But extensive connection *is* the relation between the members of every pairing. The "failure" of extensive connection isn't physical, but it is real—i.e., it is a real deviation or novelty introduced by a difference we can discover in the structure of possibility. It is slow work to discover the "structure" of possibility, but it depends upon operations that analyze patterns appearing in the primes and in their relationship to the continuum.

When we have enough primes, we will need to learn what operations can be carried out on the patterns in the primes. Projected patterns of these "cuts" onto a series of virtual plane would likely reveal some patterns, especially if the projections are done carefully (surely involving some conjecture/abduction). The alternative to producing patterns would be the denial that there is an intelligible relationship between possibility and genuine temporal flow, which seems unlikely. We should be able to make some sense of possibility-structures without falling into vicious abstractions.

There is some kind of relationship between division and divisibility and the realities in the physical world. A prehension (divisible only in analysis), and what can be actually divided are related, but prehension and possibility share the characteristic of being divisible only in analysis and thus, our thinking *about* them is always intensional—we cannot do a logic of possibility that sets individual possibilities into an atomic and mobile class of universals that then pick out particulars. The logic of possibility cannot be extensional.

Additional Tools

We have noted that when taking on a project that is formal in character, the choice of analytical tools is driven by considerations that go beyond formal capacities. Recall that in standard first order logic, Hilbert-style techniques can *formally* prove everything that can be proven by other approaches such as the tableau methods derived from Gentzen's work. However, there are significant *practical* differences in these methods that appear most dramatically when one attempts to automate formal proof. Hilbert techniques are nearly unusable, creating infinite loops on even the simplest proofs, whereas methods of the sequent calculi enable automatic methods to progress in a direct manner.[9] Pedagogical and intuitive issues are differently addressed by various analytical instruments, so the tools we choose outrun merely formal issues. How we analyze the structure of negative prehension requires care.

We need to set out some heuristic principles to clarify what we are doing and why. These principles can be summarized:

> The *analysis* of possibility is nonmonotonic
> The *analysis* of possibility is intensional
> The *analysis* of possibility is perspectival

The comprehensive explication of these heuristic principles must wait for another work. But we can gloss here a "story" that will lend credibility to these choices, unavoidable in any attempt to deal seriously with Whitehead's approach to possibility.

First, please note the emphasis upon *analysis*. The structural claims we make here with respect to possibility fall within the purview of the quantum of *explanation*, and do not embrace ontological extravagances.

The analysis of possibility must be *nonmonotonic*, because *inquiry* is nonmonotonic. Even though eternal objects are neither created nor destroyed, their ingression and egress creates relevance as a reality that must be described. We argue that the analysis of possibility is *intensional* because possibilities are relations, and not set-theoretic points patiently awaiting our notice in some set-theoretic caricature of a "possible world." The requirement that the analysis of possibility be *perspectival* is explicitly built into the actual entity, and the fact that it has its world only from a perspective.

On the basis of 1–3, the necessetarianism governing most modal systems of the past 100+ years is illegitimately ontological; logically justifiable "necessity"—that is, "necessity" that falls within a *logically* intelligible quantum of explanation—is short-hand for the *local* requirements (as expressed in numbers 1 and 3 above) of *particular* inquiries (numbers 2 and 3 above). Also on the basis of 1–3, we can see that a rough handling of possibility "itself"—as a thing-in-itself—is a gratuitously ontological *idea* that invites necessitarian *concepts* which serve only to block the road of inquiry and *explanation*.

Our concrete suggestions for the formal tools for such an analysis might be approached are also three in number: model theory, non-standard analysis, and modal logic. The most important of these is model theory, as the instrument within which the other two operate. Model theory enables one to characterize the relations among structures that are "almost the same," and weigh that "almost" along a variety of gradations. Moreover, there are model-theoretic tools, such as Paul Cohen's technique of "forcing," (and non-standard analysis itself) that make it possible to develop subtleties of structure that set theory is unable to represent.[10]

A caution about the term "model": we have repeatedly appealed to "models"—Neil Young's railroad, representations of Fenway Park—as essential aspects of our narrative. The appeal now to model-*theory* is the type of move that a non-expert might find perfectly acceptable. But certain experts with strong backgrounds in science or mathematics might find this segue suspicious. As the experts are well aware, the use of the concept of "model" in empirical science is radically different from the uses of that term in mathematics and formal logic. We are cognizant of those differences. However, these are differences of *degree*, and not of *kind*. While the differences are certainly real, the similarities are what are most profoundly salient here.

In *any* inquiry, of *any* kind—abstract or concrete, scientific or logical—there are important commonalities which may be abstractly represented. There will be (1) the thing inquired into (the "universe of discourse"); (2) the tools used to characterize that universe (the "language") and (3) the relations between these two (the "semantics," or that which makes things said *in* the language *about* the universe "true," or "false," or something else[11]). In a purely abstract setting, the range of choices in establishing 1–3 is significant, which has led to fascinating results in formal logic. Within more concrete inquiries (including empirical science) the arrangement of 1–3 is more constrained. Yet, as we will see in chapter 10, there are areas of contemporary physical science where the failure to respect such constraints is troubling. Our use of models and model-theory is neither arbitrary nor unconsidered. We insist that the connections between the two are the formal basis for characterizing logic—in line with Peirce, Dewey, and Hintikka—as the theory of inquiry.

The next formal instrument that provides some purchase on the complexities of possibility is non-standard analysis. Regarding the notion of "almost the same," non-standard analysis is the formal study of the concept of the infinitesimal, in its various guises. The idea of the infinitesimal, so central to the development of the calculus and Newton's revolution in physics, was long thought to be self-contradictory. Almost a century after Cauchy and Weierstrass had "eliminated" the infinitesimal from any further discussion, the logician Abraham Robinson demonstrated—using model theory—that the infinitesimal was a consistent and formally respectable notion.[12] Non-standard analysis will provide us with the tool for modeling differences that are too small to matter and yet grow to dominate a relational structure.

Finally, the appeal to modal logic is different from the "orthodox" philosophical conceptions of necessity and possibility The concept of "intensional" that C.I. Lewis brought to bear on the subject in 1918 is much too desiccated to play the role in the analysis of possibility that Lewis intended. The failures of Lewis's "strict implication" to capture the concept of "implies" are well known, and need not be rehearsed here. No small part of the problems dogging the analysis of possibility in the last century are due to the move away from the rich notion of intensionality found in Leibniz, for extensionally based (in the "bad," non-Whiteheadian sense of "extension") concepts. It is understandable why Lewis went this direction: he thought the formal development of intensional concepts was a failure. *Philosophical* logic has followed in Lewis's footsteps.[13] We believe that contemporary developments—constructions and discoveries that occurred after Lewis's day—permit us to gain at least a toehold on a genuinely intensional formulation of possibility within a functional quantum of explanation.

Thus, regarding modal logics, we reject as pointless such ontological notions as "possible worlds." We embrace formal machinery such as accessibility relations and frames, enabling us to represent models that differ in a real, albeit infinitesimal degree, and which function within a perspectivally framed system. Additionally, tools of modal logic have been used to model the paraconsistent overlapping of otherwise contradictory models, recently given concrete form by Bryson Brown and Graham Priest.[14]

Model theory is by far the most "strategic" choice of the three tools mentioned above. The techniques of model theory open the door to the study of quantifiers in ways that proof-theoretic methods cannot.[15] *How* things are collected into a totality tells us much about *what* kind of totality we are working with. This "how" of the collection is where the strategic character of the inquiry expresses itself in the particular form and content of the quantifiers. Also, model theory is directly rooted in *algebraic* approaches to formalization, as opposed to the proof-theoretic methods, more common amongst philosophers.[16] Algebraic approaches examine collections (even constellations) of *functionally* equivalent structures, instead of listing proof transformations in a sequence.

The application of model-theoretic tools enables one to formalize conditions of truth highlighting the role of negation. For example, the collection of propositions within the theory of a model designated (i.e. "true") and anti-designated ("false") need not form a complete, non-intersecting partition of the constellation of all propositions in the theory. Thus, there can be gaps and/or overlaps amongst the designated and anti-designated propositions, different ways in which "true," "false," "not-true" and "not-false" manifest within the model.[17]

The idea of truth/falsity gaps is of interest in the case of negative prehension. In one sense a negatively prehended actuality "p" will "positively" present itself as "not-p," but under certain circumstances it can *physically* disappear from relevance to the extent that it is presented as *neither* "p" nor "not-p." Yet there remains a kind of "thereness" to the actuality in question, even when all that is left is the shadow of the "scars of its birth," such that it is never quite *no longer there*.

Model theory also allows us to engage the shifting strangeness of the negatively prehended—at least metaphorically—targets of "non-standard analysis." This is the technical term for the "infinitely small yet somehow still meaningful." In this case, "infinitely small" means that no collection of infinitesimals (either finite or infinite) is capable of having an effect at the macroscopic level. Yet (according to the theory), infinitesimals manage to have *just such effects*. This is a curious state of affairs. Yet, Abraham Robinson demonstrated that the idea of the infinitesimal is perfectly intelligible, i.e., a logically consistent *model* (in the explicitly model-theoretic sense) can be built, successfully fulfilling all the requisite assumptions of the theory.

This concept of the infinitesimal gives us insight into the way that objects in relational systems might have considerable structure of their own, but that structure might be invisible for some purposes of the over-arching model and hence disregarded in those instances. Thus, for the purposes of some model M, object α is a single "thing," yet when viewed on its own account α might be "infinitely" nuanced in ways that M is incapable of showing.

Next, recall the distribution diagram that Crawford built based on Peirce's discussion. Multiple objects $\alpha_1, \alpha_2, \ldots$ in structure A and $\beta_1, \beta_2, \ldots$ in structure B are assigned a "1" or a "0" depending on whether they are properly characterized by feature Γ.[18] This assignment is provided by what is now known as the "characteristic function" of Γ. The characteristic function is typically written with the Greek "χ"; thus, if Γ assigns "1" to α_1, we would write this as, "$\chi(\Gamma[\alpha_1]) = 1$."

Let us change the perspective: rather than looking at the characteristic function of a single feature Γ on lists of objects in multiple structures, let us examine the long list of features—call these features "eternal objects"—on some single entity α in a single model M. Bear in mind that this "list" of eternal objects (call it "Γ_i") will be uncountably infinite in length.[19] Consequently, the list generated by the characteristic functions of all those eternal objects will itself be unboundedly long.

Thus we assert: $\{\chi(\Gamma_i[\alpha])\}$. The braces indicate that this is now an entire collection; the content of our assertion will be something like {1,0,1,1,1, 0,0,0,0,0,1,1, . . .}. The resemblance to Crawford's diagram is evident, especially *once we consider comparing not every entity in two different models according to a single eternal object, but two entities in a single model according to their respective characteristic functions of every eternal object*. Whitehead calls this difference in the models a "diversity of status."[20] In fact, Whitehead has this entire argument laid out informally:

> This diversity of status, combined with the real unity of the components, means that the real synthesis of two component elements in the objective datum of feeling must be infected with the individual particularities of each of the relata. Thus, the synthesis in its completeness expresses the joint particularities of that pair of relata, and can relate no others. A complex entity with this individual definiteness, arising out of

determinateness of eternal objects will be termed a contrast. A contrast cannot be abstracted from the contrasted relata.[21]

It is now clear why we distinguished definite from determinate order. If we were less careful, we could be deceived about whether our account of the contrasts, achieved in the elimination of eternal objects, was based on illicit abstractions. The kinds of contrasts that help us grasp determinate order among constellations of possibilities cannot be further abstracted than the formalization above allows. We can extrapolate, but we must be aware that our suppositions are only educated guesses. The possible *qua* possible is pretty elusive, but not wholly beyond our reach. Exploiting concepts such as the infinitesimal opens the door to seeing how differences too small to make a difference in any given M are still operative at a structural level as contrasts. Real synthesis is our clue to the elimination of possibilities, but possibilities are relational through and through. One remains uncertain as to their status *except* within the context of their "diversity of status."

The ordered model we are now considering essentially represents the historical route of an actual entity. However, the preceding formalization is still not quite nuanced enough. Recalling that in *physical* prehension an eternal object can be either actively negated valuated downwards or just passively dropped from relevance,[22] our list above must now be expanded to include these latter cases into something like this:

$$\{1,0,\_,\_,1,1,\_,1,0,0,0,\_,\_,0,0,1,1,\ldots\}$$

where the underscore "_" is a placeholder for some Γ_j that has gone beyond simple negative prehension and physically vanished from relevance. Again, keep in mind that the characteristics in the braces refer to a *single* entity.[23]

What does it mean for some entity α to make the *transition* (not concrescence) into its successor entity α_1? It means that for some $\Gamma_j \subset \Gamma_i$, $\chi(\Gamma_j[\alpha]) \neq \chi(\Gamma_j[\alpha_1])$. Notice first that Γ_j will always be a *proper* sub-collection of Γ_i, since it is unimaginable that an entity should change in every possible characteristic in its immediately succeeding concrescence.[24] Also, even if Γ_j is an infinite collection, it can still be vanishingly small with respect to Γ_i. Indeed, Γ_j might be so inconsequential compared to Γ_i that the differences between $\Gamma_i[\alpha]$ and $\Gamma_i[\alpha_1]$ are effectively non-existent, especially when considered by finite minds. "Effectively non-existent" is a longer way of saying "infinitesimal"; hence, the tools of non-standard analysis provide us with viable purchase on this issue. One must still wonder how infinitesimal differences can translate into macroscopic realities for finite minds to discover and utilize.

The puzzle of the infinitesimal originally presented itself to Newton and Leibniz, and their followers. But they did not have a thoroughly worked out theory of infinitesimals at hand to explain their own ideas. They had to fumble their way through and tolerate the apparently contradictory notions of infinitesimal differences with no possible macroscopic effects on the one hand, that led to exactly such macroscopic effects, with their validating consequences for Newtonian physics, on the other. Physicists and mathematicians somehow survived for 150 years with this active, self-contradictory position.

Even with the inconsistency removed in the nineteenth century by the Cauchy-Weierstrass's ε/δ techniques, puzzles about the infinitesimal remained. Robinson's work from 1959 (expanded in 1970) showed that the *idea* of the infinitesimal was *internally* consistent, but left contradictory the matter of how it could work effectively with *external* ideas. This last step—providing formal validity for the operational fact of human inquirers working with on outward contradiction—is the content of the paper by Brown and Priest (cited earlier). What is of immediate interest for us is that among the tools employed by those authors was a technique exploiting the model-theoretic character of "possible worlds" modal logic.

The "possible worlds" modal logic is fraught with problematic metaphysical assumptions. A better way of framing the subject would be as "alternative models," particularly since the latter has the advantage of being actually true. Modal "possible worlds" are nothing more than formal *models*. These models, in addition to their internal consistency, exhibit certain external relations to one another which are of substantive formal interest. The most important of these is the "accessibility relation," more of which momentarily.

One of the tools used by Brown and Priest was first presented in 1979 by Nicholas Rescher and Robert Brandom in *The Logic of Inconsistency*.[25] Using "possible world" subscripts, Rescher and Brandom indicated how contradictory propositions could be "bracketed" or, in their terminology, "quarantined" so that those propositions could be held within multiple "worlds" as true without ramifications beyond their brackets into the entire models/worlds, where the propositions could exist individually and unproblematically. Thus, one would have a schema such as $/\diamond[p_{w1}$ & $\sim p_{w2}]_{w3}/$ = True, which is to say, "the truth value of possibly [p-(in w1) and not-p-(in w2)] overlapping in world w3 is True." Rather than the existence of actual worlds, one should view this formula as capturing the overlapping (hence, accessibility) of functional models. As metaphysics, the formal conditions of physical succession in actual entities is complex enough to satisfy any logic nerd. In short, an actual entity in Whitehead's sense, i.e., the quantum of explanation, is all "the world" one needs, or can ever get, by way of explanation. Again, "model" would be better than "world." But since Rescher and Brandom used "possible-world" terminology, we use it here. It is this quarantined overlap that interests us, both for its formal qualities and its general interpretation, placing ideas in *manageable* conflict.

The Rescher-Brandom technique provides an abstract schema indicating how a contradictory pair can be managed while quarantined against wider consequences (in the formal sense). More importantly, this technique indicates how the accessibility relation can exceed conventional boundaries, including possibilities—as *connected* possibilities—that might not otherwise seem available. There are numerous acceptable treatments of modal logic that cover the basic notions, so we set that aside here.[26]

Here, the approach to modal logics found in computer science interests us. The accessibility relation remains a primary concept, but it is treated as *the transition function from one state to another*. Each "state" is a model with its own internal structure and complexities. In the computer science approach that there are the kinds and degrees of complexities that can emerge relating to the ways different models can be accessible to one another. These issues of complexity and accessibility set formal limits on the knowability of formal connections within the context of finite minds (which must deal with computably accessible relations).

Finally let us repeat that the uses of modal logic that are presented in the above eschew any connection to the standard interpretations and biases regarding necessity and possibility one finds in the philosophical literature. The tools we are drawing on have to do with "stitching together" models of actualities. Our account of the structure of the possible emerges from no such simple a pattern as $\sim\square\sim$.[27]

These topics of computability and complexity arise because of the problematic issues regarding what is real, what is formalizable, and what is effectively possible for a mind with only finite resources. In particular, tracing out the accessible—but infinitesimal—changes in state from one actual entity to its immediate successor, in the context of not merely what is actual but what is possible, raises staggeringly difficult practical problems. Whitehead recognizes these differences and provides the path to the real accessibility criteria in his idea of "feeling." The evident contradiction between the infinitesimal and the effective is due to the "disastrous confusion, more especially by Hume, of conceptual feelings with perceptual feelings."[28] The concepts of infinitesimal and macroscopically effective are resolved by the *felt* reality of the distinctions. "Our perceptual feelings feel particular existence; that is to say, a physical feeling, belonging to the percipient, feels the nexus between two other actualities, A and B."[29] We are not attempting to explain how the infinitesimal and the effective are resolved, but merely formalizing a resolution that is already achieved in experience. We explain the abstract with the concrete. Perception is more abstract than prehension, but far less abstract than concrescence and transition. Thus, the percipient actual entity, which is, after all, a genuine division of the extensive continuum, is too abstract to get at feeling, or fundamental relation, but is a fair unit for modeling concrescence and transition. Perceptual feelings are genuine relations, even if not fully concrete (as are prehensions).

These practical problems can be abstractly represented and schematized. Making sense of such schemata in a rigorously developed model-theoretic framework is more than a mere exercise in clever symbol manipulation. It illuminates the nature of science itself within a realist framework (see chapter 10); it makes explicable the notion of continuity as an emergent property of actuality in infinitesimal changes; it opens the door to a formalization of the concept of the "gestalt" as precisely such a cumulative realization of individually infinitesimal concrescences below the level of effectiveness that nevertheless concretize in a "sudden" novelty; and so on. Formal systems of relations are not shackles to the imagination, but are stepping stones to previously unimagin*able* universes of possibilities.

We are aware that we've made scarcely any real steps toward developing any such a formalism. Our gloss is only a *proposal* for a research program we will take up in the next book. For the purposes of *this* argument, it suffices to show that there are instruments at hand that suggest promising insights into the structure not only of negative prehension, but of possibility itself. Bringing these instruments into play involves no mean feat: while model theory has been variously applied to the concepts of infinitesimal and alethic modality, combining these pieces has (as far as we know) never been done. Consequently, the most we can achieve here is to crack the door, a little, upon an as yet undiscovered territory and suggest that there are credible reasons to believe that ingress to that territory can be made using those tools.

Eternal Objects in Transition

In the two previous sections we explained how eternal objects are *functionally* related to the actual entity as a concrescence. Having a procedure for discovering the immediate predecessor of an actual entity, as a concrescence, enabled us to extrapolate from a *definite* order some of the features of the *determinate* modes of order which contributed to that definite achievement (and hence could be abstracted from it in coordinate division). From a comparison and contrast of the determinate order exemplified in a concrescence with the determinate order inferred in its immediate predecessor, and so on, back through a physical route, we were able to infer some emergent structures of collections of determinate order. This functional analysis worked *with* the concrescence.

We noted that "other possibilities are real but do not belong to the route being analyzed," and these belong to the actual entity as transition. Whitehead says:

> An eternal object can only function in the concrescence of an actual entity in one of three ways: (i) it can be an element in the definiteness of some objectified nexus, or of some single actual entity, which is the datum of the feeling; (ii) it can be an element in the subjective form of some feeling; or (iii) it can be an element in the datum of a conceptual, or propositional feeling. All other modes of ingression arise from integrations which presuppose these modes.[30]

These three modes of ingression taken together are exhaustive of the role of eternal objects (possibilities) in the coordinate whole presupposed by the genetic account in *Process and Reality*. When one combines the role of the eternal object as concrescence with its role as transition, they exhaust what we can expect to learn about possibility, i.e., determinate order (including definite order), in our immediate epoch. Our analysis above shows in some detail how the first and second of Whitehead's functions can be more thoroughly presented. One reason we have held so closely to concrescence (as distinct from transition) is because we are not yet in a position to identify what was eliminated in concrescence *with* the determinate order of transition. We move into different territory, however, when we consider the third function of eternal objects in concrescence. Whitehead says that "the third mode is merely the *conceptual valuation* of the *potential ingression* in one of the other two modes."[31] This an ingression of constellations that *might* or *might not* be exemplified in the satisfaction, depending upon the conceptual valuation of the eternal objects by the concrescence.

We do not know in coordinate division whether the ingression *has* "occurred" (in the sense of exemplification), and if so, whether it owes its relevance to the objective or subjective mode of ingression. What we do know in coordinate division is that any difference that appears between the determinate order of a concrescence and that of its immediate predecessor, as inferred, will be inexplicable except by reference to some pattern of determinate order which is uncertain, perhaps even conjectural. But there *must be* a determinate order, and there is no reason to think it will be unintelligible to us, so long as we don't stray too far from the relevance functions we have discussed above.[32]

However, we will never be in a position to assert a final identity between the concrescence and the transition of the "same" actual entity on the weight of a coordinate division. Thus, to the extent that the genetic division presupposes the coordinate division, the distortions introduced by the negations involved in abstracting will affect the assertibility of the results. In a word, this isn't pure guesswork, but it isn't science either. The most we could achieve is a plausible assertion of the functional equivalence of an actual entity as transition with its concrescence. We will assert this equivalence until and unless we can be shown some empirical hypothesis that is advanced by denying it.

Whitehead continues: "It [the third mode of ingression] is a real ingression into actuality; but it is a restricted ingression with *mere potentiality*, withholding the immediate realization of its function of conferring definiteness."[33] In more ordinary language, we (and all other actual things) experience possibility *as such*, but only as a *determinate* order of possibilities left over from withheld potentiality, not as anything definite. We call this "the egress of possibility." It is as if one has made room for the feeling of what *almost* happened but at the last moment withheld itself. This is a kind of conceptual feeling. It reminds us of the "ever not quite" of William James. The real potentiality of this "ever not quite" is relinquished into the *might-have-been* in the moment of transition—it becomes, as Whitehead says, "mere potentiality."

That feeling of the might-have-been comes pretty close to the whole of what we know about the actual entity as transition. Might-have-beens are like an ontological "event horizon," immanent in the valuation of the *definite* order as a contrast—no longer a contrast with a predecessor, but with a contemporary that will never become a percipient actual entity. Only as a present becomes a past does a reason emerge as a connecting sinew with what has come to be on the one hand, what has just barely failed to be on the other hand. We feel the latter conceptually, but not physically. It becomes what Whitehead calls in the higher phases of experience an "intuitive judgment."[34]

If we had no such conceptual feeling, it would be impossible to explain things like sudden insights and inspirations that seem to come from the pure blue.[35] Without this sort of structure associated with possibility, everything actual would be wholly necessary—i.e., could not possibly have been otherwise. We mentioned earlier that Whitehead's theory of negative prehension does much the same work for him that Peirce's notion of abduction does. But Whitehead is bolder, theorizing the *possible* qua *possible* as *conceptually felt and evaluated*.

The difference between the first and second modes of ingression of eternal objects, on one hand, and the third mode, on the other hand, is that the former two are "unrestricted"; both have a traceable relationship to the *definiteness* of the concrescence. The variance in the two unrestricted modes of ingression is that one is objective and the other is subjective. If a constellation of eternal objects is "inherited," so to speak, from the act of the immediate predecessor or from the achievement of any other actual occasion or nexus in the actual world of the concrescence, the eternal objects in that constellation are objective, which is to say that they are felt by the concrescence as part of its *data* and will therefore be exemplified in the definite *datum* achieved by the concrescence. Whitehead says that "the solidarity of the world rests upon the incurable objectivity of this species of eternal objects."[36]

In short, once a constellation (including the ingressing collection physically felt in concrescence) of eternal objects has been incorporated into the achievement of an actual entity, those possibilities are "mere" possibilities no longer. Everything actual must deal with them, and were it not so, there would be no fact that *is* the world. If there were no such objective demand in concrescence, possibilities might disappear even though they had been realized, and if that could happen, no ground could be given for the *togetherness* of the world as a fact. In such a situation, any explanation for experience, as *had*, would be as good as any other; the absence of even a single ingressing eternal object in the actual world of an actual entity indicates that temporal passage has only a contingent effect on actuality. In such a situation, where the capricious disappearance of possibilities shatters experience of the world as solidarity, there can be no such thing as a quantum of explanation. Possibilities as felt by the actual entity, as transition, cannot have a merely arbitrary arrangement. What belongs to determinate order, insofar as we know it (and even though it *has not* and *will not* contribute to the definiteness of any actual entity) is nevertheless bounded by a kind of logic of possibility, the limits of which we learn, in part, from the way(s) that certain possibilities fall *just short* of being realized.

There is a clear difference between what you *can* see, and what you *could have seen*, had it come to pass. One might assert: "you never did that; if you had, I should have seen it." Such statements come very close to being either simply true or false, although the criteria that would enable us to evaluate them *in the way we would evaluate claims based on the actual entity as concrescence* are unavailable. To make such an assertion, we need to draw upon the *order* of transition, particularly the way that the elimination of unselected eternal objects still belongs to the determinate order of the immediate epoch. If our assertion is compelling, that is due to our common recognition that what *does* conform to the requirements of definite fact has a strong relationship with what *conforms with* but does *not* contribute to definite fact.

Hence, the study of the prehending actual entity, as transition, is the study of how we appropriately generalize the order immanent in those possibilities that just barely failed to contribute to the definiteness of the satisfaction of the actual entity as concrescence. This is an abstract study. But it enables us, for example, to imagine worlds in which slight variations of natural law give rise to different experiential values, such as those we find in science fiction writing, and in contrast to which we come to understand our own experience as an *example* of *possible* experience rather than as a necessitated total order.[37]

What does Whitehead mean when he says that the third mode of ingression is "restricted," while the objective and subjective modes of ingression are "unrestricted"? We understand it by comparing the objective and subjective modes. The *objective* constellations of eternal objects can make no contribution to the *subjective* form of the concrescence, Whitehead says. Those objective constellations are mathematical Platonic forms. But some unrestricted possibilities are *not* felt objectively: "A member of the subjective species [of eternal objects] is, in its primary character, an element in the *definiteness* of the subjective form of a feeling. It is a *determinate* way in which the feeling *can* feel."[38] In short, in the second mode of ingression, there is an *order* in subjective feeling, and considered in the mode of possibility, this feeling reveals a *determinate* order, pointing to its own possibility as subjective feeling. Whitehead continues, "it is an emotion, or intensity, or an adversion, or an aversion, or a pleasure, or a pain. It *defines* the subjective form of feeling of one actual entity."[39]

We have stressed the importance of the physical as we learned about the determinate order of the cosmic epoch. The unrestricted presence of eternal objects in the actual world is a part of what *defines* the world, but it doesn't follow that the world *is* just the way it *must be*. Our subjective feeling of the world in emotion, intensity, adversion, or aversion (for example) point to possibilities unavailable as data or datum, wholly peculiar to the subjective form of the prehending actual entity and in the world in no other way. Yet possibilities of the subjective species are wholly actual, and we would say that the meaning of "unrestricted" draws exclusively upon what is wholly actual for our appropriation of the eternal objects in these constellations. These eternal objects are "inherited" from the actual world of the actual entity.

But the third mode of ingression belongs neither to the objective nor to the subjective species. Possibilities considered in light of transition that are *not* part of the physical world are still felt, but in a *restricted* way. It isn't easy to learn about the structure of feelings available *only* in the mental pole. Whitehead says that "the subjective forms of feelings are only explicable by the categoreal demands arising from the unity of the subject,"[40] but those feelings that are *not* physical at all are accessible only to coordinate division (the kind of contrasting procedure we explained above).

We can illustrate this subtle relationship by considering the "Impossible Snowflake." Imagine that one snowflake, unique in its configuration, is called "Plato." Plato is hexagonal, and might never have formed or fallen in just the way it did, but having fallen, it illustrates in definiteness a determinate order in the immediate epoch. Now contrast Plato with a near neighbor, whom we will call "Octy." Octy seems incapable, in principle, of forming in this immediate epoch. Its unique pattern is octagonal, making it a *merely* possible snowflake due to the molecular structure of H_2O in our immediate epoch. Whether Octy might fall in our cosmic epoch isn't something we can know. Yet, we don't find the possibility unintelligible.

If there were an entire theory of snowflakes that never fell, but might have, their snowdrift would be a fair metaphor of the theory of the actual entity as transition. Can we ski on that drift? In imagination, yes. In concrescence, well, best stay in the lodge for some hot buttered rum.

The octagonal snowflake is a might-have-been dominated by the objective mode of ingression. The third mode of ingression can be dominated by either its objective or subjective aspect. Hence there can be might-have-beens dominated by the emotive, or advertive, or aversive character. For example, imagine a car has just sped past as you were about to step off the curb. Your health has been preserved by the difference of a fraction of a second, and you feel, as the car disappears in the distance, that you almost died. The fact of the world, in its solidarity, will not allow this feeling to be any part of its definition. Yet, there is a determinate order that includes your physical death which just barely failed to be part of the world in its definiteness. You *do* feel that determinate order, and the feeling is subjective in the same sense as the second mode of ingression, but this is a conceptual feeling *only*, bringing on a positive value, i.e., a sense of physical relief, precisely because it *isn't* part of the world. You will never be hit by just that car at just that moment. But the determinate order to which such an accident belongs just barely failed to be actual.

Such conceptual feelings are closely akin to the kind of intensely aesthetic experience had by mathematicians in the presence of ideas they have not yet been able to prove, but of whose demonstrability they are convinced.[41] Such mathematicians are dealing with *futural* possibilities that evade them, whereas determinate order as it characterizes transition depicts the past. Yet, it is impossible, according to Whitehead, to be certain about what is contemporary with the prehending actual entity, and thus, matters we take to be might-be's are sometimes might-have-beens. Such a line is never sharply drawn.

Compared with such ephemeral matters, we can easily understand the sense in which some aspects of that *physical* order are operationally unavailable to a percipient actual entity when analyzed into prehensions. This is the genetic sense of the term "negative prehension."

> The possibility of finite truths depends on the fact that the satisfaction of an actual entity is divisible into a variety of determinate operations. The operations are 'prehensions.' But the negative prehensions which consist of exclusions from contribution to the concrescence can be treated in their subordination to the positive prehensions.[42]

Determinate negative prehension transforms initial data (multiplicity) into objective datum (nexus, proposition, entity). "There is a transition from the initial data to the objective datum effected by the elimination."[43] Knowing the effect of an elimination is far easier than knowing *what* has been eliminated. The ontological event horizon is outlined by what is *not* available in prehension. But the relations between these effects and the forms, both objective and subjective, that have been eliminated are creative relations. They can be formalized in algebraic terms, and even if we have not fully carried that demand through, we have shown its determinate outlines.

10   The Nature of Naturalism

> At this moment scientists and sceptics are the leading dogmatists. Advance in detail is admitted: fundamental novelty is barred. This dogmatic common sense is the death of philosophic adventure. The universe is vast.
>
> —Whitehead, *Essays in Science and Philosophy* (92)

Model-Theory Redux

As we have noted, model-theory is an important instrument for philosophical analysis. This tool is a late addition to the methods of formal logic, unavailable to Whitehead in his work on extensive abstraction. Trapped under the shadow of the *Principia*, Whitehead turned away from the techniques of proof-theory for the more appropriate algebraic framework which dominated his earlier mathematical work, and richly informed his philosophical thinking. Model-theory is not reducible to algebraic approaches to logic, but it is highly algebraic *in its approach*, especially in comparison to proof-theoretic methods. Much in model theory, especially the interpretation of the quantifiers, is about the comparison of equivalence classes of formulas (recall the quote from Paul Halmos). Hence, the close relationship between model-theoretic and algebraic methods is readily visible. The absence of model-theory cost Whitehead an *efficient* tool for analyzing and characterizing the structural features of relational abstractions. We have sketched such a model-theoretic approach above, closely approximating Whitehead's actual analysis.

First, model theory deals with abstractions. That which is genuinely concrete is experienced in its wholeness and totality. Anything less than that totality is at best abstract and at worst recondite. Second, the worst thing about moving away from the totality of experience is not *falsehood*. As Whitehead's theory of propositions makes clear, falsehood is oftentimes the most creative of deviations. Rather, the worst thing is to be recondite. Being recondite means being not only narrowly conceived, but dubiously applicable: worse than wrong, it is irrelevant.

Regarding the second point, an abstraction, bringing a reduced picture of concrete actuality, renders the model "false" in the sense of partial, but such abstractions are *not* falsehoods *merely* on account of being partial. (See below our account of Whitehead's radical realism.) A model must drop out many details of the reality it is modeling to achieve a manageable abstraction *of* that reality. Only *some* of the totality of relational connections are highlighted by any model. (Recall our earlier model of Fenway Park). But if the model is built well, if it is constructed carefully through logically rigorous inquiry—then the relational connections so highlighted will still be *real* relations, even if not the *totality* of real relations.

Thus, Neil Young's model trains are not falsifications of "real" locomotives, even though the loads they are carrying are not measured in metric tons. Young's fascination with the details of scaling down the work needed to haul a "real" load so that his model trains accurately reflect a scaled version of the effects in sound, in acceleration and deceleration, in hill climbing ability, in the rattle and boom, of the loads on "real" trains might seem eccentric to the non-enthusiast. Perhaps, but it is *only* eccentricity. Such "modeling" would be altogether *meaningless* if there was not a concrete reality that all these factors were genuinely *modeling*.

Following these considerations, a model-theoretic approach to inquiry is entirely compatible with a robust *realism*. If a model is not a model of some-"thing" really "out there," then in what sense can it be called a "model"? Yet *how* a model relates to that some-"thing" can*not* be taken for granted if we are serious about our realism. Consider the Boeing 747. There is the scale plastic replica of the exterior of the aircraft that one might find in the lobby of Boeing's headquarters. That plastic representation is certainly *a* model of a 747. But then, the set of technical manuals used by the maintenance staff at the hangars where these vessels are housed are also a model of a 747. Unlike the plastic representation, the technical manuals will bear little visual analogy to the systems aboard the plane.[1] Neither the craft in the lobby nor the manuals actually fly, but the plane they model does. So which "model" is more "real"?

Also consider the high-resolution color photographs from the Voyager spacecraft. These images are a stunning display of visual detail and color.[2] What is interesting for us is that all of this color imagery is a fabrication—it is not "really" there, but has been digitally added to the pictures by computer enhancement. Bluntly, these pictures are not exact representations of the astronomical bodies themselves, but only visual *models* of them, in which coloration that enhances the real details has been super-added to the original images to enable the human eye to take in the real details with more ease. But that is exactly what a good model does. It allows us to cognize real and salient facts that might otherwise be lost to a finite mind in the overwhelming minutiae of the concrete totality.

Thus, models serve purposes—*our* purposes—which are "local" to the topic and intentions that motivate the making of the model in a given inquiry. But a model is not less realistic for being intentional than for being abstract. These two (intention and abstraction) are flip-sides of the same coin. One cannot abstract out relevant details without first intentionally focusing inquiry in an appropriate way, while the intentionality-driven focus is already an abstraction away from the totality of the concrete fact of the world. This is not a retreat from reality, it is the logically mandated emphasis upon salient relational connections.

In the case of model-theory, the thing it is an abstraction *from* is not so much the totality of experience, but rather of cognitive structures—often themselves quite formal—which are, in turn, abstractions from that fullness of experience. *Formal* model-theory is an abstraction from an abstraction, a model of modeling; tools for refining tools. A real tool is a reflection of the reality it is intended to work upon, even when that reality is a few steps removed, by processes of abstraction. We would argue that model-theoretic approaches come with an implicit, but essentially "built-in" implicit realist "metaphysics." This implicit realism is an expression of how model-theory understands the logical quantifiers as expressing an interrogative "game" of fitting a theory to the structure being modeled.[3]

This "game-theoretic" approach to the model-theoretic understanding of the quantifiers is quite literal. One imagines two players who are testing out a model of some kind. For our purposes we'll refer to them as "Verifier" and "Falsifier." Regardless of the terms we use, the purpose of the players is the same: one wants to show the model is "true," the other that it is "false," these evaluations being defined upon the game itself, not some context-free semantic analysis. Falsifier might present an object from the universe of discourse, challenging Verifier to show that the model can match this object properly within the model's theory. Verifier might present a structure within the model and demand that Falsifier provide any example from the universe of discourse which stands outside the theory's entailments. If Verifier has a winning strategy (one that wins on any possible play by Falsifier), then the model is true. If Falsifier has a winning strategy, then the model is false. If neither has a winning strategy, then the model is neither true nor false. (This latter possibility occurs only in games of infinite length.)

Regarding our model of Fenway Park, we could say that Falsifier has the winning strategy, due to the impossibility of modeling every aspect of the stadium under the stated purpose (i.e., the relevant universe of discourse), since certain features of the task are irresolvably recalcitrant. Yet, if these limits are incorporated explicitly into the *purpose* of the model, with hedges and caveats, then Falsifier is robbed of a winning strategy. With Neil Young's trains, we have something closer to an infinite game. Falsifier says, "Hey, Neil, bet you can't . . ." to which he replies, "Dude, let me think about it . . . and I wish I hadn't given up weed." This is not relativism, but it *is* a kind of contextualism. This is how finite minds (some more finite than others) situate models in a vast universe.

Where the quantifiers come in to "play" is with the idea of a winning strategy, using a universal ("for all") quantifier, which says that any move the other player can make will fail to trump the first player's move. An existential ("for some") quantifier says that there is something the current player *can* do that will match or beat the other player's previous move. Note that which player is asserting which quantifier depends on whether the particular move is to show the model to be true or not, and on whether the move is to trump the other player's efforts in this regard.

This game-theoretic approach to the quantifiers makes model-theory the most appropriate methodology for justifying the centrality of formal logic for good reasoning; it centers the formalism directly upon the process of inquiry itself.[4] "Verifier" represents the inquirer or community of inquiry; "Falsifier" represents that which is being investigated. Scientists who support a theory are the Verifiers, while nature (and those scientists challenging the paradigm) play the role of the Falsifier. In the absence of an objectively real nature, the role of Falsifier *makes no sense*.

This realist interpretation of the quantifiers is in line with our position throughout this book. "Logic" as the theory of inquiry; model-theory then becomes the principal formal tool for mapping key abstract relationships *within* the process of inquiry. This mapping shows how model-theory is an essential auxiliary to science, since scientific inquiry is "just" the most highly refined form of inquiry, with nature and natural processes playing the most uncompromising Falsifier we can encounter.[5]

This introduction contextualizes what follows. We return to Whitehead's philosophy of science and nature, informed now by our account of his relationalism in the context of the quantum of explanation and the intricate algebra of negative prehension. Contemporary physical science has wandered from its supposedly empirical commitments. Some blame for this drift can be laid at the feet of the philosophers. Philosophy as a public *conversation*, vigorously pursued *with the public*, has disappeared in favor of the *perceived* safety of academic isolation.

Yet scientists are not innocent. Understanding our criticisms of these attitudes amongst scientists applies basic ideas of model-*theory* in order to understand the failings of "model-centrism." Model-theoretic ideas open up a powerful line of interpretation, enabling us to formalize a significant part of Whitehead's philosophy. We already mentioned the deep connections between model-theoretic and algebraic approaches, the latter being Whiteheadian to the marrow. When seen in the light of the failures of set and proof theories, the model-theoretic approach stands out as a substantial advance in formal methods.[6] We will argue that these factors make the model-theoretic approach the most appropriate one to the thoroughly *radical* form of *realism* that is the flip-side of Whitehead's radical *empiricism*. This needs some explanation, since realism and empiricism are often treated, since Hume, as opposites. But that is only because empiricism and realism are not handled in a genuinely *radical* way.

We will look at two attempts to do an "end-around" on the careful metaphysics of nature. The first is the *ex fiat* declaration of the supposed end of philosophy by scientists who are doing philosophy badly. The second will be an examination of Bas Van Fraassen and his old(ish)-school empiricist application, using many of the model-theoretic tools in *his* project that we have been advocating here. Van Fraassen's project is ultimately untenable on account of its Humean commitments. Each of these positions illustrates something about the model-theoretic approach that serves our argument about Whitehead's "radical realism." Our starting point is the bad metaphysics of contemporary model-centric physics. The model-*centric* and the model-*theoretic* approaches are as different in spirit as the distinction between medieval philosophy and genuine science. The appeal to medieval philosophy is not entirely rhetorical.

Bad Metaphysics

Is it possible to do science without philosophy? Stephen Hawking—among others—is convinced it can be done. In Hawking's late forays into popular writing, one of the first moves is a back-handed dismissal of philosophy.[7] Hawking has sparred with philosophers in the past with no constructive intent, and even less constructive effect. The sheer vacuity of Hawking's broadside is notable. One cannot call it a "criticism" because nothing he says might be mistaken for a *reasoned* argument. But more importantly, the overtly *philosophical* purpose of Hawking's enterprise throughout *The Grand Design* renders his initial attack disingenuous; he can avoid the charge of overt hypocrisy only on the grounds that he was unaware of the blatantly philosophical method and intent of his book. Ignorance is seldom a good excuse for sloppy hyperbole. Contrary to the view expressed by Hawking, Whitehead asserts the following:

> The claim of science that it can produce an understanding of its procedures within the limits of its own categories, or that those categories themselves are understandable without reference to their status within the widest categories under exploration by the speculative Reason— that claim is entirely unfounded. Insofar as philosophers have failed, scientists do not know what they are talking about when they pursue their own methods, and insofar as philosophers have succeeded, to that extent scientists can attain an understanding of science.[8]

The problems with Hawking's view are indicative of larger of problems endemic with contemporary physics in particular, and science in general. Many of these problems are the effects of bad metaphysics. Contemporary science has been hamstrung by metaphysical commitments that are not recognized as such, and consequently cannot be addressed through the normal scientific channels. Such commitments cannot be addressed even through extraordinary channels—largely philosophical ones—until it is recognized that those channels exist and play an important role.

The willful blindness engendered by this absence—indeed, aggressive abandonment—of philosophical inquiry into basic principles has shackled physical science in particular to a clowder of assumptions for which there is little justification. Whitehead published *Science in the Modern World* in 1925, so it is not as though fundamental criticisms of hidebound physcialism are only now appearing. Yet these criticisms have been ignored by the scientific community. As a consequence we will not shy away from polemics in what follows.

Turning to Hawking's bromide, after listing certain ideas and pursuits, Hawking declares that, "traditionally these are questions for philosophy, but philosophy is dead. Philosophy has not kept up with modern developments in science, particularly physics. Scientists have become the bearers of the torch of discovery in our quest for knowledge."[9] This is a throw-away remark, unaccompanied by any argument, meant only to upset philosophers.[10] Hawking labors under the delusion that philosophy is only "real" when it is doing physics, as though the philosophical assumptions Hawking takes for granted brooked no critique. But aside from its failure in critical thinking, Hawking's dismissive swipe comes with significant history of its own. In his earlier *Black Holes and Baby Universes*, Hawking lamented the poor reception of his previous *ipse dixit*isms amongst philosophers:

> There is a real problem here. The people who ought to study and argue such questions, the philosophers, have mostly not had enough mathematical background to keep up with modern developments in theoretical physics. There is a subspecies called philosophers of science who ought to be better equipped. But many of them are failed physicists who found it too hard to invent new theories and so took to writing about the philosophy of physics instead. They are still arguing about the scientific theories of the early years of this century, like relativity and quantum mechanics. They are not in touch with the present frontier of physics. . . . Maybe I'm being harsh on philosophers, but they have not been very kind to me.[11]

It is very difficult to put a charitable spin on Hawking's remark. We are not too proud to offer criticisms of our discipline for failures on the part of some—indeed, many—to achieve an adequate grasp of mathematical ideas. But those criticisms are predicated upon a principled understanding of what mathematics is and how it can and ought to be employed. Hawking, by contrast, provides no actual reasons for his claims beyond a failure by philosophers to accept contemporary physics without question. Quite aside from his utter disdain for the very real philosophical issues that continue mercilessly to hound the developments of physics *since* the early twentieth century (still unresolved, even *within* physics) he evidently considers his *ad hominem* remarks sufficient.

Hawking fails to consider that the rough handling he complains about may have been deserved. Anyone who troubled to compare Hawking's *Brief History of Time* to Sherover's *The Human Experience of Time*, might be struck by the fact that Hawking's volume was so brief was because it lacked any *history*.[12] Realistically, Hawking's responses are sheer petulance. Some latitude must be given for the fact that Hawking's works were popular books. But some responsibility ought to be taken for the fact that they were written from a significant bully pulpit and some effort expended by Hawking to achieve at least a marginal level of scholarship. So Hawking's argument that "they hurt my feelings" scarcely seems worthy of someone who (until recently) held the Lucasian Chair at Cambridge, as he often reminds us.

This isn't the worst of it. The worst is Hawking's defense of "model-dependent realism." The problem with this proposal is not that it is bad philosophy—although it certainly is that. The problem is whether it is *any sort of philosophy at all*. Having declared the philosophical enterprise to be dead—because it is not sufficiently akin to physics—Hawking then blithely engages in a patently philosophical project, apparently in ignorance of the fact that he is doing philosophy:

> Until the advent of modern physics it was generally thought that all knowledge of the world could be obtained through direct observation. . . . [This] naïve view of reality . . . is not compatible with modern physics. To deal with such paradoxes we shall adopt an approach that we call model-dependent realism. It is based on the idea that our brains interpret the input from our sensory organs by making a model of the world. . . . But there may be different ways in which one could model the same physical situation, with each employing different fundamental elements and concepts. If two such physical theories or models accurately predict the same events, one cannot be said to be more real than the other; rather, we are free to use whichever model is most convenient.[13]

That this is a philosophical rather than a scientific position requires no commentary, yet Hawking considers himself to be functioning as a scientist. But there is nothing here that is even marginally testable; it is a program for interpreting—and a framework in which to *interpret*—the results of physics. The complete catalogue of the history of such ideas and their problems would take days to produce, although Quine's work on the underdetermination of theories leaps to mind. The fact that Hawking wants to argue for a "realism" in which it is "dealer's choice" (our phrase) as to which model is "real" whenever it conveniently "fits" one's "observations" hardly stands out as an exemplar of logical coherence. Even the word "observations" must be scare-quoted since the actual data supporting—and necessitating!—Hawking's multiple models is so thin it almost does not exist.

Hawking's "model-dependent realism" is a prime example of what we call "model-centrism." For model-centrists, theory trumps all else. In Hawking's case, rather than take from the failure of a coherent model to emerge from the scattered shards of evidence that there is a fundamental problem with theories of physics, he cavorts triumphantly, legislating that *all* the equivalent models are true and real. Another example of this theory-above-all attitude is found in the writings of Brian Greene. Greene, with an innocent charm, casually declares that, "*The* overarching lesson that has emerged from scientific inquiry over the last century is that human experience is often a misleading guide to the true nature of reality. Lying just beneath the surface of the everyday *is* a world we'd hardly recognize. . . . assessing life through the lens of everyday experience is like gazing at a van Gogh through an empty Coke bottle"[14] The solution for Greene is not merely "extraordinary" experience (since the everyday variety cannot be trusted), but experience *corrected* by theory. The question that model-centrists never ask is, "What, then, corrects theory?" Indeed, *quis custodiet ipsos custodes?*

With Hawking, apart from the lack of historical understanding, his dogmatic defense of patently inconsistent models is supposed to justify theory in the absence of observational evidence. Hawking's drive to have things both ways, real yet model-dependent, is a consequence of various contradictory beliefs Hawking holds but refuses to subject to serious, *philosophical* criticism. Hawking's commitment to realism is almost the *sine qua non* of any working scientist. Whitehead spoke to this sort of view with his usual compelling logic and grace regarding the discovery of Pluto. Initially, the idea was that Pluto had been discovered because there were deviations from the mathematically predicted paths of Uranus and Neptune:

> There is, however, no difficulty in producing a mathematical formula which *describes* the observed deviation. Such a formula will be of the most elementary mathematical character. . . . Altogether a description of charming simplicity which would have delighted Plato by its exemplification of his most daring speculations as to the future of mathematics. Every Positivist must have been completely satisfied. . . . *But astronomers were not satisfied*. They remembered the Law of Gravitation.[15]

Note the ease with which scientists could have formally "tweaked" the mathematics in order to make them "fit" the observations. Observations conflict, but simply manufacturing an *ad hoc* mathematical formula to "take care of things" enjoys (in Russell's phrase) all the advantages of theft over honest toil. This is simply fabricating parameters from the blue and inserting them into the theory and brushing aside any embarrassing contact with empirical facts. This adding of parameters is a favorite move of the model-centrists, fundamentally at odds with the realism that is a driving force with working scientists. Hawking insists that the emerging mathematical structures that he and others have developed are *real*. But these mathematical constructions do not comprise a univocal theory *of* the real; worse yet, as many physicists have pointed out, these abstract creations are so devoid of empirical content that they lack genuine *scientific* standing at all.[16]

The most notorious form of these theories—which Hawking enthusiastically embraces and uses as the justification for his "model dependent realism"—is the "M-theory" family of string-theory based mathematical constructs. "M-theory," Hawking observes, "is not a theory in the usual sense. It is a whole family of different theories, each of which is a good description of observations only in some range of physical situations."[17] Hawking profoundly mischaracterizes the situation. The problem is not just that M-theory is not "*a*" theory; the problem is that it is not a *theory* in any scientifically robust sense at all.[18] Lacking even the abstract possibility of empirical test, string-based proposals are so inherently vacuous that they do not even qualify as hypotheses. A hypothesis is something that at least *might* be tested; string-theories require such outlandish experimental set-ups—nuclear accelerators whose diameters are equal to that of our own galaxy, for instance—that they make a mockery of the distinction between "practical" and "in principle" impossibility.

Regarding Hawking "model-dependent realism" does not develop the topic with any nuance. He might have noticed the absence of empirical content in his mathematical constructs and why that was significant if he knew any philosophy. He might have discovered that the logical tools to render coherent his model-dependencies were developed by philosophers like Brandom and Rescher, and deployed by folks like Brown and Priest. But Hawking didn't bother. His dismissal of the philosophy of science included the declaration, without evidence, that many philosophers of science were simply persons who lacked the ability to do physics. It seems legitimate to wonder about the talents and proclivities of mathematicians[19] who would legislate to the entire history of human thought, rather than take the trouble actually to learn and engage that history.

Model-centrism is bald-facedly medieval in its methods and attitudes. This is not rhetorical extravagance; it is a direct analogy. Many important questions were engaged by the medievals: the relation of universals and particulars and the nature of possibility were two we noted above. But the medievals were also severely limited in their lines of inquiry: whatever question they asked and whatever answer they proposed, it had to be made to fit within the confines of the *standard model* handed to them by the Church. Some variation was permitted, but any fundamental challenges were variously extirpated by force or logical trickery. There was even an equivalent of the model-centrist's endless use of parameters, in the form of casuistry. When the orthodox deemed it appropriate, it was always possible to make some action justifiable within the scope of supposedly "absolute" dogma.

The model-centrists' unbridled use of parameters, forcing the world to fit their theory, is a reduction of reality to epicycles all the way down. We noted in chapter 6, the numerous scientific criticisms of ΛCDM within the community, over and above Whitehead's critique of the measurement problem. Many of these criticisms—including Whitehead's—came in the context of workable alternatives whose real measure of viability is ignored. If Einstein's initial presentation of GR had been left in its original form, given no more attention or modification than Whitehead's theory has received over the decades, then GR would be every bit as much an abject "failure" as Whitehead's theory is declared to be by the model-centrists. GR is the "heart" of ΛCDM, but that heart is only beating because of the endless addition of stints, bypasses, and even a pacemaker, all in the forms of model-centric parameters that do not so much "show" the theory to be true as *manufacture* the desired fit with observation.

In recent science news, the Planck satellite, which is providing the most refined images so far of the Cosmic Microwave Background Radiation ("CMBR") is triumphantly hailed as providing the most exacting confirmation of the standard model. But in the same breath, enthusiastic model-centrists also state that it shows features that cannot be explained by the standard model and may require an entirely "new physics."[20] In other words, the Planck satellite has decisively *falsified* ΛCDM . . .

We can do better than the self-contradiction of declaring the theory true while calling for a fundamentally new physics. Even though the model-centrists are faced with their own overtly self-contradictory claims; even though they are presented with (largely ignored) alternatives to ΛCDM in the peer-reviewed literature; even though such evidence as is actually available is compromised by the systematic assumptions underlying the interpretation of that evidence; and even though all of that evidence only "fits" because the overwhelming majority of gravitational cosmologists has, for the past century, devoted their careers to inserting and tweaking parameters in the standard model so as to guarantee that fit, their triumphalism remains unabated. But at least there *is* some evidence supporting the standard model of cosmology; string theory is altogether devoid of empirical content.[21]

From a model-theoretic perspective, model-centrist triumphalism has two main props. First, model-centrists have effectively denied the proper function of the quantifiers. Our game-theoretical approach to the quantifiers requires that scientists play only *one side* of the game, *nature* takes the other side. The model-centrists have usurped both sides of the play, legislating that only those moves that "verify" the preferred model are permitted. Genuine testing by nature is not allowed, either by changing the rules of the "game" with *ad hoc* parameters, or by disregarding falsifying evidence. Second, model-*theorists* realize that within the scope of any finite amount of evidence—or, equivalently, a finite number of moves of the semantical quantifier "game"—it is always possible to construct an equally valid model to "fit" evidence/moves up to that point. Yet even when viable alternatives are present, their existence is seldom even acknowledged much less given serious consideration. So the standard model is again insulated from serious test by avoiding substantive comparison that might challenge orthodoxy. Finally, on those rare occasions when a comparison is made, the model-centrists often use an older version of the alternative model, no attempt being made to consider whether the same boundless introduction of new parameters, which the standard model enjoys, might serve the alternative model to an equal or greater degree.

Hawking's move in *The Grand Design*—declaring philosophy otiose, while practicing it—is simply the latest step of the hide-bound philosophy that Whitehead so thoroughly criticized in *Science and The Modern World*. Hawking is trapped in an inconsistent triad: his theory must be about reality, it must be true, and yet it is not even fully coherent with itself. His *philosophical* position binds him because he cannot even admit he is *doing* philosophy—and doing it badly. He has bound himself to a metaphysical perspective he cannot acknowledge as such, and hence cannot defend, criticize, or modify, much less reject.

No Metaphysics?

Hawking's model-dependent realism could be defended, but the steps necessary to make it credible would not please him, since it requires relinquishing some part or parts of the above mentioned triad. It is worth our while to glance briefly at one such alternative that has been advanced by a respected philosopher of science, i.e., Bas Van Fraassen and his "constructive empiricism."

Van Fraassen avoids Hawking's inconsistent triad by denying realism altogether and moderating Hawking's truth claim. According to Van Fraassen, scientific realism holds that, "Science aims to give us, in its theories, a literally true story of what the world is like; and acceptance of a scientific theory involves the belief that it is true."[22] In lieu of this sort of realism, he turns toward a kind of anti-realist empiricism that has not been especially popular for some decades, even within analytical circles. Van Fraassen notes, "according to the anti-realist, the proposer (of a theory) does not assert the theory to be true; he *displays* it and claims certain virtues for it."[23] While noting that the idea of "literal construal" requires unpacking, he expands his comparison with realism:

> The idea of a literally true account has two aspects: the language is to be literally construed; and so construed, the account is true. This divides the anti-realists into two sorts. The first sort holds that science is or aims to be true, properly (but not literally) construed. The second holds that the language of science should be literally construed, but its theories need not be true to be good. The anti-realism I shall advocate belongs to the second sort.[24]

Hawking would be extremely unhappy with any form of anti-realism, though Van Fraassen certainly has his reasons. Van Fraassen shifts the truth claim to one of "empirical adequacy," a move necessitated by his empiricism, yet which Hawking would find extremely distasteful. However, Van

Fraassen's empirical adequacy criterion is far less of a shift from Hawking's truth claim than the shift from realism to an anti-realist empiricism, and might be less vexatious for Hawking.

Van Fraassen's empiricism is not so much a systematic doctrine or definitional ideology, as it is a "stance."[25] Eschewing a false precision, Van Fraassen argues for a broadly construed collection of notions that do not reduce to a simplistically consistent account. This move is an attempt to capture the core ideas of traditional empiricism[26] without committing any of the reductionistic fallacies that open that tradition to criticism. Van Fraasen hopes to capture a principled agnosticism toward the unjustifiable and grandiose claims of analytical metaphysics. Van Fraassen's "type 2" anti-realism also allows him to take the language of science seriously (to be construed literally) and incorporate such language into his constructive view of empiricism.

Van Fraassen's work has long shown a predilection toward model-theoretic forms of logic. Throughout *The Scientific Image*, he carefully exploits model-theoretic ideas, grounding the "constructive" aspects of his empiricism and developing his account of the scientific enterprise. This disposition ideally situates him to develop in a formally respectable manner a collection (or family) of ideas that do not meld into a simplistically "consistent" set of first-order propositions. Simplistic melding is very much what Hawking wants to achieve with his model-dependency with respect to M-theory. However, Van Fraassen is logician and philosopher enough to pull it off.[27] As he shows in *The Scientific Image*, the structure of explanation itself cannot be reduced to a single set of principles, and a multiplicity of irreducible models will typically be required in even the most elementary situations. Van Fraassen's model-theoretic approach is substantive and robustly non-reductionistic. Hawking's discussion in contrast, seems more than a trifle blunt.

However, Van Fraassen's empiricism remains problematic. He is obligated to a normative claim about scientific realism/anti-realism. Clearly his account is a failure if it is intended to be merely descriptive. Scientists are almost universally realists. So, Van Fraassen's anti-realist account has little if any historical or sociological content. That is not a problem *per se*, since philosophy of science is about the underlying logic of inquiry rather than the recorded methodologies of practice. But as a normative proposal, Van Fraassen's position provides us with no tools for criticizing the model-centric position. After all, what basis can be offered to argue against simply adding parameters whenever convenient in order to save one's cherished standard model? Since *truth* and *reality* are no longer the bases upon which we extol the "virtues" of our theories, why not appeal to the triumphalism of the orthodoxy? Following the above quote from *Adventures of Ideas*, fixing the formulae to fit the available observations is easy. The only "virtue" one might appeal to for not doing so in the case of Pluto is that Pluto was really there. But then one comes right back to the fact that Pluto was *really* there.

Furthermore, the model-theoretic issues of quantifier interpretation snap even harder at Van Fraassen's heels than at Hawking's and the model-centrists'. Van Fraassen does not appeal to the game-theoretical approach to the quantifiers, despite his otherwise robust model-theoretic orientation. This is probably due to the fact that widespread interest in the game-theoretical approach to quantifier interpretation is a relative late-comer to the literature.[28] But the game-theoretical interpretation poses problems for Van Fraassen, because the interpretation of the quantifiers in the game-theoretical approach works from the fact that there are *two* players in the "game." The model-centrists want to shove the second player aside in order to guarantee that they always win; Van Fraassen cannot entirely admit *that there is* a second player, since otherwise his anti-realist position would contradict itself. The game-theoretic approach is not just a formal way of "tricking out" the quantifiers. It is the first genuinely concrete methodology for establishing a meaningful connection with one's universe of discourse. It is the only method available that enables one to get any leverage on characterizing finite models.[29] (Recall that at any given time, a finite model is all we have validated.) While one can always be a "traditionalist" with the quantifiers, and just pack them in under the heading of the traditional IIO syllogism (*It Is* Obvious), this is not an especially satisfactory philosophical move. Yet it is the only move that remains to Van Fraassen if he insists that he is not "really" talking about the *real*. Either the universe of discourse is *there*, and the quantifiers quantify against the challenging play of reality, or we are left with the inward spiraling maelstrom of our own imaginative makings that ultimately connects to nothing.

Finally, despite his efforts at a constructive approach in *The Scientific Image*, Van Fraassen offers us no real ground or context within which to understand—much less *rationally develop*—the *speculative* elements of the scientific (or any other) enterprise. This is a serious flaw in his approach, revealed early on in *The Scientific Image* with a purely throw-away remark. While laying out the basics of his own constructive empiricism, Van Fraassen says, "It is often not at all obvious whether a theoretical term refers to a concrete entity or a mathematical entity."[30] This is Van Fraassen's old-school empiricism speaking; no *radical* empiricist would make such a comparison. For radical empiricists—such as ourselves or Whitehead—mathematical "entities" can be *concretely real relations that are objectively present in nature and experience*. Recognizing this fact is the first step in dissolving the antipathy between realism and empiricism. But for that dissolution to occur, both must be genuinely *radical*.

This inability to acknowledge real relations leaves Van Fraassen with no way of accounting for, or explaining the effectiveness of, speculative thought. That there are speculative metaphysical ideas both *in* and *undergirding* science is indisputable. Quite aside from Whitehead's critiques, consider the "multiple and multiplying universes" approach that Hawking advocates. He would have us believe that the universe instantaneously splits itself into almost identical copies over infinitesimal differences—for example, whether or not an electron jumps to another state at this immediate instant rather than another instant that differs only by an immeasurably small interval, and that such copies—*of entire universes!*—copy themselves without bound.[31]

The only possible excuse for this outlandishness is that the scientific and philosophical communities, with their limited orthodox interpretations of the mathematical tools, have so boxed themselves in with their metaphysical assumptions that they are unable to take *relational possibilities* seriously. Hawking's "bad metaphysics" approach requires that we pretend such empirically vacuous foolishness is "good science," devoid of any metaphysical commitments. On the other hand, Van Fraassen's "no metaphysics" approach leave us to make up mathematical theories that are "empirically adequate" while requiring that we avoid interpreting those theories in ways that demand something better, something *true*. We are left without speculation that *could actually motivate constructing such theories in the first place*. While it is certainly true that we must exercise the greatest care and rigor when we engage in speculative thought, it seems to us that he has become over-worried about being wrong and has abandoned any effort at being right.

Speculative Philosophy

Whitehead presented a careful defense of the nature, role, and purpose of speculative philosophy in the opening chapters of *Process and Reality*. However, our initial point of engagement comes later in that work: "But in the real world, it is more important that a proposition be *interesting* than that it be true. The importance of truth is, that it adds to interest."[32] No small part of that interest is the enormously *constructive* contribution that boldly rigorous speculation provides to all aspects of human action and inquiry, including science. An erroneous proposition can be interesting for its constructive effect, an effect it can have only because it stands in *contrast* with what is *true*. This contrast is not only important on cognitive grounds, as a direction and inspiration to inquiry, but also as a fundamental moment in the creative advance of the world. For our purposes *here,* the cognitive aspect is the only one of importance.

Science is not self-contained. This is not peculiar to science, but a consequence of being an activity of persons in the world. Oftentimes, science can get on without philosophy, and at such times philosophers may content themselves with studying and interpreting the results of science. But at times of crisis and reorganization—times which scientists like Hawking may not recognize—philosophical critique becomes more than an adjunct, but is a fundamental contributor to the process of inquiry. As Whitehead phrased it, "philosophy then appears as a criticism and a corrective, and what is now to the purpose as an additional source of evidence in times of fundamental reorganisation."[33] We think it likely that Van Fraassen would agree with

Whitehead, but Whitehead also enjoyed a background as a working scientist during much of his career, and appreciated the necessity of the realist commitment. It is that commitment that both licenses and limits speculative thought.

Speculation for its own sake is frivolous. Even writers of fiction must compose their tales within boundaries that are not only recognizable as *a* world, but sufficiently close to *this* world that the situations and characters can be recognized by the reader as meaningfully akin to situations and characters *in this* world. But that connection requires a robust commitment to the idea *of* this *world*. Speculative philosophy—that is, "metaphysics"—must be bound even more closely to reality than is fiction, since the standards by which it is evaluated include not only logical coherence and narrative intelligibility (by which fiction is judged) but Van Fraassen's own "empirical adequacy" as well.

None of those things means anything unless there is a world against which they are compared. Van Fraassen's empiricism is ultimately too Humean to allow him to escape his self-imposed skepticism and rigorously engage in speculative thought. We suspect that Van Fraassen would say that his position exemplifies the highest qualities of rigor and discipline of thought by demurring from engaging in empty mind experiments not anchored in experience. But the experience he speaks of is not the *fullness* of human experience: being Humean it fails to be *radical*. This criticism works with greater force against the model-centrists. For while they *do* engage in bold speculations—M-theories and infinitely multiplying universes—they offer nothing beyond a lip-service connection to reality. There are no tests in their speculations, no rigorous comparisons to equally viable alternatives, no serious effort to challenge their own cherished standard model. There is *only* the speculation, and the triumphalist announcements about theories declared true *ex fiat* in the effective absence of empirical facts, effusively cheered on by an under-informed media. Scientific journalism has gone the way of all journalism in the last quarter century.

Whitehead's relationship to the philosophy of science—to say nothing of experience—isn't simple. We noted how Whitehead's radical empiricism is pervasive in all of his works. Thus, for example, Whitehead consistently argues that "nature" is what is revealed in our sensory experience. But the "characters"—his term, whose meaning we will explore momentarily—of that reality are themselves things that, while intrinsically a part of our experience, only unpack themselves in our cognitive knowledge *of* that experience as expressed through thoroughly relational symbolisms.[34] The amount of reality we take in through experience outpaces anything that we can shoe-horn into our cognitive "bottles," so the Humean reductionism *of* experience is a grotesque over-simplification.

How are we to account for this tsunami of reality pouring into us from our senses? We wish to introduce what we call "radical realism." This phrase captures the fact that, especially when applied to nature, radical *empiricism* of Whitehead's variety *is* a kind of *realism*: reality is really *there*. It is the content of experience, but at the same time also the *radical* content, which is the immediacy of reality in its relationally concrete *totality*.

Consider again the efforts by Hawking and others to account for relational possibilities by appealing to the empirically vacuous notion of "multiple universes." The least of the problems with this approach is that it lacks a metaphysical framework in which *relational possibilities* can be taken seriously on their own account.[35] As a result, these would-be metaphysicians must fall back upon the dubious *metaphor* of infinitely multiplying universes, in a desperate attempt to interpret their own mathematics. On the other hand, Van Fraassen's "Humean" approach offers us no real insight into this matter, but encourages that we not talk about such things. This does not solve the problem, but merely strangles it.[36]

These issues are traceable back to bad metaphysics: bad regardless of whether it is produced by ignorance or denial. The solution, then, is good metaphysics. Such a program, as Whitehead asserts, is driven by the desire, "to frame a coherent, logical, necessary system of general ideas in terms of which every element of our experience can be interpreted."[37] In Whitehead's scheme, the development of relational wholes, including the structures of possibility, is central: his "eternal objects," which he characterizes as "Pure Potentials for the Specific Determination of Fact,"[38] are there precisely to capture such relational possibilities within an explanatory framework without dismissing them or trivializing them as parasitic on their *relata*.

This business is one of speculative *philosophy*; it is not that of the scientist because its content—while it cannot fail of exemplification[39]—also transcends the subject matter of science; it goes beyond nature (that which is given in sense experience) to include the entirety of the real, including the bases of intelligible possibility itself. Solidly anchored by its commitment to the real, speculative philosophy becomes an active, constructive participant *with* science, without attempting to *be* science. At a time when theoreticians, dazzled by their mathematical cleverness, seem to dominate physics, it is well to remind ourselves:

> Philosophy, in one of its functions, is the critic of cosmologies. It is its function to harmonise, refashion, and justify divergent intuitions as to the nature of things. It has to insist on the scrutiny of the ultimate ideas, and on the retention of *the whole of the evidence* in shaping our cosmological scheme. Its business is to render explicit, and—so far as may be—efficient, *a process which otherwise is unconsciously performed without rational tests.*[40]

This role is not ended because someone like Hawking declares it dead. As Hawking himself so woefully demonstrates, this task is all the more necessary precisely because scientists who enjoy gatekeeper status are altogether unable to see their own activities with a critical eye.

Radical Realism

Part of the difficulty we face here is that Whitehead's philosophy of science has rarely received the attention it merits, even within the Whitehead community; outside of that community it is not even known. With only a few exceptions, when Whitehead's work is applied to matters of science, it is his *metaphysics* that is invoked, and almost exclusively in the context of quantum mechanics.[41] But these *applications* of Whitehead's thought are focused upon those metaphysical developments that *logically* precede natural science. We have seen how the attempt forcibly to re-interpret Whitehead's discussion of the quantum as a *physical* notion goes astray. Meanwhile, Whitehead's work on the philosophies of science and nature (1919–1925)—which *temporally* precede his metaphysics—is seldom consulted, even though it is the most relevant source.

The general disregard for the triptych is an unfortunate inheritance from decades of Ford-influenced readings of Whitehead. Yet in these three works, Whitehead lays out his vision of nature, science, and their connection. This development does not occur in his metaphysical works, *nor can it*. Whitehead states on multiple occasions that the philosophies of nature and science eschew any comment on the relation of mind and the world, which is the province of metaphysics. Metaphysics, for its part, provides the *logical* "back story" (our phrase) for nature and the relevant methods of inquiry into it (i.e., science). But it cannot explain either nature or science *per se*, since these are dependent upon, and specific to, the facts about the particular cosmic epoch in which they are found. Metaphysics provides the logical framework for explanation, but it does not provide the particular explanations of those abstractions that need to be explained. For its part, science does not burst forth Athena-like from the Zeus of metaphysics. Rather, science (and, of course, nature) must be inquired into independently, and we must adapt our inquiries by bringing to bear the relevant specifics of the subject matter and the world as we find them. The *natural* details of this cosmic epoch do not flow of necessity from the metaphysical characters of that epoch—those characters only provide the generic order of possibility. So science and nature must be engaged as they are found, and on their own terms.[42]

In addition to the above nuances, Whitehead's philosophy of science challenges the traditional classifying schemes. The two broadest thematic strands in the philosophy of science are (1) the demarcation problem—what marks off a genuinely scientific theory from one that is not?—and (2) the realism/anti-realism problem. These two themes are not disconnected, but for brevity's sake we can treat them so. Whitehead is a realist, but what *kind* of relation to reality is involved? What are the essential features of that reality? Underneath it all is Whitehead's *radically* different notion from that of mainstream philosophy of science about the "nature of Nature," and its relation(s) to science. By this point in the book, the reader should be alerted to the significance and importance of the term "radical" without much additional explanation. We will now explain Whitehead's scientific *realism* with his already noted Jamesian radical *empiricism*.

We raise three questions: What is science? What is nature? What is explanation? There is no alternative to starting in the middle. The appearance of arbitrariness derives from the *coherence* of Whitehead's thought on this subject (as that term is specifically explained in *Process and Reality*). For historical and contextual reasons, we will approach the questions in the above order.

What Is Science?

Whitehead entered into the philosophy of science from the position of a working scientist. But unlike many of the philosophically inclined scientists who followed, Whitehead was not one to valorize theory above the experiential bases of scientific inquiry. That is, Whitehead was *not* a model-centrist. He was a peculiarly "amphibious" creature: a mathematician who was suspicious of mathematics as a route to fundamental truth, even though all of his thinking was informed by his mathematical background. This is not surprising. The best critic of a subject will be someone who actually *knows* that subject.

First, consider the context of the philosophy of science in which Whitehead found himself. Science was being transformed by the architectonic shifts caused by the general theory of relativity and the early stages of the quantum revolution. These demanded revisions of fundamental theory at both a scientific and a philosophical level. Second, philosophical inquiry was on the verge of splitting into the "analytical" and "continental."[43] The analytical/empiricist thread that inherited the philosophy of science was still dogged by the ghost of Hume. This haunting caused the philosophy of science to break into two further camps. The realists provided a robust account of how science "worked," but could not provide a logically coherent account of *why* its methods of inquiry could capture essential characteristics of reality. There were also positivists whose anti-realism grasped the Humean nettle without qualification. *Within* their empiricist assumptions, they offered a superficially coherent account of scientific activity; but any connection with reality vanished, leaving only desiccated vision of merely *descriptive* "adequacy." By the 1920's, the positivists had gained dominance in the philosophy of science, not broken until the early 1960's.

Whitehead found both philosophical alternatives unacceptable; but the anti-realist approach of the positivists was downright anathema. The purpose of science was to discover the *truth* about nature, to find the *real characters* of the natural world. Thus, speaking of space and time, Whitehead said, "This spatio-temporal framework is not an arbitrary convention. Classification is merely an indication of characteristics which are already there."[44] Speaking of a more technical part of the development of his theory regarding scientific objects, Whitehead says, "the object is more than [a] logical group; it is the recognizable *permanent character* of its various situations."[45] This permanence is not something scientists are making up, or which only approximates the reality of nature for our descriptive purposes. The intention of science is to achieve a theory that is true, not one that merely enjoys some nebulous "virtue" of empirical adequacy. On the latter attitude, Whitehead comments, "I do not think that it is any answer . . . to say that our propositions are only a little wrong, any more than it is a consolation to his friends to say that a man is only a little dead."[46]

Most explicitly, in *The Concept of Nature*, referring to atoms and molecules, which the positivists described has being nothing more than mere "logical constructs," Whitehead states:

> I assume as an axiom that science is not a fairy tale. It is not engaged in decking out unknowable entities with arbitrary and fantastic properties. What then is science doing, granting that it is effecting something of importance? My answer is that it is determining the character of things known. . . .[47]

He continues:

> Now if there are no such entities, I fail to see how any statements about them can apply to nature. For example, the assertion that there is green cheese in the moon cannot be a premise in any deduction of scientific importance, unless indeed the presence of green cheese in the moon has been verified by experiment. The current answer to these objections is that, though atoms are merely conceptual, yet they are an interesting and picturesque way of saying something else which is true of nature. But surely if it is something else that you mean, for heaven's sake say it. Do away with this elaborate machinery of a conceptual nature which consists of assertions about things which don't exist in order to convey truths about things which do exist. I am maintaining the obvious position that scientific laws, if they are true, are statements about entities which we obtain knowledge of as being in nature; and that, if the entities to which the statements refer are not to be found in nature, the statements about them have no relevance to any purely natural occurrence.[48]

These statements leave no doubt as to the *fact* of Whitehead's realism. The task Whitehead has set before us is a bold one, but the *fact* of his realism does not clarify the *nature* of that realism. We must not imagine Whitehead's realism is akin to the kind which insinuated itself into the contemporary philosophy of science, especially naïve form of realism one finds in such popular treatments as Hawking's.

Reality cannot be boxed off to the side as standing against or opposed to experience, thence to be magically "mirrored" in our theories. With

Whitehead we must resist the tendency to assume there is a simplistic point-for-point correspondence between the "elements" of our scientific theories and the real world. This might appear to violate Van Fraassen's requirement that the language of a "realist" theory be literal. However, a mistake here treats the concept of "literal" as applying only to substantive "things." Relations are also *real*. But how is one "literally" to interpret the language of *any* theory? A little attention to hermeneutics will show it is unclear in what sense any rendering of any text can be called "literal."[49]

Problems immediately present themselves regarding the interpretation of the mathematical portions of scientific theories. Here the previously mentioned irony of Whitehead's mathematical "suspicions" appears. Whitehead, the mathematician, rejected attempts to treat the mathematical elements of scientific theories as providing foundational insight into the nature of the world. "There can be no true physical science which looks first to mathematics for the provision of a conceptual model."[50] Similarly, Whitehead rejected the spatial *metaphors* infusing the interpretation of the tensor symbolisms in contemporary cosmology,[51] and he emphatically rejected any attempt to valorize the theoretical aspects of scientific inquiry over the practical ones.[52] Whitehead does not view the Einsteinian, i.e., the standard model, interpretation of the mathematics as a "literal" reading of the *concepts*; the mathematics allows for a variety of interpretations, including those merely metaphorical ones that fail to characterize the reality.

The problem is that mathematics is not self-interpreting, as Whitehead recognized. Even in so "simple" a mathematical theory as arithmetic the concept of number is rife with what Whitehead called, "queer little contradictions which thoughtful people disregarded."[53] Similarly, Whitehead notes that the theory of types developed in the *Principia* does not rescue the concept of number from these contradictions, but merely presents a "rule of safety" to avoid committing the more glaring errors. "It follows that . . . the complete explanation of number awaits an understanding of their relevance of the notion of the varieties of multiplicity to the infinitude of things. Even in arithmetic you cannot get rid of a subconscious reference to the unbounded universe. You are abstracting details from a totality."[54]

So the "literal" language of a theory is still an abstraction, but a *true* one (*if* it is true) because it highlights the *relations* that are *really* there. It is not enough to have a refined mathematical theory if we cannot even say how the most elementary parts of that structure should be interpreted. But no simplistic reference to the practical effectiveness of a theory as a guide to action will suffice either. While any simplistically direct correlation between the mathematical forms of a theory and the reality that theory is about cannot be justified, Whitehead also rejects any valorization of the merely practical over the theoretical. "In practice," he tells us, "exactness vanishes: the sole problem is, 'Does it Work?' But the aim of practice can only be defined by the use of theory; so the question 'Does it Work?' is a reference to theory."[55]

To grasp Whitehead's program fully, one must favor neither abstract theory nor practical exigency. One must "jump in" to the fullness of the scientific enterprise, not abstracting away little bits of it, hoping to find its essential or literal "nature." Essential "nature" is nothing less than *nature* itself. Superficially, such a statement might seem too obvious to mention: what else would natural science be about? But *tepid* forms of realism that try to push nature aside wherever it can be "mirrored" by abstract theories, cannot work. But rejections of realism falsify the activity of science itself. We must change our focus. *Nature* is the subject of natural science, but we need an effective notion of *nature* prior to developing a refined scientific picture of the relational structures *within* nature: we must get clearer on the nature of nature.

What Is Nature?

Whitehead offers many statements of his definition. A few examples should suffice here. In the preface of *Enquiry*, Whitehead characterizes nature as, "the object of perceptual knowledge."[56] In *The Concept of Nature*, Whitehead states:

> Nature is that which we observe in perception through the senses. In this sense-perception we are aware of something which is not thought and which is self-contained for thought. This property of being self-contained for thought lies at the base of natural science. It means that nature can be thought of as a closed system whose mutual relations do not require the expression of the fact that they are thought about.[57]

Continuing this theme, Whitehead says, "Nature is what is observed,"[58] and "Nature is the system of factors apprehended in sense-awareness,"[59] and so on.

While such statements seem clear enough, and can be easily multiplied, they also may appear to be in tension our earlier assertions of Whitehead's realism. The above statements seem to articulate an "empiricist" position, not a "realist" one. But we must move beyond the *traditional* oppositions of those classifications. We must appreciate the genuinely *radical* nature of Whitehead's realism. It is certainly the case that Whitehead is a Jamesian radical empiricist (as we have been emphasizing). Yet, this latter label might conceal the *radically* objective "there-ness" of nature and its relational structures in Whitehead's robust *version* of radical empiricism. Thus, we speak of the radical *realism* of Whitehead's philosophy.

With respect to radical empiricism, let us revisit the original texts. James says that,

> To be radical, an empiricism must neither admit into its constructions any element that is not directly experienced, nor exclude from them any

element that is directly experienced. For such a philosophy, *the relations that connect experiences must themselves be experienced relations, and any kind of relation experienced must be accounted as 'real' as anything else in the system.*[60]

To be counted as *real*, a thing must be experienced; and to be counted as valid, a philosophy must not overlook *anything* experienced. But being experienced, it is genuinely *real*, not something we are merely constructing "in our heads." Note how emphatic James, insisting upon the inclusion of relations and relational structures. Whitehead shares this emphasis; it is the centerpiece of his radicalness.

Significantly, Whitehead named the second chapter of *The Principle of Relativity* "The Relatedness of Nature." Although the particular form of relatedness that concerned Whitehead in that volume was the kind of uniformity to be found among important sets of relations (those modes of relatedness known as "space" and "time"), the general fact of nature, as a related whole whose relations are themselves natural components of nature, is equally present.

We will discuss "fact," shortly in a technical way, noting here only that for Whitehead, "fact is the undifferentiated terminus of sense-awareness,"[61] that is, nature taken as a whole and all at once.[62] Whitehead's position is sufficiently clear when he states that "fact enters consciousness in a way peculiar to itself. It is *not* the sum of factors; it is rather the concreteness (or embeddedness) of factors, and the concreteness of an *inexhaustible relatedness* among *inexhaustible relata*."[63] These *relata* are *there* for experience, not the products of some additional act of cognition—i.e., all, the relatedness of *nature* that Whitehead talks about. "[T]he relations holding between natural entities are themselves natural entities, namely they are also factors of fact, there for sense-awareness."[64] These relations and *relata* are *experienced*, and are thus *real*, not merely logical constructs or "merely instrumental," constructive adjuncts *to* that experience. Emphasizing this *realism*, Whitehead also is committed to the idea that science aims at *truth*, not merely some diluted notion of adequacy.

This realism about relations poses a different problem for Whitehead's philosophy of nature. The relatedness of nature is "inexhaustible." In such a densely relational universe one cannot know a part merely as a part; its relatedness to the whole is an essential aspect of its character. Whitehead is thrust into acknowledging no middle ground between absolute knowledge and utter ignorance. Relatedness ceases when things cease to exist in relation to one another. One must either have all of the relations or none. Thus, we must discover some meaningful sense of limitation that makes knowing possible without first having to know everything. This limitation will be a natural relation in its own right. Finding this relation, one then has an account that is an abstraction, but still a "literal" expression of relations within "fact."

The first step is recognizing how nature and natural knowledge are themselves "bracketed." The first such bracket is the limitation of nature and its philosophy: "Our experiences of the apparent world are nature itself."[65] But, "nature is an abstraction from something more concrete than itself which must also include imagination, thought and emotion."[66] The philosophy of nature "is solely engaged in determining the most general conceptions which apply to things observed by the senses. Accordingly it is not even metaphysics: it could be called pan-physics."[67] Whitehead's realism does not fret over the Humean tangles others created for themselves, attempting to resolve issues of the knower and the known *within* the theory of nature. Whitehead says:

> We leave to metaphysics the synthesis of the knower and the known. . . . [A]ny metaphysical interpretation is an illegitimate importation into the philosophy of natural science. By a metaphysical interpretation I mean any discussion of the how (beyond nature) and of the why (beyond nature) of thought and sense-awareness. In the philosophy of science we seek the general notions which apply to nature, namely, to what we are aware of in perception. It is the philosophy of the thing perceived, and it should not be confused with the metaphysics of reality of which the scope embraces both perceiver and perceived.[68]

This limitation does not imply that mind is in some sense unreal or a mere epiphenomenon; nor, Whitehead insists, should he be understood, "as implying that sense-awareness and thought are the only activities which are to be ascribed to mind."[69] Rather, "nature as disclosed in sense-perception is self-contained as against sense-awareness,[70] in addition to being self-contained as against thought." Nature, Whitehead tells us, is "closed to mind." This closure ensures that, against Berkeleyan forms of empiricism, Whitehead's view is thoroughly *realistic* in its radicalism. Knowing about nature does not alter the nature that is known. This closure also ensures that there is a primary form of limitation to the relatedness of nature such that these relations are knowable by means of humanly possible methodologies. Not everything in nature is "infinitely relevant" to everything else, even though the totality is bound into a relational unity. We, as finite minds, are able to highlight salient points and turn these into the primary focus of our inquiries, *not* because other forms of relational connectedness are not there, but because they are not *relevant*.

There is a second important form of limitation, striking at the heart of the relational structure of nature itself and our ability to know it. Not all relations are of a kind, and these distinctions are operatively present in experience—they are a *real* part of *nature*. There is a distinction *in* experience between (some types of) contingent and necessary relations (local not only to the immediate epoch, or our cosmic epoch, but to nature itself) making it possible to determine various classes of essential limitations through the global necessities of their connections. Among those necessary relational structures, Whitehead argues, are the direct experience of the previously mentioned uniformities *in* nature *of* space and time.

This claim that we have a direct experience of the uniformity of space and time can be difficult to accept. It is a bit challenging to buy into the idea that we have direct experience of relational features of nature that are in some sense universal and "necessary." Yet, taking radical empiricism seriously, one is obliged to admit to the possibility of such experience; the alternative to radical empiricism is being trapped in a Humean world where experience is but a disjointed collection of subjects qualified by predicates—so much sound and fury, signifying nothing. In such a Humean world, "you must fail to give a coherent account of experience."[71] But even taken radically, what evidence might we find in experience to confirm the claim that the uniformities of space and time are immediately present features of experience? Whitehead's response is that there is no other evidence than that actually provided *in and by experience itself*.

Consider the example of a yardstick, both wherever you are and in New York City.[72] What "logical" reason is there for thinking that the yardstick means the same thing in both places? There is an unbounded range of viable mathematical relations that could hold such that, for example, the yardstick in one place would be no more than an inch long in another. No one seriously entertains such a possibility, and not because we've investigated the matter and discovered it to be false. Rather, the uniformity of space is such an immediate and necessary factor of our experience that only from the æther of *recondite* formal thought could any alternative occur to us.[73]

Distilling such relational differences as those between the "necessary" uniformities of space and time and contingent factors of physical nature,[74] the mind casts a sufficiently robust net over reality to capture a systematic totality of basic relational characters of nature such that genuine knowledge is possible. A *natural* limit is discovered in the separation among relational forms so that the essential structures can be discerned, from which the *naturally* contingent ones can be teased out through further inquiry.

This relational picture reiterates Whitehead's numerous criticisms of the natural philosophy of his day, a view that still operates in many quarters (often implicitly); we are speaking of hide-bound physicalism.

"Physicalism" in this sense is not merely the claim that nature consists of nothing more than point-like bundles of matter in billiard-ball style collisions. Rather, it is the generic presupposition that the fundamental characters and qualities of nature come in punctiform bundles (such as infinitesimal points of time *and* space, not just bundles of matter) that are just and only the things they are without any further reference to the rest of nature; as a corollary, all relations are ultimately parasitic upon these punctiform bundles. These bundles, being ultimate (according to hidebound physicalism), neither need nor admit of any further explanation, and themselves explain all of the relational structures of nature.[75] These relational structures are themselves merely parasitic abstractions of the supposedly "real" punctiform "things." This position abandons any discussion of real possibilities within physics, leaving them to multiply universes. A considerable part of contemporary science rejects this model of nature, but it is unclear to what extent an alternative is consciously operating amongst those who make the rejection of the above naïve form of materialism explicit.

Hidebound physicalism is implicated in the general failure to take relational structures seriously. Once the fundamental structure of reality is conceived as concentrated in discrete points, then any relations amongst such points can exist only as external and parasitic interpretations we ourselves have made and imposed upon the "true" realities. The only "reality" that remains within this physicalist worldview is not that of experience and nature in their fullness, but rather of subjects (the punctiform "realities") qualified by predicates. But as Whitehead observed, "If you once conceive fundamental fact as a multiplicity of subjects qualified by predicates, you must fail to give a coherent account of experience."[76] The fallacy of materialism, in forms which, by materialism's essential nature, reject the fundamental relatedness *of* nature—is the root source of two of Whitehead's more famous charges against orthodox scientific and philosophical thinking: the fallacy of simple location and the fallacy of misplaced concreteness.

For Whitehead, while any particular set of equations is a poor source for a conceptual framework of nature, the generic habit of algebraic thinking is different. Algebraic thinking is not about some specific mathematical model, nor is it about something as broadly conceived as a conceptual frame. Algebraic thinking is about taking relations seriously on their own account by disciplining the mind to such thinking *through* the engagement with abstract algebra itself. This is the resolution of the seeming paradox of Whitehead the mathematician rejecting mathematical models as a source of *conceptual* models. The habit of algebraic thinking will *open* many doors, but it will not (by itself) provide us with the conceptual model necessary to tell us *which* door we should go through.

In order to know nature, we must know it as it is: a relational whole, closed to mind but penetrable by inquiry. Yet nature is not an *unlimited* whole. Nature is a system of relationships that variously break themselves out between the "essential" and the "inessential," the "necessary" and the "contingent." *How* these relational structures break out will be relevant to some inquirential purpose. But relational realities are not less real or less objective simply because human purposes are selectively brought to bear upon the experiential totality of nature. We choose which parts of reality to focus our attention upon. If we have chosen well, those parts are objectively, *radically*, real features of nature and experience.

Our subject is beginning to shift, and parts of it are "bleeding into" one another like patches of paint that keep running on our canvas. Our topic seems to have slipped from the radically realist "nature of Nature" into the more epistemological issue of the nature of *knowing about* Nature. This is where we began our discussion of the nature of science. So it is now time to close our circle, with a newly informed perspective on our starting point.

What Is Explanation?

Whitehead's theory of science is a theory of *explanation*, and not merely one of description. Any attempt to leave science at a purely descriptive stage while claiming it could be an adequate philosophy of science is wrecked upon the shoals of scientific practice. Describing observed phenomena is only a starting point for scientific inquiry. As the discovery of Pluto shows, tweaking parameters does not suffice to learn what is *true*, only to show how *clever* we are with our mathematics. Empirical adequacy (Van Fraassen's criterion) fails us because the only real "virtue" relevant to a scientific theory, qua theory, is whether it is *true*. Truth will ensure empirical adequacy, but we should not take for granted that a cleverly parameterized model pushes its empirical adequacy any closer to scientific *truth*—the lesson from Hawking and the model-centrists. It is not enough for a scientific theory to *be* an adequate description; it must provide us with a correct *explanation*. This is Whitehead's realism; a description can be good or bad, but an explanation is supposed to be *true*.

Van Fraassen makes some good points regarding the context sensitivity of any explanatory framework.[77] But Whitehead's work addresses those concerns by establishing the context of *Nature*, understood in his radically real way. Context sensitivity does not undermine the realist claim. The number of true things claimed *within* an inquiry might be quite large, and those true things might not say "the same thing." All forms of inquiry are automatically limited by considerations of *relevance* that cannot be eliminated in the name of a "more real" truth. We must abstract from the totality of nature to address the purpose(s) of inquiry, and this abstraction will take account of the real lines of relevance-relatedness, else our inquiry spins off into purposeless and recondite tangents.

The question of explanation does not admit of simple, or even complicated but straight-line, approaches. The work of Hempel is generally agreed to fall short of its aim in capturing the scientific notion of explanation. Carnap's work—both before and after Hempel's—focused on probabilistic theories from within a positivistic framework. Yet Carnap's statistical models were philosophically precluded from giving a metaphysically adequate explanation of themselves.[78] Peter Achinstein commented that, amongst those who do stress that hypotheses should *explain* evidence, "most frequently, defenders of such a view *do not define the concept of explanation involved*," citing William Whewell, Gilbert Harman, and Peter Lipton.[79]

Whitehead has very little to say on the subject of explanation, and per Achinstein, does not "define" the concept of explanation at all. The idea that explanation was in need of philosophical analysis did not emerge until after Whitehead's major works. There are some questions begged in Achinstein's observation about whether explanation can or even ought to be "defined." The nature of Nature requires a centering of our focus such that the totality of natural relatedness is limited to structures that finite minds can encompass. Such situated relevance might preclude anything a simplistic "definition." And Whitehead *does* offer us an opening into an "explanation of explanation," not admitting of definitional reductions. We have already introduced some of the tools needed for discussion. In discussing *what* is to be explained—nature, for example—Whitehead also showed *how* that explanation is built.

We make explicit the connection between some specific *act* of explanation and the appropriate *act* of abstraction. This act of abstraction must cull irrelevant data from the masses of undifferentiated fact, imposing themselves upon our senses and thought, selecting just those factors that can legitimately functioning as evidence within the specific inquiry. The terms "fact" and "factor" are chosen from Whitehead's use in *The Concept of Nature*. Whitehead says:

> [T]here are three components in our knowledge of nature, namely, fact, factors and entities. Fact is the undifferentiated terminus of sense-awareness [that is, the totality of nature as it is given]; factors are termini of sense-awareness, differentiated as elements of fact [that is, the first stage abstractions taken from the whole but still solidly rooted in the immediate contents of experience]; entities are factors in their function as the termini of thought [that is, second and higher order functional units of cognition.][80]

We have emphasized this aspect of Whitehead's theory throughout; Whitehead says a few lines later that, "evidently, the relations holding between natural entities are themselves natural entities, namely they are also factors of fact, there for sense-awareness."[81]

The relational structures distilled by inquiry and cognized in a logically rigorous manner *both* support the *relevant* explanatory hypothesis (where relevance will be local to the featured abstractions driving the inquiry), and these factors should be explained *by* that hypothesis.[82] If the abstractive selections are to serve the explanatory purposes, then it is not enough that these selections provide descriptive parsimony or (naïvely) pragmatic uses; they must also be *true*. Selections made for epistemic purposes must correspond to the *characters* of natural realities that are objectively present.

> The characters which science discerns in nature are subtle characters, not obvious at first sight. They are relations of relations and characters of characters. But for all their subtlety they are stamped with a certain simplicity which makes their consideration essential in unraveling the complex relations between characters of more perceptive insistence.[83]

This special insistence—including the kind of simplicity that indicates one is dealing with the *real* heart of the matter—is a central feature of *relevance*.

We have already observed that while experience does not come pigeonholed for our passive inspection, neither does it come undifferentiated and unlimited. The relational features of nature do not all stand on the same footing, and this difference is also a manifestly *present* aspect of experience.

Much has been glossed over regarding relevance and abstractive selection here.[84] We have made an initial gesture toward formalizing these ideas in chapter 9.

A context of highest relevance amounts to a system of models—basically, a Kripke frame—with the transition and relational connections between models representable along lines akin to those employed in modal logics, especially as these are used and interpreted within computer science. Explanation is possible only within contextually limited frames of intentionally selected abstractions that highlight real relational connections in the world. The context, the frame of explanation, is the structure of relevant connections that localize the *characters* of inquiry to make true statements about the world possible for finite minds.

In model theory, the model will be represented *by* a theory, and this will "satisfy" certain statements and claims. This structure of satisfaction is symbolized with the "double turnstile," or "$\vDash$." So if "T" is a theory, and "α" a sentence, then "T $\vDash \alpha$" says that α is true according to the theory T. The "$\vDash$" relation asserts an *explanatory connection* between T and α: that is, T is a symbolic representation of some model M—an explanatory framework—and serves to explain the "fact" (in Whiteheadian terms) expressed by α within the context of the model M.

What can we say about the "$\vDash$" relation? "$\vDash$" within any particular context/model amounts to an *act* of explanation. But what about the generic characteristics of that act, in *any* rather than just *some* context? The unpacking of "$\vDash$" is a problem for metaphysics,[85] and brings this part of our story back around full circle. We close this chapter by noting that the explanation of explanation, the grounding of "$\vDash$" within the limitations of *this* cosmic epoch, is nothing other than the quantum of *explanation* itself. Any account of the nature of explanation must provide a complete relational framing of the connections between cognition and the real. Limiting our attention to just nature allows us to make partial inroads on the subject, but will never permit those efforts to rise above fragmentary analogies. The fullness of such an account requires a complete metaphysics. Those who abandon metaphysics—and worse still, those who abandon philosophy—toss the possibility of explanation aside as well.

11 Synoptic Pluralism and the Problem of Whiteheadian "Theology"

> The use of the word 'nature' is more fitting to the limits of human reason and more modest than an expression indicating a providence unknown to us.
>
> —Kant, *Perpetual Peace*

In this passage, and many others, Kant expresses a profound humility before the power of what he calls "ideas." These are inexhaustible (also a Whiteheadian term, as we mentioned above) and serve as the ground for an indefinite number of "concepts" which determine our experience (including the "concept" of nature)—for cognition, for action, and via feeling, for reflection. Whitehead shares the humility, if not the precise distinction between "ideas" and "concepts." Yet, we may attribute to Whitehead a distinction between the *logic of concepts* and the *order of ideas*. Ideas, for Whitehead are the sorts of philosophical entities that govern history. In *Adventures of Ideas* he specifies seven such. The book is not entitled *Adventures of Concepts*. Nor could it be. Conversely, *The Concept of Nature* is not entitled *The Idea of Nature*. Nor could it be. We will draw on a distinction between "idea" and "concept" that follows Whitehead's usage, but also does not depart very far from Kant's views.

Religious and theological Whiteheadians have exercised tremendous influence over Whitehead's legacy in the last half-century. Much good has been done by extending Whitehead's thought in this way. Yet, we think these projects are different in spirit from Whitehead's cosmology and philosophy of culture. Whitehead's God is not, as Richard Ely pointed out, the God of religion, or indeed, a God capable of bearing human religious feeling.[1] Charles Hartshorne's God, by contrast, most definitely *is* the God of Western religion, as William T. Myers has argued persuasively.[2] Hartshorne's influence over Whitehead's interpreters is greater than most people think, to be discussed in subsequent chapters. Here we want to address the relationship between Whitehead and religious and theological thought by applying what we have said about "nature" above to the problems of religious and theological natural*ism*.[3]

A number of interpreters of Whitehead, including sympathetic voices, such as Hartshorne, George Allan and Donald Sherburne, along with critics such as Donald Crosby, Robert C. Neville, and Robert Corrington, have associated Whitehead's naturalism with an untenable interpretation of his cosmology. Most have conceded the field to Lewis Ford and then attacked the weaknesses of Ford's views as if these were *Whitehead's*. Ford's interpretation of Whitehead is manifestly wrong. We have corrected those errors. What remains is to explain what Whiteheadian naturalism really *is*, addressing this result to the tradition of religious and theological naturalism that has grown alongside process philosophy. The discussion should return to philosophy of science. If something can invigorate religious experience genuinely in keeping with Whitehead's thought, we favor it—and we have some suggestions of our own.

Philosophical, Religious, and Theological Naturalism

What sort of *concept* of "nature" is appropriate to philosophical and/or religious naturalism? There are at least two different questions here. *Philosophical* naturalism depends upon *concepts*, in the sense that conceptual determination is a condition for philosophical discourse. If philosophy exists apart from the discourse of reflective experience, we do not know what it would be or how it would be related to the reflective discipline of our tradition, The disciplining of reflection is a necessary condition of philosophizing well, although we do not say that unreflective or non-reflective experience has no *place* in philosophy.

Religious naturalism *may* depend upon concepts for its articulate *expression*, but there is a precognitive domain of praxis that grounds religious experience as a mode of human experiencing; such praxis does not depend upon linguistic or conceptual articulation, nor upon the disciplining of reflection. One could *be* a religious naturalist without knowing it, which is not true of philosophical naturalism. The inclusion of reflective experience, disciplined or not, is also compatible with religious naturalism. A "power of concepts" (to use Kant's language) *is* a precondition for being a religious naturalist; we doubt whether any animal lacking the ability to conceptualize experience could be a religious naturalist. It is a serious question whether dolphins, whales, and higher primates might possibly be religious naturalists, since it seems possible that they have religious feelings and experiences and also a power of concepts. That question is beyond our scope.

Being a *philosophical* naturalist depends upon conscious adoption of a position in a way that religious naturalism does not. Our view raises questions about the status of *theology*—what it is and what it does. Systematic theology seeks to employ the tools of philosophy, because it seeks to be thoroughly reflective. Such theology can endorse naturalism *only* in a philosophical way, and hence, theology is not the creator of a view of nature; it borrows a view of nature from philosophy (or social or natural science operating philosophically), asking what happens to religious issues, investigated in a thoroughly rational way. A "natural theology" is thus thoroughly dependent on whatever view of nature it appropriates from philosophy (including social and natural sciences). The varieties of theological naturalism are further elaborations of the respective versions of philosophical naturalism so appropriated.

There is no *purely* theological view of nature; the question of what nature is, even including God in one's cosmic elements, even where all of nature *dependent upon* that God (e.g., as "creation"), is a philosophical rather than theological question. Theology owns only one question: "*what* is God?" To ask *whether* there is a God is a philosophical question (despite the assertions of atheologians like Altizer and Taylor). It follows that one cannot be a theological naturalist without knowing it. Religious naturalism thus neither implies nor depends upon theological naturalism; a religious naturalist is free to reject the legitimacy or authority of *any* reflective and conceptual assessment of natural, religious experience. Religious experience requires no theology. Practitioners and intelligent disciples of any religion can be seekers of certain kinds of experiences *of* nature. The practice of philosophy, however, *requires* reflection and must maintain authority over its own organization of concepts and ideas. Theological naturalism presupposes religious naturalism, otherwise it would have no content (becoming an empty logic of ultimates[4]).

Thus, we should treat philosophical and religious naturalism as separate fields of experience that sometimes overlap, incidentally and interestingly. Yet, they have a few crucial common features. When religious naturalism allows that human beings naturally *express* their experience, and to the extent that such expression is mediated by concepts, the question "what is nature?" will arise for religious naturalists just as for philosophers. Philosophical naturalists may believe that the question "what is nature?" is presupposed by the activity of philosophizing, but it is not. One could have a systematic, subtle metaphysics, epistemology, logic, aesthetics, and ethics without *ever* considering the concept of nature, or even the *idea* of nature. Like all philosophical concepts, the *concept* of nature has a history and did not always exist (philosophy is older than any clear *idea* of "nature," let alone a concept thereof), and the concepts of nature we articulate in our philosophies have changed over thousands of years. Presently, the question "what is nature?" is, by ubiquitous convention, a required piece of any complete philosophical inquiry.

Philosophy and the Idea of Nature

Whether there is a fundamental "idea" of nature, distinct from the many "concepts" of it, is a serious question. We leave aside whether a single *concept* of nature exists adequate to every possible scheme of thought and human activity. Such questions for the doctrinaire (model-centrists being a prime example). Kant decided that the *idea* of nature was not as inevitable as were the ideas of God, freedom, and immortality, but he thought that nature was the closest thing to an "idea of reason" in the strict sense, lying below the three necessary postulates of pure practical reason. The idea of nature is *almost* necessary for the determinate interpretation of our experience. If the idea of an orderly, self-organizing system of all possible sense-experience were, in the infinitely distant future, wholly consistent with the three necessary postulates (especially freedom), then "nature" would be an "idea of reason." But a tension exists between all versions of the idea of nature and some aspects of our experience of acting freely.[5] It is unwise, and unphilosophical, to treat as necessary *in thinking* what is contingent in *experience*. We are not certain that we *have* experiences that go beyond nature, but we are also not certain we *do not*.

Part of the problem with philosophical naturalism in the past two centuries is that it has been forgetful both of the *history* of the concept of nature, and of the *contingency* of the concepts that have successively been framed in the history of philosophy. Even Richard Rorty's emphasis on contingency misses this point. If there is one thing he affirms dogmatically without apology or doubt, it is evolutionary naturalism (in his sense[6]). But Rorty sets the agenda on this question. We cannot accept his rejection of philosophy *as* inquiry (as opposed to edification), although we cannot ignore Rorty's critique of the pretenses of philosophy, setting itself up as the arbiter of the *idea* of nature. Rorty says:

> I have been insisting that we should not try to have a successor subject to epistemology, but rather try to free ourselves from the notion that philosophy must center around the discovery of a permanent framework for inquiry. In particular, we should free ourselves from the notion that philosophy can explain what science leaves unexplained.[7]

Rorty is right that a *permanent* framework for inquiry is not the desideratum, but there is a world of territory between admitting that philosophical inquiry is contingent and pluralistic, and rejecting the claim that philosophy picks up where scientific explanation leaves off. Philosophers can be inquirers without claiming to be authoritative meta-scientists (see chapter 10). But the widespread error Rorty is pointing out *does* turn on this: philosophers have assumed that when they speak of nature, they are speaking of the same thing science studies. That is an understandable assumption, but it requires very close scrutiny. The cost of denying it would be very high. Our position is that a philosophical *idea* of nature includes but is not exhausted by the "nature" that is studied by the sciences. We take this to be consistent with Aristotle's and Whitehead's views.

As a remedy Rorty presents some unpopular ideas, because the dogma of narrow naturalism is so closely held by philosophers as to make anyone who calls it out a kind of heretic. Whitehead was also such a heretic, and we follow him to the stake, if need be. But radical empiricists cannot sit by idly while philosophical dogmas are being used as empirical, cultural currency. To restore to philosophy a robust idea of nature requires that we avoid the mistakes Rorty dragged into the sunlight. But leaving nature entirely to the natural sciences, is not the answer, as we have argued. Philosophical inquiry not only includes nature, it is almost always *about* nature, and no gain comes from simply allowing current science to tell us what we mean by the word.

Reflective and Unreflective Experience

Let us return to the topic of philosophical and religious naturalism. What religious and philosophical naturalism really *do* share is a dependence upon *experience* (not the *idea* or *concept* of experience, the *having* of experience). Royce argued (in ways that will never be refuted) that the practice of philosophy has an inexpungeable religious aspect.[8] The activity of philosophizing, even when it deals with doubt conceptually rather than existentially, requires that practitioners have a kind of faith in the orderliness of the universe and thought's relation to it. Religious persons need not be philosophical, but a philosophical person must be at least slightly religious.[9] The difference between philosophers and other religious people turns whether the issue of "nature" arises in an *insistent* way.

For unphilosophical religious naturalists, God, freedom, and immortality remain *ideas* that have great power, but religious naturalists often deny that these ideas yield to a single system of conceptualizations. The idea of nature is taken in tow behind the other three ideas. If the idea of nature is given equal treatment with those other three, we move into the jurisdiction of philosophy (even if we are only doing theology). Insistence upon this comparison of nature with the other three ideas of reason arises not from reflection upon the relationship between those three "unavoidable" ideas and the idea of nature. We cannot help thinking that Rorty's half-hearted efforts in latter days, to make room for religious experience in his worldview was motivated by a guilty sense of intellectual irresponsibility about excluding something very real in human experience we have usually called "religious."

Consider: the *onset* of reflective experience seems to be coeval with the development of an idea (not necessarily a concept) of nature. If we take God, freedom, and immortality to be the fundamental and irreducible "ideas of reason," we might, by analogy, assert that nature is *the* fundamental "idea of reflection," only in the sense of reflect*ing*, i.e., the idea is active in transforming unreflective experience *into* reflective experience. We do not assert that in every act of cognition we also think "nature" explicitly, only that the whole of the reflective order of our thought is designated by the word "nature."

The reason one could have a philosophy without an explicit *concept* of nature is because the *idea* of nature connects *the rest of experience* with the *experience* of reflecting. That is the most important claim in this chapter; we will reiterate it. Reflection, as an activity, is, for human beings, as far as we know, a natural experience—it requires nothing beyond the natural order for its existence or employment. Yet, it is a "reversion" of the temporal flow, to use Whitehead's term.[10] Like all the other things that belong in and with "nature," human reflection is counted, as based upon the natural *experience* of reflecting. Perhaps there might be *more* to reflection than the natural experience (we don't know), or perhaps all that the experience consists in, naturally, is *available to* our reflections (the product of reflecting). This point merely restates the twin theses of radical empiricism and radical realism we have defended. There is probably more to the *experience* of reflecting than can be held in any reflective *idea* of reflection. Here we discover a maxim: From a Whiteheadian point of view, there is *always* more to experiencing x than is contained, implied, or even suggested by an *account* of that experience of x. Reflecting reduces experience, and hence its products, call them "individual thoughts," are static reductions ("reversions"). It is unwise to mistake a concept, a thought, or even an idea of x for the experience that gave rise to it.

Thus, *naturalizing* reflective experience does not require the denial of reality beyond "nature" (see chapter 10), whether as a concept, idea, or experience, and it does not imply that reflection upon the *experience of* reflecting is perfectly transparent—the latter involves a kind of judgment in which we seek to find generalities and universals adequate to our particular experiences, and when we fail to find adequate ones, we fail to know very much about our own experience of reflecting. We are claiming that the *idea* of nature connects the rest of experience with the experience of reflecting. We do not believe that the idea of nature connects reflecting with the rest of experience, *per se*. Moving from reflecting toward unreflective experience, *freedom* is the relevant idea.

Kant allowed that of the three ideas of reason, freedom is the only one of which we have concrete experience—in the experience of moving our bodies in accordance with the actions we autonomously prescribe to it. When the body is energized, enlivened, Kant says, this is the function of reason in nature. Let us reserve the idea of "freedom" for the active processes by which our reflective activity becomes unreflective bodily action. The idea of nature is appropriate to the other process—whereby unreflective experience becomes available for reflecting, and then becomes the product, that is, reflection, or individual thoughts and their order.

Nature, as an *idea*, is implicit in all philosophizing, but whether it should be made explicit—in extent, and likelihood of error—is also a part of the task of philosophizing. Anyone who ponders the relationship between reflective and unreflective experience encounters the *idea* of nature and may be prone to conceptualizing it—in one way for monists, two ways for dualists, and many ways for pluralists, such as Whitehead. The germ of a *philosophical* concept of nature is evinced whenever we offer a solution to the metaphysical problem of the problem of the One and the Many, rejecting the other paths. It seems that the idea of nature is relevant to *all* the ways the many can be joined and increased, whereas the idea of freedom seems to govern (and, we do mean "govern") the ways that the unifications of experience become diffuse elements in the world.[11] Thus, the idea of freedom, so far as we experience it in our bodily movements, is an ethical solution to the problem of the one and the many. Choosing a course of action, forsaking all others, we solve that problem in practice. *E pluribus unum*.

The idea of nature is grounded in a relation between unreflective and reflective experience; both types of experience are tied to the question of *existence*. The reason these two kinds of experience are joined in the broader question of existence is that both unreflective and reflective modes of experience do *exist* (even when existence is devoid of novelty), but together they fall short of full-fledged action. The problem of action *includes* the problem of existence, not vice-versa, and that is what makes us process philosophers. The idea of freedom is more comprehensive (and therefore more pressing, more philosophically urgent) than the idea of nature. So, stating the issue in terms of the classical problem of act and being, we assert that the problem of freedom includes the problem of experience and existence, and it exhausts the problem of act and being. This is not a chapter on freedom, however, and while we hold that any discussion of nature already implies a reciprocal discussion of freedom, we now set these complements aside and pursue our main objective.

Experience and existence are different enough that they not only engender the idea of nature, as a contrast of unreflective and reflective experience, but they also engender that contrast *as a problem for reflection*. One does not solve the mind-body riddle by asserting that one is a philosophical naturalist; in making such an assertion, one simply *states* the problem in the broadest way. If naturalism has a philosophical meaning, the mind-body problem (that is, the problem of the contrast between reflective and unreflective experience) isn't the crux. How naturalism acknowledges and approaches the very real differences between those two modes of experience is the heart of the matter. Self-aware philosophical naturalists know that unreflective and reflective experience do not correspond, and do not attempt to get a theory of truth to do any heavy lifting. Hence we can have *models* of concepts but not of ideas.

In Whiteheadian language, if there were no mentality in the universe, it would not be a cosmos. The mental pole of the actual entity elevates the mere *abstract possibility* of intelligibility to a *real potentiality*. Without mentality, there might still be a flux, but there would be no "time" (no rational order) because "before" and "after" would have no stable relations existing in anything like "order." The mental pole not only reverses the flux, it does so in ways creative of novel order; that order is raised to a level of intelligibility in its sustaining of a compound collection of relations prehended. Reflection tears apart that whole in activity but not in action. This analysis called "reflection" partializes without destroying the cosmic order. It is not a model, per se, but it grounds the real potentiality for modeling the real. We discussed this process under the names genetic and coordinate division. But as Whitehead says, "there is no holding nature still and looking at it."[12]

The contrast between unreflective and reflective experience is a productive contrast; the former becomes the latter, but is also a reductive activity for the purposes of action. Where action fails to fulfill the promise of that-for-the-sake-of-which we reduce unreflective to reflective experience, we have "missed the mark," or, in traditional language, "sinned." Sin, here, means leaving something out of experience for the sake of an *illusory* greater good, projected in some plan of action, and carried out. This is missing the mark—due at least in part to the gap between the goal as projected and the real aim. There is no such thing as "evil," but evil is a fair word to apply to grievous error.

The reduction of unreflective to reflective experience is the price we pay for *choosing* our actions. The idea of nature is the limit of the reduction—it provides the "musts" that we cannot afford to ignore in reducing unreflective to reflective experience. The aim of action upon a reflective product is only normative adequation—this adequation can fairly be called "modeling." We don't care *why* we have acted successfully, we only care to have been justified in our thinking once the consequences of our actions are available. Adequacy of genuine action to its *plan* of action is the justification of selection and reduction in the transformation unreflective to reflective experience.

Yet, we must ask, *why* are the reflective and unreflective modes of experience so different? Why do they even seem to obey different laws or rules (since the norms of logic are identical with the laws of physical nature, as far as we know)? Is there any third realm that unites them? (We have suggested that freedom answers this demand; that is another story.) This sort of doubt which gives rise to our various philosophical *concepts* of nature. "Existence" is that realm to which both unreflective and reflective experience belongs, but whether they are there united in an identity is a metaphysical issue beyond our ken. We are pluralists. Reflective and unreflective experience could be united in different and perhaps incommensurable ways, but they could not fail to be constructively related, even if we are unable to discover all the ways they *are* so related. Similarly, Whitehead says that "space as pure extension, dissociated from process, and time as pure serial process, are correlative abstractions which can be made in different ways, each way representing a real property of nature."[13]

We take the view articulated here to be Whiteheadian in character, emphasizing that we read Whitehead as a radical empiricist. We also think that radical empiricism, as a fundamental orientation on what philosophy is and how it ought to be done, is most compatible with the idea of nature we have just suggested. But what sort of view is this?

Synoptic Pluralism and Whiteheadian Theology 229

Varieties of Naturalism

John Shook has recently offered a compact, systematic schematization of the varieties of philosophical naturalism we want to endorse, as far as it goes, without qualification or modification.[14] He rightly sees that Whitehead, William James, Peirce, Santayana, Paul Weiss, Stephen Pepper, Nelson Goodman, and—most importantly—Richard Rorty in *The Linguistic Turn* (but not later writings), endorse a version of naturalism that he calls "synoptic pluralism." It is bold to group these thinkers together. Shook describes this view as "the most open and flexible naturalism, defining reality most generously."[15] "Synoptic pluralism" holds that "reality has a variety of aspects or modes as known by the many sciences, and also has aspects or modes known by experience and perhaps pure reason as well, that the sciences are incompetent to describe or explain."[16]

Shook's schema of naturalisms is based upon the positions philosophers take regarding science's contribution to our understanding of "reality." This is the best approach to sorting the types of philosophical naturalism, but the explicit limitations must be noted. Shook does *not* claim to be sorting out types of naturalism that take the *concept* of nature to be dispensable, i.e., non-philosophical naturalisms—such as religious naturalisms are free to do. Various romanticisms, nativisms, atavisms, aestheticisms, asceticisms, pyrrhonisms, cynicisms, hedonisms, ritualisms, and any version of a major or minor religion that rejects the exclusive authority of theology, while embracing religious experience, could be called "non-philosophical" and a variety of naturalism without being *un*philosophical or *anti*-philosophical, but not strictly "a philosophy."

Shook's schema takes for granted a certain positive view of the relation between philosophy and the enterprise of scientific knowing. That is a position not only about science, but also about what philosophy is, what it does, and what it *can* do. Those who take a different view of philosophy, less optimistic and more deflationary, are not likely to endorse naturalism for *philosophical* reasons.[17] The difference between Rorty in *The Linguistic Turn* and his later writings is that his naturalism is still *philosophical* in the earlier writing. When he wrote *Philosophy and the Mirror of Nature*, his naturalism had become non-philosophical. Still later, he accepts the authority of science, but not for philosophical reasons; he did reject the authority of philosophy to deal with the *concept* of nature for historical *and* philosophical reasons. Rorty accepts the authority of science *in spite of* the philosophical bankruptcy of all efforts to justify its knowledge claims. And *that* is a problem for philosophy. Is there a philosophical *concept* of nature that avoids the narrowness of most philosophical naturalisms and is also of some use to reflective religious naturalists?

Radical empiricism, as an orientation on the scope and limits of philosophy, can underwrite a synoptic pluralism that frames a concept of nature that is non-dogmatic, philosophical, and useful to reflective religious naturalists. Indeed, that is how we read Whitehead's concept of nature. A religious naturalist does not require a *concept* of nature, only an *idea* of nature, and that idea is already functional in the relation between his/her unreflective and reflective experience. We do not claim to address religious naturalists except insofar as some of them *become* philosophical in their doubts and responses to those doubts (Neville, Corrington, and Hartshorne loom largest). In the remainder of this chapter, we will examine the question of nature, and philosophical and religious naturalism, by examining the relation between *experience* and *existence*, as radical empiricists.

Experience and Existence

Both unreflective and reflective experience have immediacy.[18] The immediacy of reflective experience can be expressed in the "what is it like to be a bat" question, but applied more narrowly to the experience of reflective beings.[19] There is a something "it is like," i.e., an aesthetic and unreflective experience, *of* reflecting *on* one's experience. This immediate experience of reflecting has been chased by philosophers for a long time, but Bergson's "reflective intuition" or Husserl's "categorical intuition," or better, Whitehead's "conceptual feeling"; even our most mediated modes of experience possess some immediacy.

The *general* problem of immediacy is how to think about and describe the point at which the *experience* of x is indistinguishable from the *existence* of x; perhaps the two are "identical," but what does that mean? The absence of discernible difference does not imply ontological identity for radical empiricists. Still, there is no reason to postulate that the problem of the immediate experience *of reflecting* is *different* in kind from any other immediate experience. What empirical inquiry is advanced by such a hypothesis? What *experience* could legitimize such a claim? (These are the questions Whitehead always asks when dealing with the limits of experience.) We cannot come up with a thoroughly empirical reason to postulate a dualism of that sort (what Whitehead calls the "bifurcation of nature"). The pervasiveness of mediation in the experience of reflecting may be the dominant character of its very existence, but immediacy cannot be wholly absent in reflecting.[20]

If there is an *act* of thinking that does not imply an *experience* of thinking, we don't know how we could include it in a philosophy, except as a negation, and here we follow Aquinas. We also believe we have covered this in our discussions of negative prehension. Thinking is one kind of reflecting, and we can generalize this point to all types of reflecting—however many types there may be. Whatever the immediate experience of reflecting may be, unless we find a reason to think of it as utterly and irreconcilably different from other kinds of experience, we postulate that it is of the same kind as other immediate experience, and the difference, where it comes in, lies in the presence, the degree, and the modality of mediation.

The question of connecting experience to existence should be treated as the *same* question in both reflective and unreflective experience. If we have overlooked some *reason* to be ontological dualists about reflective and unreflective experience, we await correction. But we have just suggested that whatever "mind" may be, it is compatible with bodily experiencing at least as far as the immediacy of experiencing goes. "Nature," then, is the idea of a *possible* unity of existence and experience where the unreflective aspect of experience is becoming reflective. Such existence is a dynamism (not static being), and its process of becoming reflective is a reduction that has the effect of slowing down temporal passage in intelligible ways.

If we assume that the idea (and the various philosophical concepts) of "nature" can do the work of connecting experience (both unreflective and reflective) and existence, the question is *how*? Many philosophical naturalists hold that a concept of nature *can* do the needed work. There are many varieties of naturalism, each crossing this bridge in a slightly different way. Let us endorse again Shook's schema and his conclusion that there are seven viable types of philosophical naturalism. He points out that all seven viable types still suffer from "unresolved problems requiring further intense philosophical work."[21] We offer some of that work here.

Challenges for Synoptic Pluralism

Shook says, "Unless synoptic pluralism can develop its own compelling naturalistic ontology, its enthusiasm for multiple modes of reality can easily amount to ontological dualisms and pluralisms that entirely depart from naturalism."[22] In Shook's account, the main problem with all seven viable naturalisms (as with the *non*-viable types) is that, under pressure, they morph into a different type of naturalism. When that happens, naturalists are inconsistent in maintaining their respective positions. We naturalists equivocate, Shook believes, and either fail to stick to a clear concept, or keep shifting our view (consciously or not) from one type of naturalism to another, rejecting the earlier by implication.

In addition, synoptic pluralism has the extra problem of morphing into a non-naturalistic position, and also, we add, a stance too friendly to non-philosophical naturalisms. That is the move Rorty made after *The Linguistic Turn*. He did not defend his dogmatic naturalism because it wasn't a philosophical position.

But a Whiteheadian concept of nature goes a long way toward justifying a synoptic pluralism; it is clear and unequivocal, and does not morph under our noses. We think most people will allow that Whitehead's naturalism is thoroughly philosophical, but some believe that his view of God or of eternal objects removes his system from the realm of philosophical naturalism. Interpreters such as George Allen and Donald Sherburne have sought to "naturalize" Whitehead's thought by re-working or eliminating some of the basic elements. We believe such interpreters are misreading Whitehead, not grasping his method or his commitment to radical empiricism. Whitehead's naturalism is so "generous" (to use Shook's term) as to let in apparently non-natural entities.

Whitehead aims to "lay the basis of a Natural Philosophy which is the necessary presupposition of a reorganized speculative physics."[23] He calls this "cosmology," and the ontology internal to that project is in *Process and Reality*. It is a type of naturalism. In discussing a radically empirical Whiteheadian naturalism, remember the qualification he introduces in framing his concept of nature. Noting that not all philosophies and religions employ the idea of nature philosophically, he says:

> The mention of these vast systems and of the age-long controversies from which they spring, warns us to concentrate. Our task is the simpler one of the *philosophy* of the sciences. Now a science already has a certain unity which is the very reason why that body of knowledge has been *instinctively* [unreflectively] recognized as forming a science. The *philosophy* of a science is the endeavor to express explicitly those unifying characteristics which pervade the complex of thoughts and make it to be a science. The philosophy of the science*s*—conceived as one subject—is the endeavor to exhibit all sciences as one science, or—in the case of defeat—the disproof of such a possibility. *Again I will make a simplification, and confine attention to the natural sciences, that is, to the sciences whose subject-matter is nature. By* **postulating** *a common subject matter for this group of sciences a unifying* **philosophy** *of natural science has been thereby* **presupposed**.[24]

Whitehead is postulating a concept of nature that will enable him to show how a *philosophy* of science presupposes a common subject-matter for all the sciences. Thus, the philosophy of science presupposes what we have called "nature" above, or, the slowing down of temporal passage that has the effect of emphasizing the reflective character of existence, transforming primarily unreflective experience *into* reflective experience. It is not that we know for certain that "process" is organized this way; we *presuppose* such an organization when we postulate that the sciences have a common subject matter, and that "nature" includes that subject matter.

Whitehead does not draw knowledge from the sciences, nor demonstrates the unity of individual sciences, nor the unity of *all* the sciences; he shows only how the *philosophy* of science presupposes that unity. The unity itself, in the sense of "existence" we have used above, i.e., immediacy of experience and existence, is the practical possession of individual sciences, granted to them by our *instinct* that a whole group of thoughts is aimed at the same *mode* of existence. Whitehead avoids slipping into non-philosophy or non-naturalism by *simplifying the investigation*. He famously uttered of the dictum "seek simplicity, and mistrust it." The position is a synoptic pluralism that seeks to be radically empirical. If we have Whitehead wrong, our aim has been to articulate the position, not exclusively to explain Whitehead. Yet, this is Whitehead's project, even if it is possible to catch him in occasional "lapses of attention," as he puts it.

Narrow Naturalism

We finish by pointing to some of the perils associated with the types of naturalism that are not based on radical empiricism. One in particular is reductionist Darwinism.[25] These views are damaging the reputation of philosophical naturalism and, being too narrow, they give religious fanatics an advantage in public debate, in legislation, and in the courts they ought not to enjoy. These views bear a strong analogy to the model-centrism, but are biological, or, rather, bio-historical. Not only are these positions philosophically inferior to synoptic pluralist naturalism, there are good practical reasons to avoid them. If Richard Dawkins and his ilk are to be regarded as spokespersons for naturalism, scientific and philosophical, then naturalism is in trouble.

First, in addition to being naturalists, we are also evolutionists. We follow Whitehead in this regard. But the *idea* of evolution is being over-stretched these days. The undeniable operation of natural selection within genetically informed biochemistry and cell biology, and its long-term statistical results upon both genetic configurations and, to a more limited degree, the phenotypes of organic forms, is beyond serious questioning. "Natural selection" is clear in that context. We have 60 years of evidence since the culmination modern synthesis. Whitehead did not live to see the discovery of DNA, and so we are obliged to extrapolate a bit regarding this important character in nature. How to generalize and apply our observations since DNA was discovered to longer spans of time remains a serious question. The principle of evolution certainly applies in "nature," at least as an interpretive, narrative concept, and quite probably more, but how *much* more is yet to be determined. To give one's unconditional commitment to a single version of evolution, as Darwinians do, is unwarranted and unphilosophical. It is a scientism that threatens the future of more thoroughly philosophical naturalisms.[26]

Generalizing the mechanism of natural selection beyond what can be genuinely observed under experimental conditions requires tremendous care—it is bio-history, subject not only to errors that characterize even careful science, but also those that haunt the historiographer. Sober history is difficult enough, but history that was neither observed nor recorded by reflective agents at all is beyond speculative. We have known about DNA for sixty years. We became able to document alterations in DNA about fifty years ago, and have thoroughly understood the replication process and *some* of its variables for some thirty years. We have been gathering reliable data for that long. We have only a few generations of actually observed higher organisms with which to generalize the operation of natural selection and genetic variation, and due to their complexity and the difficult-to-control variables of environing circumstance, the *meaning* of the term "natural selection" becomes both vague and general, placing it into genuine scientific dispute. And that is what *should* happen.

Good science recognizes its own vagueness, the limits of its observation and their safe generalization, and works to clarify what remains poorly understood. Thus, we cannot reliably say that higher organisms "evolve" according to the documented patterns of simpler organisms. Our dislike of religious fanatics who claim absurd things is irrelevant to our science. Present social upheavals aside, we must not let responsible science devolve into irresponsible scientism, *slowing* our general progress in both science and in the public judgment. We avoid these consequences through a collective commitment to propositional and conceptual sobriety. Even if our errors are productive, we must remain fallibilists.

We *can* say that there is no compelling reason to think that complex higher organisms *do not* evolve. If they evolve, it is reasonable to look for order in that process, and the well-documented occurrence of natural selection in simple organisms is a good place to begin. We *can* say better hypotheses about higher organisms are scarce, and that it the term "evolution" can be applied analogously to higher organisms, with the qualification that the term "natural selection" is too broad and vague to do the same work relative to higher organisms that it now performs relative to lower ones. Increasing the orders of complexity in nature affects our capacity to generalize safely.

Creationists and intelligent design ideologues appeal to the concept of "irreducible complexity," asserting that some organic structure or other is too complex to have arisen incrementally via natural selection, since there are (they claim) no intermediate stages leading up to the final, complex structure which would give the organism a selective advantage.[27] As with so much creationist nonsense, this is an *argumentum ad ignorantiam*. It speaks to the vacuity of creationist-driven ideology according to which it is possible to make an impossibility proof regarding the development of mechanical structures, via mechanical processes, from mechanical bases. Aside from the fact that such impossibility proofs are themselves all but impossible, the fact that the structures, processes, and bases are all firmly located within the same semantic and ontological framework—i.e., all *mechanical*—is a "clue" that they are looking in the wrong place.

Issues of complexity and emergence dog reductive, natural selection biology, but they are not to be found in something as easily accounted for as bacterial flagella.[28] But these issues are more semantic and *logical* than ontological. For example, the three problematic areas that Nagel focuses on in *Mind and Cosmos* are the emergence of consciousness, cognition, and value. (The first two are, of course, generically included in what Whitehead calls the "higher phases of experience," while the third is built into the very structure of prehension.) The problem with these three is not that they are some very queer kind of mechanical "stuff," akin to bacterial flagella; the problem is that they are neither mechanical, nor any kind of "stuff," at all.[29]

Indeed, one cannot even *talk* about such things in purely mechanical terms; one must jump from the semantic realm of the mechanical into the full-blooded field of intentionality in order to speak about such things as consciousness, cognition, and value. These are among the most immediate aspects of our experience, yet reductive naturalism and narrowly construed Darwinism cannot even acknowledge their existence, much less talk about that existence in the mechanical terminology to which they have limited themselves. Indeed, reductive naturalism *requires* these terms that it cannot account for: there must be intention-rich semantics in order for *any* theory, even narrowly construed Darwinism, to be *interpreted* at all, since interpretation is an *intentional* act informed by intention-rich semantics. Nature as investigated by science is only closed to mind; it does not pretend that mind does not exist.[30]

Above the level of complex biological organisms, the broader suggestions that cultures evolve, that geological change is a kind of evolution, and that a solar system or stars or the universe "evolve" simply adopt a favored metaphor. All of these *change* according to discernible patterns. But if *evolution* is to retain the clear meaning that *good* science requires, we must acknowledge it is only a metaphor at when we generalize it beyond biological life. It may not be the best metaphor in some of these instances, although it is suggestive. We see no reason that science should not seek out some mechanism corresponding to natural selection at the level of geology or planet formation or universe expansion. We have our doubts whether any such thing will be found.

Still, what will we say about the processes whereby unreflective experience becomes reflective? It is patterned change. Is it "evolution"? We have only begun to explore that. Any philosopher who wants to claim that philosophical naturalism necessarily implies such "evolution" must be smarter than we are. It seems possible that *reflecting* is an eddy in the current of dynamic process, a temporary retrogression that has become an oxbow in the river of process, so to speak. We like this metaphor. We take this to be Whitehead's view as well. Eventually, "becoming" finds a way to pass by the eddies of reflection. Consciousness is a hard problem not just because narrow naturalism cannot address it conceptually; it is hard because consciousness might be a pathology of temporality, for all we know. If our reflective experience *is* temporal disease, the best we can hope for is an extinguishing of our suffering. No naturalist can afford to rule this out, however pessimistic it sounds. The refusal to consider whether consciousness is a retrovirus of temporal passage, operating on the double helix is irresponsible. This idea of temporal pathology is not simple nihilism, it is a live option for synoptic naturalists; it is reasonable to acknowledge it and seek alternatives with chastened hearts. We see precious little of this beatitude among scientists, less among philosophers, and none in the press.

When scientists and scientific journalists use the term "evolution" to mean something other than the reliably documented phenomena indicate, they give fundamentalists and other idiots an unearned advantage. If we use the term "nature" in a way that purports entirely to close the gap between experience and existence, we become dogmatists ourselves. What we can know is that *experience exists*. Existence is not thoroughly hostile to the experiential modality. We are the proof. If by "nature" we mean "existence," we must allow that there may be more to existence than we experience, and more to the experience *of* nature than we *can* reflect upon; but *what* we experience *does* exist, insofar as immediacy is common to all modalities. "Nature" expresses that confidence, with the well-warranted *idea* that ordering our experience rationally and scientifically is the best way to clarify and verify our confidence.

Radical Empiricism Redux

We end this inquiry where it really began. The best way to keep dogmatists and religious fundamentalists at bay is to adopt the position of radical empiricism. It is worth revisiting what James said:

> Radical empiricism consists first of a postulate, next of a statement of fact, and finally of a generalized conclusion. The postulate is that the only things that shall be debatable among philosophers shall be things definable in terms from experience. The statement of fact is that relations between things, conjunctive as well as disjunctive, are *just matters of particular experience*, neither more nor less so than the things themselves. The generalized conclusion is that therefore the parts of experience hold together from next to next by relations that are themselves parts of experience. The directly apprehended universe needs, in short, no extraneous trans-empirical connective support, but possesses in its own right a concatenated or continuous structure.[31]

What James calls "directly apprehended" we called "immediacy" above. Our position begins with a postulate. It is not that we *know* that matters beyond experience are not debatable by philosophers (clearly philosophers do debate things beyond experience all the time). We don't go that direction. This is a postulate. It is more important for naturalists to respect this postulate than philosophers who do not claim to be naturalists. Attempting to discuss, debate, and know what may or may not exist beyond experience lacks a solid method. If such a method appears, we can revise our postulate.

The second constraint is that we begin not with experience in general, but with *particular* experience; so far as we know, there *isn't* any "experience in general" that is neither yours nor mine, nor anyone's in particular. It may exist, but we do not know how to get at it. If there is knowledge, and there is, it comes from properly generalizing particular experience. Particular experience includes both continuity and discontinuity, so science cannot afford to generalize continuity to the exclusion of discontinuity in nature. Thus, pluralism is the wise stance relative to the various fields of science. Dualism is unempirical and against naturalism, as is monism.

To assume a perfectly orderly and continuous nature that lies "behind" experience, as *we* have it, is roughly the same as assuming the existence of God. We have no experience to confirm or disconfirm the notion. If "nature" becomes a substitute term doing the work of the idea of "God," the naturalist becomes unempirical. *Ideas* of reason belong to the domain of the act, not to the domain of reflection. *Concepts* of reason, on the other hand, belong to the domain of reflection as well as to that of action. Naturalism is easily seduced by the temptation of over-generalizing continuity, of ontologizing it, either locally (as Dewey and Mead do), or globally (as Peirce does). Perhaps nature "in itself" really is continuous—general relativity assumes so; but perhaps nature contains genuine discontinuity—quantum physics mandates this. Our "quantum of explanation" imposes no ontology on nature but only a condition upon explanation. The principle is that parts and wholes have a plurality of reciprocal relations, but that neither part nor whole constitutes its self-sufficient reciprocal. The many become one and are increased by one, we read somewhere. Choosing one alternative among these for all time is not a serious concern to radical empiricists, since they build explanations in both the local and global domains, not from a favored scientific theory, but from particular experience with both continuities and discontinuities. Bad metaphysics derives from privileging one over the other.

Dewey modifies radical empiricism in a useful way in adding the postulate of immediate empiricism: philosophers will take things to be, existentially, what they are experienced *as*. We do not *know* this, we *postulate* it. The general problem of experience and existence remains. But we worry over the *genuine* problem with Dewey's postulate, not a host of pseudo-problems. Deweyan naturalism is thus, as Shook rightly notes, almost synoptic. But in the end Dewey's view is unempirical, since it takes the problematic situation and the perspective of the problem-solving organism as primitive. These categories are not primitive to philosophical naturalism. They are concepts, heavily conditioned by the more or less independent history of thinking. They are not "ideas" in the same way that nature, God, freedom, and immortality are "ideas."

What is the relationship between particular experience and existence? We venture some safe generalizations:

(1) Experience exists. Thus among the modes of existence, whatever they may be, experience is one.
(2) Experience comes in a variety of modalities, including physical, biological, physiological, perceptual, and reflective. Whether these are

explainable in terms of fewer divisions is an open question, to be vigorously investigated. Whether there are more modalities than these is also an open question. However, this list reflects Whitehead's concern in *Process and Reality*.

(3) Existence, whatever it is, is not wholly hostile to experience, although, given the apparent scarcity of *reflectively conscious* experience (i.e., highly organized feeling) in the universe, would appear to be a fairly minor mode of possible existence.

(4) Existence may, for all we know, contain inexperienceable modalities, but we can make nothing of them except through the lens of experience.

(5) Particular experience is the ground of generalization, and generalization is fallible.

(6) Nature is experienced existence, from unreflective to reflective; it may be more than that, but it is *not* less.

This is enough for good science. Science, done properly, does produce knowledge of experienced existence; and because it is methodical, repeatable, inter-subjective, and systematic, it is capable of discovering and correcting some of its own errors. There is no reason a radical empiricist needs to insist that every error of science is correctable or discoverable. The method itself grows and changes. Radical empiricists only say is that science produces the most reliable measurements of experienced existence that we currently have. Therefore, insofar as we have reliable knowledge of nature, it should stand the test of science. Science should not presume to tell us what exists in itself, only *how* existence is experienced, offered in reliable generalities. There are other kinds of knowledge than science, but these are less reliable in proportion to their inability to discover and correct their own errors of generalization.

Thus, moral knowledge, for example, is more appropriate to the solution of moral problems than is scientific knowledge; not all persons who attempt to solve moral problems are equally accomplished in the requisite kind of knowledge. Scientific knowledge is not required for moral knowledge. There is surely a relationship between and among types of knowledge, but *how* they relate is, for a radical empiricist, an open question. Scientific knowledge is to be favored in all situations in which the problem to be solved is cast in terms that can be controlled by science, usually only a small part of a larger problem. There is a scientific aspect to nearly every problem, and good science should be applied to *that* aspect. But let's not imagine that solving the scientific aspect of a problem solves the entire problem. This overstretches science.

For example, if the problem is "what is the origin of human beings?" this is not *just* a scientific question. It has a dominating scientific aspect. But the question addresses multiple modalities of human experience (such as consciousness, cognition, and valuing), and while science may illuminate some of the terrain involved in these other modes, it must not be overstretched. If we ask the question for religious reasons, science can supply reliable details about the possible meanings of terms like "origin" and "human," but cannot provide an exhaustive answer to all that the question means. When science dismisses or eliminates those other modes of experience, it does a disservice to everyone, especially to science, and ceases to be, in our sense, naturalistic.

12 Possibility and God

> There is finitude—unless this were true, infinity would have no meaning. The contrast of finitude and infinity arises from the fundamental metaphysical truth that every entity involves an indefinite array of perspectives, each perspective expressing a finite characteristic of that entity.
>
> —Whitehead, *Essays in Science and Philosophy* (60)

Introduction

Our extended discussion of naturalism reclaims for the idea of nature a robust sense of possibility. Whitehead's interpreters are persistently averse to "eternal objects" (admittedly not the best term for "possibilities"). We have shown that there is nothing problematic about treating Whitehead's theory of possibility as belonging to a perfectly respectable philosophical naturalism. Indeed, narrower kinds of naturalism involve non-empirical assumptions. Radical empiricism is satisfied only with a synoptic pluralism regarding the relations of existence and experience. Possibilities, while theorized on the basis of particular experience, should not be *limited to* particular experience regarding their existence. We do not know whether possibilities can or do exist independent of particular experience, but there is more reason to suppose they might than to suppose that they cannot.

We take Whitehead's theories of "eternal objects" to satisfy the requirements of this delicate relation between existence and experience, but most of Whitehead's interpreters have difficulty understanding this part of his cosmology.[1] They commonly conclude that this feature of his cosmology is overly Platonistic and thus anti-naturalist. Such interpreters too hastily seize upon Whitehead's statements that eternal objects are "universals," and, thinking that they know what this means, they conclude that he has gone off the tracks. It seems to occur to very few to work carefully through the idea of the elimination of eternal objects from prehension so as to grasp why these "universals" are, for Whitehead, generalizations from particular experience, with certain restrictions lifted for the purpose of abstract theorizing. These are not Plato's forms, but, as Whitehead says, closer to what he imagines Plato would say if he lived in the twentieth century.

Analogously, many of Whitehead's readers attempt to expunge "God" from his cosmology, assuming they also know what *that* word means. Conversely, other interpreters try to expand Whitehead's cosmological God beyond cosmology and into all sorts of areas that are beyond the scope of the inquiry in *Process and Realty*. The divisive arguments in the literature about the consequent nature of God are carried out with little or no reference to the way Whitehead's various books are different inquiries with different purposes. Whether his concept of God in a work is or is not beyond the scope of a defensible naturalism depends both upon what "naturalism" means and what role the concept of God was playing in a given inquiry. Our account of Whitehead's naturalism covers all of his inquiries, but we allow that the most important test of that naturalism comes in *Process and Reality*. Demonstrating that his theory of possibility in that work is well within the scope of naturalism implies, in our view, that the burden of showing that *any* claim that eternal objects are non-natural now devolves upon those who claim it. Whitehead would find the charge surprising.

But in light of our account of possibility and naturalism, we still must show how to handle the vexed question of the *idea* of God, and its several conceptualizations in Whitehead's various inquiries. We insist that the conceptualization changes from one inquiry to the next, but it is incumbent upon us to show something about both the idea of God that informs Whitehead's thinking and that the most important conceptualizations align with the ways he treats possibility in his most important works.

In the previous chapter we mentioned the complex relation between Hartshorne's interpretation of Whitehead and the subsequent legacy of Whitehead interpretation, especially in the United States. What we may fairly call the "Claremont School" of Whitehead interpretation is more indebted to Hartshorne's worldview than it acknowledges.[2] The result has been three generations of Whitehead scholarship that tends to affirm some un-Whiteheadian principles. Chief among these is the claim that there is no negative prehension in God. That is not Whitehead's view, but it is Hartshorne's.[3]

Yet, if there are no negative prehensions in God, then God cannot be a personal being. That point would not concern Whitehead, but it is a dilemma for those who follow Hartshorne. If there are negative prehensions in God, then it is still an open question as to whether God is a personal being, but affirming negative prehensions in God would provide a fairly firm basis for thinking of God as having personality. This is what Hartshorne should have asserted, both on behalf of Whitehead, and for himself. The issue of whether there are or are not negative prehensions in God has been widely written upon in the circles of process thought; it is not our objective to engage that literature exhaustively here.[4]

But our thesis reveals an important difficulty in Hartshorne's philosophy. This difficulty does *not* show up in Whitehead's philosophy, due to differences in the methods of Hartshorne and Whitehead. Our purpose is not to criticize Hartshorne or defend Whitehead. Rather, we aim to show why the issue of the nature of possibility is the key to understanding the issue of negative prehension (chs. 8–9). The effect of the following argument will be the liberation of Whitehead from the theological interpretation of the Claremont School. We want to make it clear that we admire these scholars, their purposes, and their accomplishments. Hence, we also aim to liberate the Claremont School from a procrustean theological bed made from eiderdown they plucked from the works of Whitehead. One's account of possibility in relation to negative prehension has profound implications for one's ideas of God, particularly whether God is or is not a personal being.

Analogies and Such

Our argument has similarities with Robert Neville's views, but rather than dealing with "creator" and "created being," and then following out the implications of the idea that the God that is "beyond the creator," we will approach the issue in a less traditional way. Neville's view, although current, has a traditional structure. He attacks the problem by rejecting the Thomistic analogy of being, offering instead an account of God that re-invigorates the medieval quest for the univocity of being. The disagreement between the sanguine partisans of univocity, such as Duns Scotus, and more cautious analogizers, such as Aquinas, was well worn even in the thirteenth century, but is novel and curious in our time.

We would call his approach "traditional" because he does not attempt to give a new theory of analogy, but defends a vanilla conceptualist account of universals (following Peirce and Scotus), and reframes the Thomistic analogy of being as a mid-level of reference. Only then does he move to a more modern, semiotic account of language. There is a difference between the theory guiding Neville's philosophy of symbols, and the account of language he advocates for doing metaphysics. The latter was adopted early in his career and he continued to rely upon it thereafter.[5]

Neville's association of the problem of God with the problem of the *unity* of God (or, the unity of *language* about God), and his assumption that the being of "person" requires or implies an *ontological* unity, are positions we do not share, and neither does Whitehead. We take unity to be a reduction or abstraction to a single modality of the societies of actual occasions that genuinely exist only in plural modalities. If, as Peirce says, it is order (rather than disorder) that calls for explanation, then unity, even complex unity, as an extreme manifestation of order, is rare and curious in the universe. The obsession with the unity of the self in Kant and post-Kantian philosophy is a red herring. Following Whitehead, we hold diversity to be a co-condition with unity (or identity, which he uses interchangeably with unity) of existence. Unity alone is not to be sought. Its attainment would be the destruction of all depth and intensity in experience. Unity *under contrast* is an achievement, but by no means is it so very unified as to supply an old-fashioned ontological "identity."

In criticizing Whitehead, Neville relies upon Lewis Ford's rendering of the Category of the Ultimate,[6] and defers to Ford on the basic interpretation of Whitehead. It should be clear by now we disagree with Ford. Thus, when Neville criticizes Whitehead, he is often speaking of a *misinterpretation* of Whitehead rather than of Whitehead's actual views.[7] Ford *created* the problem of temporal atomicity in Whitehead. This and other habits of misreading Whitehead allowed philosophers of religion and theologians to kidnap Whitehead's name and leave his actual philosophy unexamined. It is time for the real Whitehead to roam the planet again.

Obviously *Religion in the Making* is better suited for appropriation by theologians, but few of them make use of the conception of God articulated there. Arguably, even this book (and its God) is unsuitable for the purposes of theology, since it makes no claim at all about whether any God ever existed, contenting itself to examine how the idea of God changed in Western history as the civilization came to be "rationalized" (and this is a good thing in Whitehead's sense of the term).[8] A better view of Whitehead reveals a philosophy that, contrary to the wishes of many, asserts no ontological knowledge and is of little theological value. Neville's critique of Ford's Whitehead is *philosophically* unnecessary, its main value lying in its cautionary effect on theologians who might believe Ford.

However, Neville's critique of Hartshorne is almost the same as our own.[9] Unlike Whitehead, Hartshorne's God has theological value. Thus, we must wrangle with the metaphysics of negative prehension in God to see why, in effect, Whitehead *allowed* it for *philosophical* reasons, while Hartshorne *rejected* it for *theological* reasons—and then Hartshorne has to deal with the philosophical fallout. In our estimation, Hartshorne cannot free himself from the philosophical tangle he creates when he rejects negative prehension in God, although the theological gains may be almost worth the philosophical cost.

Ironically, if we are correct, Hartshorne, who is a personalist and insists upon the personal character of the divine being,[10] is the one who has undermined that very claim by holding that there are no negative prehensions in God. Meanwhile Whitehead, who explicitly denies being an unqualified personalist with respect to God,[11] has provided an account that leaves *open* the question of whether God is or is not a personal being.[12] One of the best examinations of this question is Neville's,[13] but unlike Whitehead, Neville assumes that a personal order of actual occasions requires an over-arching unity, a personal being to whom the personal history belongs. Whitehead does not assume or assert that.[14] Otherwise Neville's unimpeachable account of the relationship between the ideas of "person" and "God" would be definitive, but as it stands, there is still room to discuss the issue of whether God is best thought of as a personal being.

We will follow the excellent analysis of R.J. Connelly in his book *Whitehead vs. Hartshorne: Basic Metaphysical Issues*.[15] Connelly confronts how Whitehead and Hartshorne differ on the issue of negative prehension, arguing that while it is clear Whitehead allows no negative prehensions in the primordial nature of God (which is unsurprisingly the least personal way of understanding God), his position leaves open the possibility of negative prehensions in God's consequent nature. Hence, Connelly takes on a number of interpreters, particularly William Christian, who have argued from textual clues that God's consequent nature is all-encompassing and hence free of negative prehensions.[16] Connelly goes on to defend the claim that "Justice is done to [Whitehead's] notion of the consequent nature, and analogy between God and other entities perhaps becomes more meaningful, if we conclude simply that God's physical experience compared with that of other entities is the most inclusive and the least exclusive. Negative prehension may be the exception in God's experience and transformation through intellectual feelings the rule."[17] We think Connelly hits upon the interpretation of Whitehead that is least hostile to Hartshorne's typical gestures, but still, the most inclusive idea of God will be the primordial nature, and it is the least personal precisely *because* it is the most inclusive. Still it seems consistent with Whitehead's temperament to grant to his concept of God inclusiveness to the extent that it doesn't interfere, conceptually, with the irreducible eternal objects and actual entities.

We will build upon Connelly's account here, not as a contribution to the understanding of Whitehead, but to the understanding of how God ought to be conceived by process philosophers who think in the *Whiteheadian* line. We do not think that Whitehead pursues this line very far himself, but there are better and worse ways to proceed upon the basis of what he *did* say. Hence, this is a reform of Whiteheadian philosophy of religion, and if theology is needed (which we question), also of Whiteheadian theology.

Regarding Hartshorne's view of negative prehension, Connelly cannot make it fully self-consistent regarding whether the subjective immediacy in present experience is ours, God's, or both, and this problem is connected to the issue of the objective immortality of past experience, and the full, active presence of past experience in the present (however "present" is defined). Auxier has addressed this issue with Hartshorne's philosophy elsewhere, and rectifying the ambiguities in Hartshorne's view, to make it consistent.[18] This view builds upon and supports Connelly's view. Hartshorne also *should* have endorsed the idea of negative prehension in God, even though he endorsed the opposing view.[19]

Connelly concludes his analysis of Hartshorne saying that "there is no systematic reason for saying that God has to feel everything. It is enough to realize that God's experience is more perfect than that of any other actuality."[20] We add to Connelly's claim a systematic reason for saying that God does *not* feel or prehend everything (positively at least), and suggest that this characteristic goes to the very heart of the issue of the *personhood* of God, of actual entities, and of their freedom, when we interpret Whitehead. We do not necessarily express our own views in what follows, but rather we aim to provide a roadmap for understanding a Whiteheadian God—a philosophical God-function that squares with Whitehead's view of possibility and actuality, and with the quantum of explanation. This project is most easily carried out in contrast to Neville's views.

Some Terms

Neville sets out his schema of basic terms concisely in chapter five of *Eternity and Time's Flow*.[21] He defends a particular view of relationalism and its implications for time. For Neville, all three "modes of time" (past, present, future) have their specific essential/necessary characteristics. He draws upon necessity as a springboard for possibility. Whitehead neither affirms nor denies ontological necessity, and he never employs it as an ungrounded principle. But no radical empiricist can rule out its reality. After all, when we genuinely don't know whether something *is* the case, we should not assume it *either* is or is not the case, in some final way. The suspension of traditional metaphysical claims is for the sake of doing better philosophy, and this move does not aim at excluding completely the traditional uses of philosophical necessity. The habit of excluding radically empirical approaches, by neglect, hostility, or ignorance, is a weakness in the traditional approaches.

Happily, Neville's method and his habits of inquiry rise above such complaints. He says traditional metaphysics is inadequate for both theological and religious purposes, and we concur. We think that even if Whitehead does not do so, there is a version of personalism that answers both theological and religious needs without having to rely upon a *philosophical* conception of God. That account falls beyond our present scope, but following out the consequences of our account of extension and negative prehension does require a discussion of *Whitehead's* God.

Preliminarily, we need to explain some terms. In Western thought it has been commonplace to accept Aristotle's claim that necessity (*ananke*) is the most basic modal notion, that we derive the idea of impossibility *from* necessity, by conversion—what is necessary *cannot fail to be*, and therefore it is *impossible* that it *not be*—and the idea of *possibility* is then derived from impossibility.[22] Aristotle thus defines possibility in terms of necessity.[23] What is clearest, Aristotle says, is that " 'necessary' and 'impossible' do signify the same but . . . when applied conversely."[24] Also, historically, possibility and necessity have been understood not only as *logical* principles, but *metaphysical* principles as well.[25] We find the logical sense of possibility and necessity most clearly manifest in the law of non-contradiction, "it is not the case both that A and not A," at the same time and in the same respect. This is only as an abstract formula which is determinative of truth conditions for other propositional forms, but it is normative for our *thinking*.

To think *well*, as an ideal of inquiry, we must conform to the law of non-contradiction (among others). It keeps us from getting confused about our various levels of generality, and we ought not contradict ourselves at a *more* abstract level of generality when we have found a self-consistent generic contrast at a more concrete level of the analysis. When we *do* find ourselves locked into a contradiction at a higher level of abstraction, we can be assured that our thinking is confused. Hence, non-contradiction is very important, but it has no *a priori* ontological weight.

The metaphysical significance of necessity and possibility is not altogether clear, although it is often thought to be best expressed in Aristotle's formula (above). This is not just a norm for clear thinking, but supposedly a universal claim about what can and cannot *exist*. From the above, we get the following conditional norm: if we wish to think about what *is*, rather than what *is not*, we should conform our thinking to the metaphysical claim that "grounds" the logical norm. (We might well ask *why* we would wish to restrict our thinking in that way, but that is getting ahead of the story.)

Possibility, therefore, is habitually considered by philosophers from the standpoint of the principle that nothing is possible *except* upon the basis of some metaphysically prior actuality, whether that actuality is contingent or necessary in its own way of existing. They assume that we *want to* think about *something* rather than *nothing*. Neville's arguments about created being and God are examples—the determinations of God are dependent upon a prior undetermined being, called "being itself," and further determinations, whether contingent or necessary, are *possible* only upon the basis of a prior (positive) actuality. For Neville, the one creates the many, and only having *done so* is the creator created *qua* creator. In others words, unity (in a strong, metaphysical sense) precedes plurality.[26] Thus, traditionally, the possible is derivative, while the actual (and necessary) is prior.

Process metaphysics does not fully accept this view, and certainly Whitehead does not. While it is true that everything that *comes to be* actual becomes actual through the agency of some prior actuality (God's contribution to creative synthesis), it does not follow that possibility is derivative of actuality. Indeed, it is equally true that everything that becomes actual must first be possible, and insofar as God "becomes," *whatever* God actually becomes was first possible. Process philosophers, while not holding to the absolute independence of possibilities from actualities, nevertheless explore and explain the nature of possibility as a fundamental metaphysical notion, at least coeval with actuality.

Possibility can be distinguished from actuality in numerous ways. Whitehead's way in *Process and Reality* renders possibilities lacking in actuality but still existing. What the character of that existence is remains unclear, but the account is adjusted to the demands of cosmology in the "modern style." This view denies to actual occasions any *immediate* experience of past possibilities, or "might-have-beens." Our analysis of this mediation in chapters 8 and 9 explains this point in detail. *We* think humans *do* have an immediate experience of the might-have-been, although only internal to a single durational epoch, but then, that's what "immediacy" is—the indistinguishability of existence and experience. That is not Whitehead's *explicit* position, although we have argued (chapter 11) that a Whiteheadian naturalism *requires* such a view. As far as we can tell, Whitehead wasn't very interested in the question, since he treats a subject-superject (or a transition-concrescence) as something with no duration or endurance. The actual entity is an explanatory hypothesis, or more strongly, an irreducible quantum of explanation. The "actual entity" is not the outcome of a series of events.

Whitehead's position is superior to Hartshorne's, because, contrary to his own intentions, Hartshorne sets up a structure in which God cannot be a personal being—in spite of Hartshorne's avowed personalism—while there is a way of seeing Whitehead's God as at least arguably personal, despite, as we have noted, Whitehead's explicit disavowal of personalism.[27] Although it is an open question Whitehead does not assert whether God *is or is not* a personal being. If Whitehead did assert personality in God it would look a lot like E.S. Brightman's position in *Person and Reality* (1958). We disagree with Brightman's position too, which cannot handle the question of immediacy, especially the immediate interaction of what he calls "purposes" in the "illuminating absent."[28] In Whiteheadian language, if you imagined that eternal objects were active (as Brightman and Leibniz do), then what sorts of interactions would they have beyond our *experience* of them? Leibniz says they would all "strive to be." Would those interactions be mediated? What would be the "order" in it? Brightman cannot answer this question without positing that monads have "one window," which is the power to form purposes. Leibniz's answer was more consistent, i.e., no windows, pre-established harmony.

Whitehead handles this question about the order among possibilities by hypothesizing that possibilities are utterly *in*active—they are not purposes (God's or anyone else's). Their *order* is not active. It follows from this limitation that possibilities for Whitehead are uncreative and uncreated. God just *deals with them*, and with their "order" as given modes of limitation. That is why Whitehead's God is not the God of religion nor necessarily personal. But depending on how the question of immediate experience is handled, God *could* still be a personal being of a sort. For example, God could be in the process of becoming more personal, or less personal, as God struggles with the non-rational Given (this is how Brightman characterizes it). We find neither position satisfactory, but of the three—Whitehead, Hartshorne, Brightman—Whitehead is the one who makes no actual mistakes. He never asserts any ontological knowledge, whereas Hartshorne and Brightman do. We can adopt Whitehead's hypothesis about the utter indifference and inactivity of the possible for the purposes of cosmological "explanation" (in Whitehead's sense, that the abstract is *explained by* the concrete). When a different set of abstractions needs to be explained, we can drop that stipulation about possibilities being inactive (e.g., when cosmology is not our main topic). Hartshorne and Brightman do not have this kind of flexibility—they are not radical empiricists.

Whether we could improve on Whitehead's cosmology by postulating active purposes as the character of the possible is another matter. We are inclined to think that such a change would *not* improve the cosmology, because of the character of cosmology; it does not try to answer questions such as whether possibilities are "purposes in nature." Yet, such a hypothesis might improve our ethics or theology. As Kant demonstrated in the *Critique of Judgment*, such questions as purpose in nature are relevant to our moral knowledge, but get in the way of science. Still, we should not flush all intentional structure from scientific inquiry, and Whitehead guards against that tendency. We can't see how a philosophy of science can be improved by building in teleology at the level of possibility. Peirce also rejects this move. Bergson most definitely rejects it. James rejects it, because he denies independence to possibility. Dewey waffled on the question, for decades, before coming around to the same view as Peirce in 1938. Dewey's waffled because he doesn't want to divorce possibility, even by hypothesis, from the world of active purposes. It seemed to him that allotting to possibility any independence *at all* invites the vicious abstractionist into the conversation. But finally he recognized that one *can* hypothesize true universals (deploy universal propositions in logic) without believing they are "out there" somewhere. Dewey had conflated the process of fixing belief with the process of hypothesizing; he had pinned hypothesizing to the problematic situation in ways that Peirce found inappropriate.[29]

For Whitehead's cosmology, the "higher phases of experience" are treated uniformly and he does not seek to discover their *origins* or to contrast those higher phases in *our* immediate epoch with more primitive versions of the same. He recognizes that the "rationalization of the world" is both the cause and the effect of civilization, and he sees tragedy as well as progress, but he takes it for granted in *Process and Reality* that we wouldn't choose to revert to a less rationalized condition.

In *Religion in the Making*, the higher phases are a central concern. Whitehead replaces the idea of "eternal objects" with "ideal forms," allowing us to think of possibilities as active purposes at various stages of civilization along the way, but Whitehead has no praise for that way of thinking, being clearly glad that the human race outgrew it—even if the cost was unbounded "solitariness." There is no nostalgia for the old God of religion, but our religious instinct will treat possibilities as active purposes. This is the implicit basis of the Claremont School of interpretation: they treat possibilities as purposes, as do most religious people. Whitehead says very little about the relationship between immediate experience and the might-have-been when he is doing philosophical anthropology, but his position could be different from his cosmological view of possibility; if his view of possibility is different, why not also the specific case of the might-have-been?

Considering the contribution of eternal objects to creative synthesis, one sees their ingression as a formative element in every act. More plainly, even God has to work with possibilities and ordered constellations of possibilities to say "let there be light." Thus far, Whitehead, Hartshorne, and Neville agree. But for those who follow Whitehead's radical empiricism, there remains the other main option—that God might not act. A God who *must* create is not free, i.e., is a machine at bottom, and certainly not a personal being. Neville handles this contingency by distinguishing between God the creator and the transcendent God that *is* being itself. But we do not think the latter God, if there is one, can be conceived of as anything more than a *possibility* pointed out by a specific line of *thinking*. Regardless of whether God as "being itself" actually exists, the *most* we get from Neville's argument is a possibility for *thinking about* God. As far as we know, no argument or mode of thinking *necessarily* commands or demonstrates the existence of the intentional object of the thought. Thought is freer than that in our experience, and need not imply the existence of the object of thought. "Explanation" does not mean commanding thought in some particular direction, forsaking all others; rather, explanation means replacing the abstract idea at hand with the more concrete combination of actualities and possibilities that gave rise to just *that* abstraction (and not some other).

We may also entertain the idea that the first creative act of God might have been something other than "let there be light" (which, admittedly is the product of human mythic imagination aiming for the infinite act). Whatever the original act is, we must confront the question whether it was the *only* possible act God could have enacted, or whether refraining from acting (which is also an act, perhaps) was a genuine *possibility* for God. Process philosophers do not agree on these points, and we hope to show that even so, at least the implications are clear regarding the *options* for self-consistent thinking left open to them.

Our position is that we get the best *philosophical* account of God by hypothesizing, in keeping with radical empiricism, that God creates freely from among genuine *uncreated* possibilities, including the option not to create at all, because possibility and act are *genuinely related* but *independent* modes of existence.[30] We do not think that we can concretely conceive of a God who refrains from acting altogether, but we think we can imagine it through negative analogy to ourselves. When *we* don't act, we have at least a vague idea of what that means and what the structure of the alternatives to acting-as-we-have *would be*. This is the "cash value" of our analysis of negative prehension in chapters 8 and 9. We do not mean that a naïve analogy or anthropomorphism is proper to thinking about this kind of problem.

Given the above, one *ought* to treat possibilities as ontologically coeval with actualities. Prior to the first act (ontologically not temporally prior), we see that the actual and the possible are not distinguishable, but that does not imply they cannot *be* distinct, only that they cannot be *known* as distinct by knowers such as we. Prior to the first act, all possibilities *could* become actual (indeed, this is where we can ground the idea of omnipotence in the primordial nature of God, as against Hartshorne). No possibility is yet excluded—there are no "might-have-beens" yet, since the first act commences separation of past from future by the mediation of the present. With the first act, infinitely many constellations of ordered possibilities become *might-have-beens*, viz., all those that were not concretely compatible with "let there be light," including the act of restraint involved in *not* creating. After "let there be light" (or whatever the first act was), there is no longer any literal sense to saying "God is omnipotent," only "God *was* omnipotent." The first division of the extensive continuum is the becoming of temporality, and it conditions all future divisions by casting the form of time over the whole *qua actual*. While these might-have-beens remain *real*, remain *possibilities*, they no longer have any potency or potentiality.[31] And here the distinction between the actual and the possible first emerges, via the mediation of the potential.

The word "potentialities" thus designates possibilities that still "might-be," and does not apply to possibilities that are now "might-have-beens." In this context the actual *makes* the difference, and we see the truth of the Aristotelian position. Every possibility that still "might be" *has* that status because it has potency, and its potency arises from something *already* actual. But there is still an important difference between possibilities that still *have* potency on the basis of present actuality, and possibilities that do not. The might-have-beens remain real (since they *once* had potency), but can never be actual. Yet, these might-have-beens still contribute something to the present as determinate but non-definite realities. The might-be's are also real, and while the vast majority will never be actual, at least one collection of them will become actual. The members of this collection are *known to be* determinate only insofar as the constellations are ordered according to principles of mutual concrete compatibility and continuity with the actual. They can be abstractly sorted from one another according to numerous proposed operators (we demonstrate the most general form of these in chapter 9), as science does, but such analysis is contingent and must be judged according to its *probability*. If there are determinate orders that are *wholly* independent of the definite, concrete (and actual) order, we have no way to know about them.

Perhaps such a determinate order is not wholly closed to our imaginations, but Whitehead was never very much impressed with our human imaginative powers. The model of Abbott's *Flatland* might provide us with a template for understanding determinate order beyond the influence of the actual, but even at this we are confined to analogizing. If we are in the position of Flatlanders trying to imagine the possibilities for determinate order, then we are unlikely to perform at a higher level than do our planar counterparts when they attempt to imagine three dimensional beings. That failure has no bearing on what is genuinely possible, only on what we can find out about, hypothesize, and theorize. The Whiteheadian theory of extension presses our capacity to think about determinate order about as far as it has ever gone, but not so far as to fall outside of the quantum of explanation. In short, this is not Rortyan deflationism about philosophy. It is rather a sort of radical realism about the relation between what might be known by some possible knower and the human capacity for knowing that relation in the present.

To put this in Whiteheadian terms, some eternal objects mutually imply one another and ingress together, while other eternal objects are excluded from a particular ingressing collection, being both logically and ontologically incompatible with the ordered collection that ingresses. We have explained the logical conditions and the practical structure of such ingression in chapters 8 and 9. But the various constellations of eternal objects are themselves partly indeterminate since infinitely many constellations have potency at any given moment, and there is infinite overlap of candidates. There are no elements to these constellations *per se* (and hence these are not "sets" as traditionally understood), in that the determinate constellations are not constituted by "elements"; membership is vague and the determinate order is prior to the enumeration of elements. No metaphysical or traditional (logical) principle of identity is needed, which is to say that there are no temporal atoms. It is true that an eternal object, considered as a member of one ingressing collection, is not precisely the *same* eternal object when it ingresses with another, since the identity of each eternal object, if we *postulate* an identity, is inferred from its relations to all the others in its constellation. But there is still an analogy between an eternal object considered as ingressing with one partly determinate ordered constellation and the alternate version of that eternal object seen as ingressing with another ordered constellation.

The best way to think of these eternal objects is as concrescence (rather than transition), as *genuine* possibilities, as might-be's, as possibilities with *some* potency, seen from a given present (a frame of reference). These might-be's are potentialities and are determinate in relation to one another precisely to the extent that some potentialities are excluded and others entailed relative to a given constellation. But these potentialities are partly indeterminate insofar as each constellation of potentialities still implies incompatible possibilities along the course of its further development, and some will *later* be excluded while some will be included in the actual. The structure of this inclusion and exclusion has been discussed.

From any given present, therefore, incompatible potentialities are equally real. This is not to say that incompatible potentialities can become actual. As an actual entity concresces, it excludes all potencies incompatible with its self-creative act and with the contributions of others, rendering one determinate collection *definite*, as we have argued, also rendering all incompatible determinate constellations mere might-have-been's. This process is the achievement of the hybrid entity Whitehead calls the "generic contrast."[32] And yet, such exclusion is a manner of prehension—that is, the actual entity *takes account* of the potencies it excludes *by excluding them*. The exclusion of physical feelings is one process and the exclusion of conceptual feelings is a different process, as we have argued. In either case there is a broad *and* a narrow sense of inclusion. Broad inclusion means including x as a possibility *only* (positive and negative prehension), while narrow inclusion means positively prehending x, i.e., taking from x a positive contribution in the achieved satisfaction of the concrescing actual entity.

This account creates the following general, provisional ontology: the cosmos at any given moment (after the alpha point) consists of: (1) actualities, i.e., the present and objectively immortal past taken together; (2) potentialities, i.e., the actual plus all constellations of possibilities not presently *in*compatible with it, which is the future (not all potencies are equally *relevant*, since some have little potency, which is what we mean by "improbable," and some are distantly removed in the future, which is what we mean by "indefinite"); and (3) the might-have-been, i.e., those *real* constellations of possibilities that once had potency but no longer have potency from the standpoint of a present (or abstractly, any definite frame of reference). Let us consider some of the things we can know, discern, or hypothesize about these modes of being, in reverse order.

The Might-Have-Been

We insist upon the *reality* of the might-have-been. Following Whitehead, we use the term "real" to designate the actual and the possible together, i.e., the actual entity (that is, the quantum of explanation) and the eternal objects (considered in their independent sense, as unchanged by actuality).[33] And here, the medieval realists were correct. This is also close to Neville's motive for affirming the transcendent God, but without the theological baggage. We do not know whether universals are "real," for Whitehead or any other radical empiricist, having never encountered anything more than a contingently exceptionless generalization, but we could not deny that true universals *might be* real (even if "merely" so), and they are certainly useful. That position is consistent with Whitehead's view, even if he does not explicitly say so. Thus, instead of treating "eternal objects" as functional universals, as Whitehead does (and that is as far as he goes), we are content to think of them as possibilities existing at various levels of generality, from exceptionless generality exemplified in *every* actuality down to the individual relation of one actual entity to an immediate predecessor or successor in its actual world (and here we restate the "two descriptions" of "reality" for Whitehead, the macroscopic process of transition and its converse microscopic counterpart, called concrescence). All such relations require some degree of generality, and hence exemplify some possibility.

Insofar as might-have-beens exist (and we are saying they do *exist*, drawing upon Hartshorne's distinction between existence and actuality, which we take to be consistent with Whitehead's view, just as Hartshorne claims; we are not saying these might-have-beens "subsist" *á la* Meinong), their reality is limited to being universals—they are intelligible to us as ideas, and they display a kind of order that follows (as far as we can think about) the order of thought and imagination. Might-have-beens are possibilities that were negatively prehended, dismissed from conceptual feeling, even at their highest point of relevance to a complex organism.[34] Without repeating the full exposition of negative prehension, suffice to say that a philosophical account of the might-have-been is dependent upon what is *not felt* conceptually, by an entity, and can easily allow that everything that exists physically in the universe is positively prehended.[35] The reality of the might-have-been is relevant when we consider those eternal objects, or possibilities, that are incompatible with the actualities and potencies felt at the *highest* grade of relevance, contributing to the subjective form of an occasion under the contrasting tension of objective unity and objective diversity.[36] We can sort out ordered constellations of might-have-beens by numerous means, the most popular of which is probably historical thinking arising from nostalgia. This kind of thinking is easier than abstract mathematics, although the formal accomplishment in human thinking required by this act of abstraction is strikingly similar.

For instance, we can take two mutually exclusive might-have-beens and follow out their order by means of their historical implications. Had General Lee taken Longstreet's advice and called off Pickett's frontal assault at Gettysburg, we can call to mind, upon that basis, numerous constellations of might-have-beens that then may be adjudged according to their plausibility (we reserve the term "probability" for potencies and "plausibility" for judging the hypothetical likelihood of might-have-beens[37]). So we might, on one hand, consider the plausibility of the constellation of might-have-beens that accompanies the assumption that Lee called off the charge, but then did not follow Longstreet's further advice to attempt a flanking march (e.g., Lee withdraws and chooses another battleground in Pennsylvania or Maryland or even Virginia); and we may, on the other hand, entertain the constellation of might-have-beens that accompanies the assumption that Lee did try the flanking march.

These two lines of thinking suggest very different lines of "research," and one line is far more determinate than another, since these scenarios display variant degrees of compatibility the actual. If we want to tell a plausible story about the flanking march, we have an advantage. We can examine the positions, condition, and state of military intelligence possessed by both sides, and consider the timing of the movement of the army. We can show plausibly which units would have been withdrawn, and by studying the habits of the generals involved, we can even form a plausible account of which units would have remained to detain the Union Army, their strength in numbers and experience in battle, and gauge perhaps the likelihood of success for either side. This would be a highly speculative analysis, but one which could be made plausible by a person who was diligent in gathering facts and figures, and who was sensitive to the flaws in his/her own reasoning, and to alternative lines of reasoning that may seem more plausible but can be answered. There would be no final resolution, but there might be a compelling story.

For the person who takes the other line—that Lee cancelled the charge but retired from the field without attempting the flanking march, the task is far more complicated. This scenario is less compatible with the actual, and the ways in which a strategic retreat and move to another location might have proceeded are amazingly varied. We could say only that his army would have to go somewhere soon to re-establish supply lines and deter any attempt by the larger Union force to threaten Richmond. Further, one could not begin to form a very plausible account of where Lee would have gone and what battles might have followed, and indeed, one could not say with much plausibility what the retreat itself would have looked like. All of this reasoning is algebraic in character, and all of it depends upon narrowing down whole constellation of variables so as to draw on more determinate and law-like regularities with as little vagueness as we can manage. Plausibility is the reward for good reasoning about the possible.

As compatibility with the actual decreases, the effort necessary to exclude these rogue types of possibilities from an imagined scenario increases. Yet, one cannot say that the might-have-been has no order or structure. It has at least (and exactly) the amount of determinateness that can be gleaned from its compatibility with the actual, albeit abstractly in every case, since it has lost the potency it once had. Still, might-have-been's do come in ordered constellations, and depending upon what principles of order one emphasizes (whether historical, or geological or ecological or mathematical or even purely imaginative), one can tell numerous different and comparatively plausible stories about the might-have-been.

All of these stories are algebraic, in the sense we have defended. We have used some narrative principles of history above, but a simpler example is the debate between uniformitarianism and catastrophism in geology, which employs a much less imaginative narrative structure, since one need not consider the wills and decisions of individual persons. The uniformitarians realized they were involved in a form of historical reasoning that emphasized gradualism and that they needed to presuppose uniform natural laws of a simple kind (e.g., linear) to have a *plausible* basis for excluding infinitely many might-have-beens. Thus, the principle of uniformity in natural law became, in and of itself, the target of philosophical controversy. Since catastrophists are gaining the upper hand again, the research to exclude completely different might-have-beens proceeds apace. What few participants in this debate seem to realize is that they have been arguing over a principle of *exclusion* in the might-have-been, not about the fundamental laws of nature. Uniformitarian thinking now proceeds *ex-hypothese* rather than from a conviction that the "laws" of nature (a suspect metaphor these days) do not themselves change. And, as a hypothesis, both the advantages and disadvantages of uniformitarianism are before us; we are aware of how the weaknesses in the hypothesis affect the plausibility of the stories we tell about geological phenomena. These weaknesses force us to recognize that creation scientists can tell alternative stories which, even if extremely implausible, cannot be finally refuted based on empirical evidence alone.[38]

The upshot is that the might-have-been admits of order—indeed of many forms of order. Our choices as to which relations to emphasize in discussing the might-have-been take direction from our interests in the present—Whitehead and Dewey both emphasized the perils and rewards of selection, and this awareness ought to chasten us.[39] We must not confuse the plausible, as such (and the interests that lead us to tell plausible stories about the might-have-been), with the probable.[40] Our orderings of the might-have-been deal with possibilities *qua* possible, not with potencies *qua* possible. We interact with the might-have-been in a fundamentally different mode of consciousness (more precisely, conceptual feeling) than we use to interact with the actual and *its* potencies in the present, in which physical feeling dominates. The same logic does not apply, nor are the standards of knowledge the same, in these situations. We experience the might-have-been abstractly and imaginatively when we experience it at all, at least from a Whiteheadian perspective—if there is a phenomenology of the might-have-been, he does not provide it. Our best efforts to concretize the might-have-been proceed by *analogy*, which depends first upon a negation of all that has become actual *since* the constellation of might-have-beens we are considering has lost its potency. The structure of determinate order in the might-have-been is algebraic, just as it is in every other description, but the variables are now permanently indeterminate as related to the actual.

Thus, for Whitehead, there can be speculative and abstract "knowledge" of the might-have-been, but no immediate experience of it—*not even in God*. And here is the rub: it follows from our treatment of the might-have-been and our prior discussion of God that God, as concrete, subsequent to the first act, *experiences* the might-have-been *differently* from the actual. Does God *also* experience the might-have-been abstractly? We cannot know this, but what we can say is that *if* there is no difference in God's experience between the might-have-been and the "was" (past actuality), then God cannot be a personal being. We will make good upon this assertion as we proceed. For now let us foreshadow the argument by indicating that we think it makes the most sense to say that God negatively prehends might-have-beens.

Potencies

Earlier we said potencies are possibilities that still *might be*, considered from a given present. They are also "merely real" and are what is left over once the might-have-been is negatively prehended. As Whitehead says, "it is the mark of a high grade organism to eliminate, by negative prehension, the irrelevant accidents in its environment, and to elicit massive attention to every variety of systematic order."[41]

Regarding the dependence of the present feeling on negative prehension, Whitehead says, "The negative prehensions, involved in the production of any one feeling, are not independent of the other feelings. The subjective forms of the feelings depend in part on the negative prehensions."[42] This is why nothing comes to consciousness without actively excluding data; to fail to exclude is to fail to take on a subjective form. As Whitehead says, "consciousness is the feeling of negation" in our perception.[43] The actual world of a high-grade organism is saturated with the past, perishing but contributing to the environment all of its present characters. Past conceptual feelings that were incompatible with the possibilities physically felt, are *unfelt* conceptually, and are conditions for any and all consciousness. If no possibilities are dismissed, no consciousness arises.

Thus, we now identify potencies, or "real potentials," by their impressive relevance and systematic order, which is so compelling as to evoke, to the exclusion of every alternative, a "massive attention." Potencies therefore also admit of analysis in terms of relevance to the actual (i.e., the present). The most relevant potencies are the most probable ones, and analysis that seeks to anticipate *which* potencies will become actual may proceed variously. Scientific prediction is an example, and other frameworks for prediction are available. The framework chosen is more reliable when it is specifically adapted to the phenomena one is addressing, not merely borrowed from some more special or more general inquiry. Hence, in quantum physics, for example, one sort of probability (almost ontological in meaning) is needful, and in the reproduction and mutation of yeast quite another model is called for (statistical models suffice). Also if one gives an account of the likely actions of complex organisms, psychological, sociological and other models of prediction are needed. The more complex the phenomena, the more difficult it becomes to forecast which potencies will become actual, at any specified level of generality. Every animal has its own ways of dealing with potencies, so we ought not get carried away with the human ways of coping.

In spite of the difficulties involved in forecasting future actualities, certain relational structures will be required of all complex organisms. These relational structures capture, in the most general way, the contrasts among various ordered collections of potencies. Some potencies definitely imply others—causally, logically, mathematically, or otherwise—in the sense that all possibilities, including potencies, exhibit order and the compatibilities and incompatibilities in such a multiplicity fall into ordered constellations, whether we wish them to or not. We cannot always know whether some possibilities are included in particular collections of potencies currently under consideration. We *can* be confident that *some* possibilities are always excluded: assuming a collection of potencies, call it A, becomes actual, other particular potencies will be excluded, and will be destined therefore to become might-have-been's, never conceptually felt by any actual entity. So, hypothetically, the differences between an outcome we expect *more* and other outcomes we expect *less*, we contemplate the more probable *as if* it were actual, and we may then ask ourselves, "if A is actual, then what is excluded by A?" Is there some mode of consciousness that will allow us to consider the relationship between A, as an imagined, anticipated actuality, and the particular potencies and collections of potencies that will probably be excluded by A? In a word, yes.

Whitehead captures this situation in a profound but difficult complex disjunction: "In awareness actuality, as a process in fact, is integrated with the potentialities which illustrate *either* what it is and might not be, or what it is not and might be."[44] Once the might-have-been is negatively prehended and "dismissed," and its physical potency conceptually "eliminated," what we have is either the contingent present (what is and might not have been), or the future (what is *not yet* and might be), and consciousness of the relation of the present to the future is dependent upon our negatively prehending the might-have-been. In short, our capacity to distinguish the present from the future depends on our ability to prehend possibility negatively, meaning that the distinction between present and future depends *in awareness* on the distinction between *the was* and the *might-have-been*. If you think about this in common sense terms, it isn't so difficult. How do you know, right now, that one collection of potencies is available for present action, while other potencies will remain regardless of which collection you act on now, yet, some collections will be eliminated? How do we explain the basic structure of "choice"? It means that some choices eliminate other choices. All choices eliminate some choices. Some choices eliminate fewer other choices than some other choices. That difference is the relevant difference for understanding our awareness of potencies, and thus we also grasp that in potencies we integrate the actual and the possible.

With Whitehead, we think there is a mode of consciousness that allows us to distinguish among constellations of possibilities *qua* possible, and it is the *same* mode of consciousness we use to separate might-have-beens from past actuality. This is another way of saying that memory and imagination are really the same function in consciousness. What differs is our temporal orientation and awareness of this function—what we *take ourselves to be doing* in forming a generic contrast between these two illustrative temporal moments. When we seek to distinguish potencies from *one another* on the basis of an imagined actuality that is *not yet*, we do so by imagining what *will be* negatively prehended and what *will be* positively prehended by our future selves. When we seek to sort out ordered constellations of might-have-beens on the basis of past actuality, we imagine first a future that never was or will be, and then ask what we *would have* positively prehended in that imagined future and what *would have* been negatively prehended. We then construct plausible stories to support one set of might-have-beens over others.

When we sort out *probabilities*, that is, possibilities that still have potency, we similarly imagine an actuality that *is not* and try to construct a probable story about why some potencies and collections of potencies *will* become might-have-beens while another collection will become actual. In both cases the operation depends upon our willingness to bracket present actuality, negate it, and imagine another actuality *as if* it were the present, then building upon the fundamental features of the actual (degrees of relevance, for instance) in order to tell either a plausible or a probable story, but not both.

This operation impoverishes the actual (the concrete present) of all its complexity and concreteness, moves it thereby, to an imagined space/time location, ignores its own act of abstraction, and then reasons by principles of inclusion and exclusion about *which* possibilities are most relevant/compatible. One might call this operation the "deactualization of the actual" and the depossibilization of the possible, constellations becoming collections by fiat of cognition. One might as easily call it what Whitehead does, i.e., abstraction. It is the potentialization of the actual, and integration of actual and possible, at the expense of both—since both suffer negation with extreme prejudice.

It requires a very complex organism to eliminate so much of what it is feeling. It is more than an act of abstraction (although it is at least that) because we not only ignore some actualities in imaginatively relocating the actual, we also reconfigure all the ordered constellations of possibilities and force them to conform themselves to one privileged collection of potencies (the one we are merely *calling* actual at the moment). But what we are calling the "actual" in these cases is not in fact *actual*. It is not "fact," to use Whitehead's favorite word for the most stubborn features of physical experience. Rather, it is either a potency in our real present (in which case we are concerned with prediction and probability), or it is a dead potency, a might-have-been whose chance for actuality is past from the standpoint of our concrete present, in which case we are interested only in plausibility to serve some present interest. We interpret ourselves by such activities, give ourselves the reasons for what we have done that supposedly explain what we shall do. In both cases we are really thinking about potencies, but the difference is that considering might-have-been's is carried out by analogy through past potencies, while considering might-be's is carried out by analogy from the standpoint of future or present potencies.

Actuality

In thinking about possibilities and potencies above, the pivot of each was its relation to actuality. But there is a distinction between concrete actuality as presently and immediately experienced, and the *concept* of actuality or *idea* of actuality as it operates in our reflective thinking *about* possibility and potency. Whitehead was fascinated by the problem of immediacy and never assumed it was the same as simplicity. Hence, actuality as an *idea* is just an operational principle we employ to get a handle upon possibilities as related to potencies. Yet, this limitation, which renders our philosophical account hypothetical, is nevertheless also part of whatever we mean by "experience."

The consideration of possibilities *qua* possible does not require a *concept* of actuality, although it is clear that it requires the assumption that something *is* actual (since only thus would there be any existing beings thinking about pure possibilities). As Peirce argued, the possible, taken alone, has no necessary connection to existence. Mathematicians, he claimed, when they operate *as* mathematicians do not care a straw for existences.[45] They explore the relations among possibilities *qua* possible, and when they become interested in applying those relations to concrete existences, to actualities or to potencies, or even to *past* potencies now perishing, or as a matter of past contingencies, utterly unactualized, they cease being mathematicians and begin being scientists. Peirce insisted upon an independence of the possible *qua* possible, and criticized Dewey, for example, for not recognizing how relations among possibilities might be considered without choosing a particular actual (e.g., historical) frame of reference.[46] This is only to say that we do need the *concept* of the actual to think about possibilities. If we wish to consider the possible in relation to anything else that exists in any other mode, we immediately must *posit* the actual, in concept, and distinguish the might-have-been from the might-be in relation to that postulated actuality. That distinction will give us the potencies for whatever actuality we postulated and enable us to analyze these potencies (as either plausible or probable, depending upon the character of the actuality we have postulated as either past, present, or future) in light of their degrees of relevance to/compatibility with this actuality.

But the *concept* of actuality (as distinct from the concretely present actual experience and its concrete relations to the future and the past) is abstract, and not only abstract, it is itself just one possibility (for thinking). We can alter our present postulate of the actual in as many ways as we like—emphasizing or ignoring certain relations, to treat our concept as past or future alternately. With each alteration of our concept of the actual we witness corresponding changes in the relations of the potencies implied to the possibilities presupposed. When we alter our postulated actuality slightly, we see some movement within constellations of possibilities in accordance with their relevance/compatibility. We are reminded that some event we have treated as a might-have-been suddenly becomes a "was" if the postulated actuality is altered in certain ways. For example, if we have postulated an actuality in the future and analyzed it to discover what potencies it contains and what possibilities will be excluded so as to become thereby might-have-been's, and then we alter that postulated actuality in some way, we see that our further analysis has removed from potency some possibilities (eternal objects) belonging to the earlier constellation; we have now made them might-have-been's, while other possibilities that *were* might-have-been's in the earlier analysis now regain their potency, at least by hypothesis. This exercise is akin to Husserl's "imaginative variation," if it is different at all. The difference, if there is one, would lie in the ontological status attributed to the process by Whitehead—e.g., this isn't just phenomenology for Whitehead. These hypotheses are realistic.

This sort of thinking pervaded the weeks following the disputed U.S. presidential election of November 2000, for example. Many commentators and ordinary people would posit a futural actuality (e.g., the recount of ballots in a given Florida county is prevented by court order), and then ask themselves, what potencies will be rendered might-have-beens by this? Then, altering this postulated future actuality a bit (the same recount is allowed to finish), they ask what difference there is between one scenario and the other. What difference will it make to some predicted or desired outcome? Some predicted, at that time, that the recount would make *no difference*. But what they really meant was that the winner would be the same, since obviously the future in which the recount is allowed and the one in which it is forbidden are quite different futures (the ordered constellations of possibilities are quite different, but they both happen to include the same outcome of the election). The variations are obvious.

This operation of postulating an actuality proceeds upon treating one complex of possibilities as actual, and we generally choose one we regard as highly probable in such circumstances. Then we rearrange the remaining potencies (might-be's) from those possibilities newly excluded (the new might-have-been's). We are free, if we like, to imagine a very improbable future or past, or even present, actuality and enlarge upon it—for example, the implausible possibility that a new nationwide election would be held in 2000. We may then say something like, "I think Ralph Nader's voters would vote for Al Gore the second time," might seem plausible to us on the ground of a new nationwide election. Some other person might interject that voter turnout would be much higher if a new nationwide election were held, and we can see this would also be plausible. But due to the improbability (back then) of the postulated actuality upon which these assertions depend, we feel inclined to call these "probabilities" merely plausible rather than saying they were ever, at any point, probable. And knowing the outcome, we are telling a story much along the lines of the Civil War story earlier. From the standpoint of the concrete present, we are almost prepared to treat highly improbable futures as might-have-been's *already*. Yet, responsible thinking requires us to resist that urge, reminding ourselves that improbable future actualities nevertheless have *some* measure of probability.

Something similar happens when we postulate past actualities that *were not*—Lee's *choices*, assuming he did not order Pickett's charge. Note, the postulated actuality in this case is really *just a possibility*, and indeed, the postulated actuality is a might-have-been. We recognize this might-have-been as "interesting" (in the Whiteheadian sense) because we regard this decision as having been once genuinely *within* Lee's power, an important decision with far-reaching consequences, and hence, as an actuality that at some moment had compelling probability. The past probability of this might-have-been makes it seem almost within our grasp, regardless of what sentiments we may have about it (we don't wish away the ultimate outcome of that war). But if the past relevance (its high probability at one point) of our imagined actuality to a past concrete actuality (what Lee really *did* decide) makes it seem lively, we may also recognize that our thinking is not *bound* by such relevance. We may as easily postulate a past actuality that was improbable *even at the time*—e.g., Lee suddenly changes his mind about his entire cause, rides off into the woods, and surrenders himself to the Northern troops. We cannot rule this out, but it will not receive much serious discussion among historians. The point is that any variation in our posited actuality may make a great difference to our subsequent judgments regarding ordered *collections* of potencies and ordered *constellations* of might-have-been's based upon the variations in our posited actualities.

But what we have learned from this exercise is that the only way we can think about actuality, so far as we know, is to subjugate it to the order and variation that *attends* possibility, rendering actuality itself an open process, which includes the concrete process of thinking, while the content and form of *what is thought* has a less open character. The move from actuality to some version of it *for thinking* is a reduction of the actual to a *concept* of the actual, but the movement of thinking itself by which this conceptualization is accomplished, is not as tragic for knowledge as it may seem, at first, since a similar reduction of the possible itself to a *concept* of possibility has already occurred, and knowledge just *is* some kind of consciousness about a relation between what is possible and what is actual.

The "negation" (as Whitehead astutely called it) that we experience in an act of knowing, the elimination of what does *not* interest us for the sake of a focused act of attention, and the exclusion by negative prehension of infinitely many conceptual feelings, is a condition *for* knowing. Thus, negative prehension provides us with a structure of possibilities that then frames every available variation in how we postulate the actual (as a concept). This is the reason we would do well to treat possibility as the most basic modal notion, or at least not derive it from necessity. To do otherwise fails to recognize the conditions imposed by the act of *feeling* that which we wish to *think about* (which is a feeling of negation or elimination) upon the product of thinking—i.e., the thought.

The Implications: Narrative Intelligibility

We might ask, what limit is there to our ability to project possibilities of possibilities of possibilities, etc.? This rephrases Whitehead's remark about the eighth category of existence, contrasts of contrasts, since we take possibilities *qua* possible to have the finality of eternal objects, while possibilities *as thought*, as generic contrasts, have an "intermediate character."[47] We have practical limits, to be sure; our interest in a topic disappears if the process of positing actualities strays too far from relevance to some concretely actual past, present, or future. And yet, the loops of positing actualities and enlarging upon their possibilities are quite broad.

For example, J.R.R. Tolkien's Middle Earth is an enormously intricate concoction of posited actualities followed by an analysis of potencies and might-have-been's, leading to further posited actualities, and so forth. Can Golem actually regret killing Deagol for the Ring of Power? Regret is attached to the might-have-been; Golem never existed at all, let alone did anything he could regret. Yet, we feel *sorry* for him in the concrete present as we read his story because such a pathetic creature *might* learn to regret something he did, if there *were* such a creature. Middle Earth in fact has a developed set of might-have-been's corresponding to its posited actualities as Tolkien expounded them at astonishing length.

So in creating a fantastic narrative of this sort, we may vaguely analogize Middle Earth to a dark or supra-ancient or even future actuality that we *could* in principle experience. But we already understand that the relevance of Middle Earth to us is comprehended by an imaginative leap, not by a concrete analysis. In fact, when we examine literature such as Tolkien's Middle Earth, we actually encounter the unity of memory and imagination, for in a sense Tolkien is *remembering* Middle Earth, and in another he is *imaginatively projecting* it. We go along with him in an amazingly complex act of consciousness called traditionally "the willing suspension of disbelief." This is well named in its recognition of the negation required. Concrete actuality withdraws in favor of this alternative "actuality," and we as readers surrender ourselves to Tolkien's care (or Twain's or even L. Ron Hubbard's), allowing him to surprise and delight us with the further determinations of the potencies and might-have-been's within his imagined actuality.

An imagined actuality is just a privileged possibility. Its fundamental character is governed by the *structure* of possibility, not by that of concrete actuality. To concretize this privileged possibility, we must retrieve the suppressed assumption of *our own* concrete actuality (not our *concept* of actuality). We cannot make the posited actuality accord with the concrete One.—the Society for Creative Anachronism *wants* to experience medieval warfare without hurting anyone (i.e., not wanting really to experience medieval warfare *at all*). Those who re-enact historical battles take upon themselves the same pathos. The exercise is still filled with nostalgia. Among the ends of "education," according to Whitehead, we find remembrance of the past in order *not* to repeat it, but that end may be better served with less celebration of warfare *per se* than a re-enactment emanates.

As far as we can determine, we have to exist *actually*, in the concrete present, in order to *read* Tolkien, and the smart-aleck who points this out to us is not very interesting. Yet, actuality, the concrete present in its immediacy, is precisely the ground upon which possibilities as a whole (might-have-been's and might-be's taken together) become intelligible. Concrete actuality cannot be exhausted in any *thought* of it, since the thought is not the thing—it is a negation of the thing in its concrete fullness and a re-creation of it as an object of consciousness, in short, an analogy in sign or symbol. We hold these same conditions to be true of God's thought, if God thinks, but we grant that God's mode of experiencing may be such as to render thought (symbolically mediated abstraction) unnecessary. Certainly the *concept* of

Possibility and God   263

God as primordial, in Whitehead, is unthinking, which is precisely why there are no negative prehensions to be found in *that* condition of deity.

This limitation, the requirement of negative prehension for thinking, is a claim about the *thought* of the actual present, not the *experience* of it, and it is one thing to recognize that no *thought* can comprehend the whole of the actual present, and quite another to claim that no *experience* can comprehend the whole. Whitehead says, and we agree, that the principle of relativity requires that we *think* of everything physically existing in the actual world of an actual entity as positively prehended, however vaguely. But can concrete actuality be exhausted in any experience?

The Implications: God, Personhood, and Negative Prehension

Here we confront the question of God. Royce believed we had no choice but to concede that there is an experience of the whole, since there could be no experience of the parts unless these experiences were parts of a total experience, and he believed we would all concede our experiences both exist and fall short of an experience of the whole.[48] Hartshorne defended the reality of an experience exhaustive of concrete existence using different arguments, but from similar logical motives.[49] Neville takes the wiser course of depending only upon the revelations of being in open moments, a sort of Heideggerian phenomenology of Being, but without the trappings of the destruction of the history of ontology. Even Neville focuses too much upon what is positively experienced or experienceable, in our view, and it leads him to deny the personhood of God, while ironically, Hartshorne affirms it when he ought not. Meanwhile Royce holds on to the personality of God by sharply distinguishing the God of religion and history from the God of philosophy, even though the latter is only a concept.[50] Thus, process philosophers have suggested a variety of responses to the problem raised by the *idea* of God's thinking and experiencing process, and what limits or conditions are appropriate to the *concept* of God.

Here, Whitehead's notion of negative prehension becomes relevant. If an actual entity may *take account of* the whole in concrescing by either including the contribution of some other present actuality (positive physical prehension), *or* by excluding its conceptual aspects (negative prehension), then the existence of the concrete whole need not imply the existence of a total experience. And here we come to the crux of the question about negative prehensions in God. Hartshorne never was persuaded by Whitehead's ideas about eternal objects (that is, possibilities), and as a result, Hartshorne built a conception of God around the idea of an experience exhaustive of all actuality.[51] Hartshorne argued that if anything were not *in* God or were not experienced *by* God, there could be no truth about that actuality.[52] For example, he argued that propositions about the past actual experiences of persons now dead would be neither true nor false *unless* God experienced them and preserved them. And since Hartshorne was confident we would agree that there is a fact of the matter, some *truth* about whether these experiences were or weren't *had*, he was also confident that we would conclude that God must be the experiencer in whom these experiences are preserved, and through this act of preservation there is still a *true* and a *false* about these experiences. Thus, if God does not experience all that is actual, then there is no final ontological distinction between the *was* and the *might-have-been*—that all past actualities would collapse into indistinguishable possibilities unless some *being* experienced them all and retains them totally.

This is a worthy assertion, but we think it is false, which is to say that it does not possess the universality and necessity Hartshorne claims for it. There need not be a total experiencer of all that is actual. Without such an experiencer there could still be a truth about the difference between the *might-have-been* and the *was*. E. S. Brightman, in arguing with Hartshorne, challenged him thus: If God has complete experience of all *my* experiences *as* I experience them, then my experiences are not really *mine*. Suppose I believe something erroneously. God will know *both* that I believe it, and that it is an error, but God will not believe it, and hence will not "contain" or "include" my *experience* of it, since I experience the belief *as true* and God does not and cannot. Brightman argues: God only experiences *that* I believe it, but not *what it is like* to believe it and not know it is an error. Hence, moving now to our own language, God negatively prehends my act of believing while positively prehending the physical content (feeling) of the belief along with its erroneousness.[53] The only other option is to say that God is not the sort of being who "knows" things, which Hartshorne cannot abide. (See chapter 13.) This is bad news also for those who follow Hartshorne in saying there is no negative prehension in God, and in attempting to read Whitehead as though he held or should have held such a view.

Brightman further pointed out that, in general, God negatively prehends not just my *beliefs* but also my *will*. We might attribute to God a complete experience of *all else* that is actual, and we might presume that all else *apart* from our individual wills is perfectly preserved in God, but if we once say that God positively prehends our wills, and indeed positively prehends *all* that is actual, we take away the ontological basis for our freedom. Also we assert that God can include contradictions, concretely, in the divine self (e.g., both believe x and know x is false). At work here is something like Whitehead's insight about how the actual entity's subjective insistence on consistency, applying the categoreal obligation of subjective harmony to the phases of concrescence, *replaces* positive feelings with negative prehensions.[54] Brightman favored saying that *we* negatively prehend God, and all else, and positively prehend only that which is given to our consciousness in a single time-span. This view is radical in the bad sense and we do not agree with it.[55] So, according to Brightman we negatively prehend our own past (and must infer or imagine it), and similarly with our bodies, other minds, the external world, etc. This makes conceptual feeling the master of physical feeling, and hence is an ungrounded kind of idealism. But Brightman is right to say that the finite self does far more negative prehending than positive.

A genuinely Whiteheadian God could be thought of as positively prehending everything but the might-have-beens and the self-active wills of individual, finite persons (including subhuman personal beings, and superhuman personal beings, if there are any). We could say, it is because God negatively prehends our wills that we are free. We think this romanticizes and exaggerates the importance of the human type of experiencing, but there are reasons beyond those sanctioned by a severe radical empiricism to give some weight to such ideas.[56]

Whether divine negative prehension is a *self-limitation* on God's part or a limitation *imposed on* God (assuming God constitutes the divine self as creator) is the precise point where our view runs up against Neville's. Neville is committed to the view that God is limited by the order and conditions of creation, once the act of creation occurs. In short, the possibilities for *our* cosmic epoch limit God, but God also *created* those possibilities. But in our view, we could never distinguish between the limitations imposed on this problem by the conditions of our own thinking (that concrete actuality has to be transmuted into a possibility and varied according to some principles of order), *and* limitations of concrete actuality *itself*—this is why we favor descriptive metaphysics and provisional ontologies. We are unlikely ever to get an answer to the question about whether God creates *all* the possibilities, and we ought to be content to recognize this limitation in our philosophizing. Everything Neville says might be right, but we see no adequately careful philosophical path to it.

Hartshorne would not allow any of this kind of talk, ours or Brightman's, and he responded to Brightman with compelling theories of mutual immanence, psychicalism and dual transcendence.[57] These concepts defend the broader idea of a complete experiencer of all that is actual, and through that, all that is possible. But Hartshorne should not have bothered. So long as everything is experience*able*, possibilities and actualities alike (as negatively prehended might-have-beens, or as actual experiences of some being), it does not matter whether they *are* actually experienced by a single being or even by all the beings taken together. Negative prehension is *experience*. It counts. It does not violate the principle of relativity. That principle governs whatever we can *say*, *know*, or *say we know* within a cosmic epoch. The principle of relativity assures us that things *are* related, but it doesn't say *how*.

We cannot know whether God does or does not experience the whole of the actual, but what we recognize that *if* God positively prehends the whole, as Hartshorne claims, *then* God cannot be a personal being, as Hartshorne also claims. In short, that view of God contradicts Hartshorne's own personalism, and that is why Brightman would not accept it. The reason: a precondition for the personal mode of existence is that a personal being must experience the actual *as actual* and the possible *as possible*. Being able to distinguish the might-have-been from the "was" *in thinking* is only one tiny example of the larger point (since thinking is only one exceptional kind of experiencing). But it is a crucial test case for *philosophy*.

Thus the following broad generalization is warranted, and exceptionless as far as we know: For all personal beings there must be a difference between the *was* and the *might-have-been*, and an implied difference between the will-be and the might-be. If there is no difference in God between the *might-have-been* and the *was*, i.e., no differing experience *of* these, then God cannot be a temporal being *at all*. There would be no difference between God's concrete nature, primordial nature and consequent nature unless the *was* and the *might-have-been* are prehended differently *in God*. Here we differ from Neville. His transcendent God, lacking in determinations, cannot be the same God as the Creator, and indeed, Being *itself* is not God at all. We do in fact worship God the Creator, for only *that* God can be prayed to or can have a will for our lives that is different from our own (and hence, sin is made possible by this differing of wills). None of this has any relation to Whitehead's conception of God. God simply *is* non-temporal for Whitehead, and the difference between the *was* and the *might-have-been* is of greater concern to us than our *ideas* about God.

The one certainty about Neville's transcendent God is that there is no negative prehension in it, which is why he is quite right to deny to it any personality or personhood. God is, on Neville's account, the kind of being described by Aristotle, Plotinus, and Spinoza, except that Neville defends not their monistic metaphysics, but the milder thesis of univocity. God would have to be a fully actual being for whom no potency is real, and hence a being that always and eternally *is* all that it can have been. But the defenders of univocity in the Middle Ages took that milder approach to avoid the charge of pantheism, which was a kind of "atheism," which is what got the Latin Averroeists in so much trouble. But the non-heretical defenders of univocity, such as Duns Scotus, desired to retain an idea that God as a personal being, and they actually believed it. One wonders why Neville doesn't simply go the whole way to Spinozism or Plotinianism, given that no one will roast him for defending extreme monism. Why does he not characterize his scheme of determinations of "being itself" as the hypostases or emanations of the One? Is it philosophical modesty or caution? If so, it seems that he is not quite cautious or modest enough, since, according to our argument, that sort of modesty requires that we leave *open* the question of whether there is negative prehension in God, and consequently, we also have to leave open whether God is a personal being.

One can create a self-consistent account of God with such ideas as Neville defends, but one cannot satisfactorily account for the reality of *thought* in a changing universe. The latter was a greater problem for Whitehead, although he was always careful not to exaggerate the efficacy of the mental pole. The perfectly actual God is complete, whether Hartshorne's or Neville's or Royce's, but bears no real relation to the conditions governing our *thought* about the changing universe, and it is thinking about the changing universe that is stamped with the determination, exceptionlessly as far as we know, of personality. In this case one is confronted with the issue of whether there *is* a being who experiences the whole, since the unmoved mover doesn't experience anything but itself, if that. It is not just a religious or theological God that *thinks*, but a philosophical one whose *concept* has to be consistent with a universe in which thinking *occurs*, especially when the thought produced *by* that thinking is a thought *about* God, taken to have a real relation to its intentional object. Previously we showed how the problem of *intention* undermines mechanistic, reductionistic, naturalist philosophies, and here we have applied the same problem in the opposite direction—to those who attempt to conceive of God as more than one formative element in the universe.

Thus, when it comes to conceiving of God, we might be thinking so badly that we effectively fail to have a viable analogy between what has been produced with the cooperation of negative prehension and what concretely exists, but we cannot be thinking about *nothing at all*. We temporal beings are worse off than all those might-have-been's if God has no real relation to the tiny part of the universe that thinks. In that case, we poor little humans are worse than illusions with no reality of our own—as Hartshorne faithfully argued against the Thomists for so many years—we are rather self-deceived illusions who cannot so much as analogize.[58] Mark Twain came to a conclusion like that in his darkest decade, going the Calvinists one better than their own worst estimate of our worth. Advocates of such views are often determinists, Calvinist or otherwise, and are not troubled at the idea that finite beings are not free. But that claim would be troubling to Hartshorne, as an ardent defender of ontological freedom. Yet, his claim that there are no negative prehensions in God seems to imply something like this unwanted conclusion. Hartshorne can have his God only at the cost of confusing possibility with potentiality, and equivocating on the personhood of God, since only the God conceived of as having negative prehensions *is* love.

Why must we conclude that there must be a single being who experiences everything? We need not, and indeed *must* not if we wish to preserve the idea of God as a personal being. We do not think God's negative prehensions are limited exclusively by our wills, although it seems God must negatively prehend our wills if we are to be free, ontologically. Additionally, Whitehead's God negatively prehends might-have-been's. The consequent nature of God must negatively prehend the might-have-been's to prevent past actuality from slipping into mere past possibility, in which case the *might-have-been* would be indistinguishable from the *was*. Unless there is a grounding for this distinction in God's mode of existence, there can be no ground for our own habit of making the distinction, and all our analogies fail.

Here we find the value of Hartshorne's suggestion that without God there is no truth. For a non-theist, what is the ontological ground for distinguishing the *was* from the *might-have-been*? Laws of Nature? Those depend upon the same distinction; alternate accounts of such Laws could account for the same present.[59] Personal memory? This implies solipsism and a transcendental unity of the self that could never be demonstrated. Transcendental ego? Even that ego needs an actual past to distinguish its temporal retending from its intending and protending. Configuration of the present? Infinitely many pasts could have produced the present; if one and only one of them actually *was*, what basis do we have for throwing out the others? We repeat: How would a non-theist support the claim that the *was* and the *might-have-been* are different? Here we find also the reason that the Whiteheadian interpreters who attempt to eliminate God from the system are on the wrong track. In short, no God, no actuality, including no actuality of the *possible* in distinction from the actuality of the concrete.

Since we can postulate that God experienced the past actuality *as* actual when it was present, and was witness to the unactualized potencies becoming *might-have-beens*, we propose that we think of God as bound by the following "principle of possibility": *God cannot positively prehend a **might-have-been** and cannot negatively prehend a **was**.*

This principle implies a fundamental distinction in God between the mode of experience that corresponds to the actual, and the mode that corresponds to the possible. That is, God has conceptual feelings. It is analogous to the same function in our finite consciousness, and this distinction renders both humans and the consequent God temporal and *personal* beings. Here we confirm Connelly's prediction that we can make the "analogy between God and other entities . . . more meaningful"[60] by saying God's consequent and concrete natures are consistent with negative prehension. We would cease to be temporal beings as soon as *our* positive and negative prehensions, in relation to the concrete actual, became interchangeable. By negatively prehending the might-have-been *we* open up the door to contemplating the past *in abstracto*, along with all other possibilities. By positively prehending the concrete present, we are given our own actuality, and for particular, finite existences, there exists also a difference always between what we *are* and what we *might-have-been*. That difference is the ground of the difference between what we will be and what we merely might-be.

We are philosophically obliged to analogize our situation to God's. If there are differences among the primordial, the concrete, and the consequent natures of God, then we may see God as a temporal being. But this difference in aspects of God presupposes the difference between the *was* and the *might-have-been*, and on that ground, the difference between the will-be and the merely might-be. On this view, God does not know what God will become because God does not know completely which actualities "He" will be obliged to prehend negatively and which "He" will be obliged to prehend positively in the future. However, God presumably *does* know which past actualities "He" is prehending positively and which might-have-beens "He" is prehending negatively. This, along with a vast amount of prehensive present actuality provides God with the best basis for knowing the relative probabilities of any given future, and anticipating adeptly what persuasive lures are most efficacious in the present. But God, so conceived, in order to *be* God, need not know what *we* will do. Rather, God, as Whitehead indicates, functions first as the original ground of the actuality of the universe, and as the sustainer of its present actuality, and as the being that loves what present actuality *still makes possible* in the future. This is not Claremont theology, but it isn't wholly without consolation for spiritual people. We are not "co-creators" with God, whether of the good, or the bad, or the ugly, but we come from dirt, pass through the fleshly coil, and return to the dirt as witnesses of an actuality we never made and over which our influence was nearly negligible, but real.

If, however, by "personal being" we mean a being that, at a minimum, has a will and prehends the universe, then there cannot be a complete identity among the will, the universe, and the act of prehending it. If these are the same, we are pantheists.[61] However, if we assume the principle of possibility we suggested—that whatever is positively prehended once is always positively prehended, while whatever is negatively prehended is always negatively prehended *by the individual beings who have excluded those possibilities*, and that the actual and the might-have-been *correspond* to these distinctions—we are free to believe the concrete and consequent God is a *personal* being. More importantly, we have the basis of a principle of individuation without subscribing to any temporal atomism. Personality becomes our best clue to genuine individuality in the universe, while time is our best clue about the character of continuity.

This does not settle the issue of whether God is a personal being. That requires further discussion of what is meant by "personal being." But it seems impossible that a non-temporal being could be "personal," in any sense we can understand. A temporal being must somehow preserve in its *mode* of being and experiencing the difference between the *was* and the *might-have-been*. This act of preserving is the very source of any being's personal identity. That includes humans.

Finally, this principle of negative prehending in God also relieves us of another problem in Hartshorne's conception of God. If God retains perfectly the entire past, that means that evil is never really overcome in God's consequent nature—it is active just as truly in God's concrete present as it was when it occurred. All Hartshorne usually says about this is that God continues to suffer that evil. If that is so, then we have little hope for its eventual defeat, no matter how much good God is able to do later on the basis of it. Hartshorne's view is unsatisfying, even unsettling. If we are right, Hartshorne is off the hook, so to speak, since God negatively prehended the evil intent of evil wills, and always continues to prehend those evil wills negatively, although God *can* prehend the positive (morally salutary, albeit unintended) effects of evil wills. Thus, the evil will is not preserved in God as a positive part of the concrete nature of God, but is taken account of and preserved negatively *only*. That would make the forgiveness of sin easier to understand, and by analogy could apply to human forgiving as well, since it is a mode of forgetting what one never knew in any case, i.e., the intentions of the other.

13 God's Mortal Soul

> At this point the discussion must be halted. It has run into exaggeration. The essence of the universe is more than process.
>
> —Whitehead, *Modes of Thought* (100)

Whether we have fulfilled our stated purposes well, others may decide. But there remains a discussion that bears upon hope and possibility that we cannot ignore. As Whitehead says, "hope and fear, joy and disillusion, obtain their meaning from the potentialities essential in the nature of things."[1] We have tried to say something clear, clearer than Whitehead himself, about what might be intended by a word like "potentialities" in such assertions. We find that it has cost us our hope, perhaps even our joy, because the God that we find in Whitehead's thought is little consolation in our solitariness. The faith of philosophers provides only a pale deity and a strained quality of mercy. Must we be so brutal? We conclude this book by answering that question in the affirmative. But there is, after all, this: "the universe is more than process."

Robert Neville denies, with no hesitation, that God has a "soul," whatever that may mean. This denial is not as worrisome as it may sound to theists, since Neville holds determinations of God, whether conditional or essential, to be a function of God's creative activity, but the *transcendent* God is beyond all those determinations, and "soul" would be one such. We departed from Neville's view because the concrete and consequent natures are the only aspects of the divine that a religious person could reasonably address. Those aspects of God concern *philosophers* only in limited ways. These aspects, if regarded as religiously important, are either to be worshipped without much analysis or analyzed within the limits of theology (see chapter 11).

We have defended a robust distinction between the God of philosophy and the God of theology and/or religious worship, both because we think that is the correct reading of Whitehead and because we think his view is true and unproblematic: we may speak of God without seeking univocity and also without failing to refer to what we speak about. Whitehead's view of symbolic reference thrives on the reality of error, error without which we would neither grow nor learn. The quest for the final and immutable (dare we say inerrant?) words about God, or anything else, is a linguistic conceit driven by the very illusion of possessing the *thing* when all we really have is a *word* for it.[2] Whitehead retained the word "God" and was right to do so. That is the word philosophers use when grappling with the essence of the Whole. It is no rigid designator, but it also cannot fail to refer, at least not if anything *has* an explanation.

We endorse a God *for philosophy* that is Whiteheadian. Functionally this concept of God is no more than a conception of a whole of *the possible and the actual taken together* (as the rational and creative whole, as Whitehead understood it). Neither is reduced to the other. A whole need not be a strict unity, let alone an identity or a condition for identity. We defend radical-empirical pluralism in our reading of Whitehead, an idea with which a few scholars do agree—although as far as we know, no one else has followed the idea through to our conclusions about the relation of the idea of God to the Whiteheadian concept of possibility. Specifically, we argue that Whitehead allows negative prehension in God, as concrete and consequent, and one of the results is that Whitehead leaves open the philosophical question not only of whether God is a personal being, but in so doing, leaves completely untouched the religious and theological questions associated with *personality* and God—questions for theologians and ordinary people to confront, with and without the aid of philosophical reflection. This is the correct position for a radical empiricist.

We think that almost no Whitehead scholar has Whitehead's God right, since most either (1) deny that Whitehead's philosophy needs a God,[3] or (2) deny that Whitehead's God prehends anything negatively (as we have described relative the Claremont school). Through a widespread aversion to Whitehead's views about "eternal objects," both groups fail to grasp Whitehead's view of possibility. Anyone who wants to avoid the idea of eternal objects in Whitehead is almost certain to confuse possibility with potentiality. Thus, many process philosophers have ended up (sometimes unwittingly) with ideas closer to Hartshorne's than to Whitehead's—ideas which are not as philosophically defensible as Whitehead's (in spite of their abundant theological and religious virtues). These weaker ideas are then either defended, or modified, or rejected, but none of this activity has very much to do with Whitehead's actual philosophy.[4] Also, without a proper understanding of the role of eternal objects in Whitehead's metaphysics, the entire theory of extension and of continuity in relation to discontinuity is misunderstood. The result is a grand mess. We have tried to clean it up.

Hartshorne's influence on Whitehead interpretation is the primary source of this widespread misreading, but it has been repeated and extended by the "Claremont School." They have an interesting God, but it is not Whitehead's. The atheists reconstructed Whitehead without God and eternal objects, while Neville and Frederick Ferré developed alternative ontologies, saving a little from Whitehead, but essentially giving up on Whitehead's project as *he* saw it.[5]

To move our project forward, Hartshorne's view of God has been pressed into two different and incompatible, non-overlapping magisteria—the personal, concrete and consequent God, on one side, and the impersonal, Absolute primordial God on the other side. Harsthorne's error comes in asserting that there is no negative prehension in the former. He is entitled to claim this only about the latter, and given his own commitments to personalism he should have allowed negative prehension in the concrete and consequent God. Whitehead has almost nothing to say about this problem.

Hartshorne's conception of God as the all-others-and-self-surpassing being is one of the keys to his philosophy, and indeed, to the overall grasp of most process philosophy beyond Hartshorne. But we think, contrary to Hartshorne, the process God must be viewed as negatively prehending some things—primarily our individual wills, and those past possibilities that no longer have any potency (might-have-beens); see chapter 12. Negative prehension in God provides a ground for saying that God is in *some respects* finite, a point Hartshorne grants, although he denies God's finitude in the crucial respects we insist upon.

There are other modes of finitude that Hartshorne acknowledges, and those who follow him (consciously or not), such as Suchocki and Griffin, affirm some of the same modes of finitude as Hartshorne.[6] The reasons for denying negative prehension in God also vary, but they seem motivated by a worry about the loss of the past *to* God or *in* God, like an assault on the objective immortality of the past. Hartshorne's argument about the relationship between truth and the preservation of the past in God (rehearsed in the last chapter) is a powerful one—so powerful that it leads people to conflate Whitehead's ideas about objective immortality and perpetual perishing with Hartshorne's claims about negative prehension.

To comfort those who fret about God's mortality (whether or not they acknowledge their fretting) that odious negative prehension is *still prehension*, and as such, *is* a "taking account of" the real existence of those supposedly "lost" possibilities and of the wills of others in the concrete nature of the deity. We are not forgotten by God just because we are not completely absorbed by God. It is akin to the old English saying "remember me to him." It means that although I am not there, I am "there by proxy." Such is anyone's negatively prehended presence.

The concrete and consequent God need not be sub-philosophical, although the operation of keeping *philosophy* within its proper limits, while also defending its fence-line from the incursions of theology or unreflective judgment of model-centric scientism, is delicate. Having a *philosophical* conception of God *as* process forbids the identification of God *with* process. In short, Whitehead isn't necessarily a panentheist. Hartshorne acknowledges this point, many times, but his followers, in Claremont and elsewhere, seem quickly to forget this caveat. We don't believe Hartshorne ever forgot it.

Thus, we might still hold, with Hartshorne, that God "preserves" or "remembers" even that which is negatively prehended in and by God, and we might do so while remaining within the proper boundaries of philosophy. This move preserves the objective immortality of the past, as Whitehead understood it. If Auxier is correct in having defended the claim that Hartshorne's metaphysics, contrary to his own statements, actually supports also a very rich notion of *subjective* immortality,[7] then there remains a third puzzle (apart from those of negative prehension and objective immortality) associated with the philosophical God: whether and in what sense God has an "immortal soul," and whether the recognition of finitude in God extends to saying God is in some respects mortal, or at least not immortal in every respect.

Having defended two modes of finitude in God, where Hartshorne allows no such contrasts (negative prehension of our wills and of the might-have-been), and having argued that such modes *are* required if we are even to *leave open* the question of whether God is a personal being, we now move to an examination of what these philosophical modes of "divine" finitude imply. We are attempting to replace a serious contradiction in Hartshorne's view with a consistent account that preserves a point he insisted upon: God's personal character. That more consistent account revives the issue of metaphysical possibility and follows from Whitehead's version of that story.

The resulting philosophical conception is not the claim that God is necessarily, or even probably, a personal being; rather, the question of God's personhood and personality is beyond the reach of Whitehead's philosophy and *all* other forms of responsible radical empiricism, at least until such time as we gain access to experiences that *might serve* as serious evidence *in philosophical reflection* for or against any claim we might make on this score. Being beyond the reach of appropriately restricted philosophical investigation, the question remains a very important religious and theological issue. The theological investigation of whether God is personal actually *is* limited by the results of philosophy, since theology presupposes the philosophical validity of the question itself while philosophy need not make any such presupposition, and theology has no methods that are entirely independent of philosophy (as we have argued, see chapter 11). But, as James rightly argued, nothing in philosophy or theology infringes our religious appropriation of a personal God because not only are the methods inadequate and the evidence indecisive, they probably always will be. And we recall here that experience *always* outruns theory.

Moving now to a discussion of a new *mode* of finitude proposed in-and-for a philosophical conception of God, i.e., that the God of philosophy is best conceived as *mortal* in some respects, we begin with a word about *why* the process God needs to be seen as having a soul at all. Hartshorne's God is a personal being, and unless God exists in a personal mode, many of Hartshorne's arguments as resolutions to old theological problems fail. It is not simple to say what it means for God to exist in a personal mode; Auxier has treated those issues in some detail elsewhere.[8] Suffice to say that treating as empirically real the experiences of consciousness, freedom, and temporality requires something corresponding to a personal mode of existence in God. We do not claim to know whether there *is* a God, let alone whether God actually exists in a personal mode, but we do know that Western religions conceive of God this way, and that many people believe such a God does exist. Those are religious and/or theological claims that derive from widespread experience across time and distance. They must be treated seriously.

Philosophy gets its subject matter from life, but also brings to bear upon it the critique available from the systematic examination of the reflective structures of thought. When we apply the limits and affordances of reflection to the range of experience that is designated by the word "God," we arrive at a more rarified set of conceptions; these answer *not* to the norms of life, but to the small part of life that conforms to the best norms of *reflective thinking*. Reflective norms and lived norms, or more traditionally, intellectual and moral virtues, are as often at odds as in harmony—so that the most intellectually virtuous people are too often the most morally challenged, while intellectually simple people often have little difficulty in achieving and maintaining high moral standards. This dissonance (if that is a fair word for it) is more than puzzling; it demands an explanation. Here, in this gulf between the order of thinking (and its excellences) and living (and its excellences) we find *philosophical* conceptions of God. After all, we worry not only about the abstract whole, but also the accretion of value in the universe, when we philosophize.

Philosophically, we are also aware that a self-consistent worldview is difficult to espouse when one takes up the contrary assumption, viz., that we ought to conceive of God as impersonal, since it is not easy to explain how personal existence, such as we humans definitely have, can *come to be* in a universe that was wholly impersonal, as far as we know, prior to animal life on our planet. That is a recent development, from what we can tell empirically. Thinking of God as impersonal becomes philosophically inconvenient, especially since, *only* a personal being has any *need* of such an account of its own ultimate origins, and *only* the personal beings we call "human" actually *possess* any need of that kind. Yet, the personality structures of that "interpreting being" pervade all of our categories of explanation, including whatever categories *enable* that being to explain (or explain *away*) its *own* personal mode of existence.[9]

It is far easier to assume a universe informed by principles of the personal, and *then* account for what is apparently *impersonal* in the universe as a degenerate or misunderstood instance of personal existence. (One can account for the mechanical with the intentional, but not vice-versa, as we have noted.) A materialist is a strange kind of human creature who wishes to *commence* his/her investigation of the nature of the universe by dismissing or denying the most pervasive characteristic of his/her own experience (i.e., that it is personal). There can be no evidence in favor of such a move because the evidence *itself* is also mediated by the assumption that it will be interpreted and *judged* by a personal being with purposes and needs. The need to profess materialism is a pathology, not the outcome of high quality reflection upon experience. It is far less empirical to begin with a universe supposedly devoid of all personality (a recondite view, since it involves negating our own fundamental mode of experiencing), and being less empirical, the view is ultimately less rational (in the only defensible sense of "reason"). Thus, it behooves us to begin with the idea that there is at least *some analogy* between our personal mode of existing and the universe, and its God, if there is a God. This view is less recondite and abstract, and is not in danger of becoming a dogmatic stipulation, if we only suppose it rather than assert it.

Hartshorne conceives of God as a maximally affective and affected being who experiences *our* experience to the maximal degree, and who retains the subjective immediacy of those experiences *without loss*. This position is necessary to Hartshorne's understanding of truth, for if a past event were lost to God, then there could be nothing knowable or extant that could *make it true* that these events ever happened (according to Hartshorne). This view is not empirical; it is a logical argument about the conditions for the truth of propositions *about* the past. Bergson, by contrast, held that the full weight of the past *exists in the present*, not in terms of its felt content (or subjective immediacy), but because the present would not be *just the present it is* had anything in the past been even slightly different from the way it *was*. This is hypothetical, but it is an *empirical* claim.

Bergson's view is therefore ontological rather than logical. This claim is rendered as empirical as it *can be* in chapter two of his book *Matter and Memory*, and the argument stands up to contemporary science. Bergson designates the active presence of the *full*-past-in-the-present with two terms: "memory" is appropriate as we consider the images we experience more in their particularity and individuality; and "matter" (the repetition of events of minimal duration whose "togetherness" is found in nested, overlapping hierarchies of duration), when we anticipate possibilities for action in the present. This range of memory to matter includes, presumably, the subjective immediacies accompanying the experiences of all experiencers of *any* duration, in addition to the objective potencies and dynamics of an impersonal sort. Bergson is not concerned with subjectivity of this kind and leaves the privacy of images out of his account. The image we habitually privilege in perception is simply a center of activity. In short, it *functions* as a quantum of explanation, or, what is the same, an actual entity.

In contrast to Hartshorne, Bergson is not committed to the recoverability or continued existence of these subjective experiences, except insofar as they affect the configuration of action in the present. Bergson offers us a version of the objective immortality of the past which allows for the possibility that experiences can be genuinely lost—in the sense that they are not *now* being experienced by any person. Yet, even lost subjective immediacy, if there is any, would remain, minimally, in permanent contribution, meaning wholly active, relative to the present and the future. Recovering the past is an empirical problem for Bergson, not primarily a philosophical problem, but he does point out that a severing of the *communication* between past and present can occur in many ways, as can the *recovery* of the past in the present. The phenomenon we *should* be trying to explain, according to Bergson, is not the survival of the past in the present (clearly the past survives), but the general characteristics of severance and recovery.

Here in Hartshorne and Bergson we have the two extremes in the process tradition: the former theistic and dedicated to preserving subjective immediacy in perpetuity (albeit only in God's experience), and the latter non-theistic and appealing only to the reality of "centers of action" *in* the present. We cannot here resolve the issue, but we want to raise a worry about Hartshorne's account. We think his view cannot answer it. We have been very frank about the fact that Whitehead's God isn't "God," but we realize many philosophers simply cannot hear or read that word with any sense of neutrality. In Whitehead studies, it is Hartshorne's influence that leads "naturalistic" interpreters to conclude that they must jettison the idea of God in order to render the philosophy of organism scientifically respectable. Not only is this move unnecessary, it also renders the Whiteheadian philosophy incoherent and *un*scientific (i.e., non-naturalistic).

Our objection to Hartshorne makes the Bergsonian approach more appealing, since it is not subject to the same criticism. It is also closer to Whitehead's view. The problem that concerns us is a version of the problem raised for the immortality of the soul in Plato's *Phaedo*; it comes up routinely in the history of philosophy. For Hartshorne it makes sense to say that God has something like a "soul," assuming we understand this as analogous to God's "personhood" (an existential feature of the divine) and God's "personality" (an actuality which *is what it is* due to the *way* God has experienced the actual unfolding of the universe). The warrant for the analogy between finite and infinite personhood is only that we are philosophically obliged to have *some* provisional idea of the "whole" to which our philosophizing refers and on which it depends. As we argued above, it is more empirical to analogize *our own* mode of experiencing to that whole than to deprive the whole of that personal modality and create, needlessly, a mystery (as materialists do) about how *we* came to experience everything in the personal mode.

Thus, we analogize a distinction between *personhood* as a functional and, evidently, an exceptionless characteristic of all possibility. Such a distinction includes *personality* as a generic feature of the actual determinations of the universe, except where there is some reason why personality has been dissolved or destroyed by some process or force, such as abstraction. Given that nothing existential or actual is ever lost to God, according to Hartshorne, we might understand God's "soul," on such an account, as having at least the following features:

(1) Personhood—God's existential mode of being as a personal being; here Hartshorne asserts that God cannot succeed in existing impersonally (not that God would attempt to, but with nature as God's body some are tempted to think of nature as impersonal, which Hartshorne denies).
(2) Personality—the way actual events in the past *have been* experienced by God as actual (not as mere possibilities). Contrary to Hartshorne, this experience, including the exclusion by negative prehension of might-have-beens and the wills of other persons, both restricts and enables God's present experiencing of the concrete universe and future experiencing of what *will be* the concrete universe.
(3) Inexhaustibility—the inexhaustible capacity of God to take in *new* experiences, each in its full uniqueness and each making its full contribution to the aesthetic unity and variety of the real, including the concrete future. Hartshorne attributes this character to God, but denies it to human persons, a point on which Auxier has argued he is inconsistent. In any case, ours is stated below, but we can indicate here the "mortality of God" is the issue. The inexhaustibility question is the central issue surrounding Whitehead's enduring concern with the accretion of value. Thus, when we speak of the inexhaustibility of cosmological "experience," i.e., the radical-empirical criterion of futurity, we mean the problem of the accretion of value, taken in the grandest terms. The problem of evil is one of the many forms in which this issue comes to a head, but we will speak more of limitation or finitude than of evil.

Each of these general categories, possibility/personhood, actuality/personality, and inexhaustibility/potentiality, poses a basic sort of problem for grasping the nature of the process God. The first is a metaphysical (and derivatively, also theological) question requiring abstract but literal language to address. Here, logic is our best (and at present only) tool for discussing the personhood of God; analogy provides the best logic, and it is at this level that God's negative prehending of the wills of other creatures is grounded. Correcting Hartshorne's interpretation depends upon getting the logic right here. In Whiteheadian language, God negatively prehends the self-valuation of every actual entity in its actual world, and thus, every actual occasion.

The second category poses all the same questions as the first, but in less abstract and more complex terms—requiring the introduction of empirical, historical, and epistemological questions. Here, discussion in analogical terms of basic concepts like "God's love" and "God's knowledge" must be added to our language about the first level, and our religious experience along with our theology become relevant. Those who would defend God's actual goodness, for example, must not ignore evidence. In this second level God's negative prehension of might-have-beens is grounded, because not everything that *could have* happened actually did—at least not if we analogize divine and human experience. This is to say, God is free, in some sense, *because* there is a difference, for God, between what was and what might have been (see chapter 12). In level two, abstract logic no longer reigns supreme, and self-consistency is only one among many virtues of discourse. Thus, God, as concrete and consequent, negatively prehends what never happened, but might have. This is contrary to Hartshorne's many assertions.

The most difficult level to discuss, by far, is the third, which expresses a complex relationship between possibility and actuality, and points to the future. Such discussions presuppose fairly settled answers to the first and second levels, and make use of both abstract, logical and concrete, analogical language. But the third level passes through the logical and empirical (evidential) restraints of discourse and launches itself into a dangerously speculative domain. It is not even clear what could count as "evidence" in a discussion of God's capacity to absorb new experiences, and it is not clear what authority "logic" (or any kind of language) could ever possess in governing such an idea. Time just *is* contingency, at least in our experience of it, and the most slippery aspect of time is *future* time, which is the temporal mode addressed by the inexhaustibility thesis above.

This extreme contingency might be reason enough to stay away from the topic, but we cannot do that if we want to correct an important misunderstanding of Whitehead's philosophy. Having expended much effort in the interpretation of possibility, actuality, and potentiality, we should not now become coy about the portents of such an analysis—even as it bears upon religion, civilization, or hope and joy.

We notice the Abrahamic religions *require* such an account of God's inexhaustibility just as much as the process philosophers do, which is *why theology exists*. We finite beings do seem curious about matters beyond our ken. The concept of God can never make systematic sense unless we can show how God is related to the temporal world, and in what sense God surpasses that world. To this extent, Hartshorne surely stated the right problem, even if his solutions are not the same, or as plausible, as Whitehead's.[10] If the God of the Abrahamic religions is *unaware* of the events in the temporal world, then the humans know something God does *not* know, i.e., what it is like to exist finitely and mortally in time. That we should have such knowledge, while God lacks it, would be contradictory to the classical idea of God, as well as to Hartshorne's. The God of philosophy, we have argued, has to be conscious (this is the same as saying there is negative prehension in God), and *is* conscious (at least after "let there be light"—i.e., the primordial nature of God would not be conscious in any important sense, but neither would it make sense to assert the *un*consciousness of the primordial nature[11]).

The classical answer to the riddle about God and time is: God knows as an eternal present all that happens "in time" (from the limited, human point of view). But this idea *also* places a limit on God—*if* God experiences as an eternal present what *we* experience as unfolding in time, *then* God lacks the experience of having "His" experiences accumulate, clarify, and fill one another out in the course of time. This experience has great value to finite souls. This development is what makes it possible for us to become moral, virtuous, refined, appreciative, reflective, wise, and many other things. If God's experiences come all at once, in an eternal present, then God *lacks* an experience of the accretion of value-in-becoming that is basic to finite persons. Such a God is not superhuman but rather subhuman, in our view—a view asserted by Hartshorne, Brightman, and others.

Yet, God's experience is, traditionally speaking, *all* the experience there is, and all equally present at every moment. This raises uncomfortable theological questions about the difference between finite, individual experience and God's. Since nothing changes for God, on such a view, from a finite standpoint, this would make it boring to be God. We must grant that perhaps boredom springs from a deficiency in the human mode of existence that God does not possess or endure. However, it is not only hard to prove the supposition that there is no change in God, it is impossible to do so, because to prove it we have to leave the categories of God's personhood and personality and enter upon a discussion of divine inexhaustibility. Only at this level does the matter of divine boredom become a concern.

Is God's future already over, and if so are we are obliged to affirm that God's inexhaustibility is actually exhausted? The only analogy to this in our experience is to a dead person. Obviously the death of God has been discussed for well over a century, and philosophers and theologians still struggle with it. The close association of the God of philosophy and the God of religion and theology has been a stumbling block to all concerned, regardless of whether the religious people or the philosophers had the upper hand at one time or another. But the death of God doesn't have to be bad news for religious people, or perhaps even philosophers. Rorty and Gianni Vattimo have argued that getting over the old God opens up some positive space for discussing a God that might play a less destructive role in our cultural future.[12]

Be that as it may, one can see why the assertion of an eternal present in which nothing changes can become an uncomfortable accompaniment to theism when examined critically. Such problems motivate Hartshorne's theses of dual transcendence. God is inexhaustible, he argues, not only because God surpasses the created world, but also is self-surpassing. That is, Hartshorne's God has a future.

The inexhaustibility thesis *is in play* regardless of whether one conceives of God's experience as being contained in a single eternal present, or in an on-going development. This is why philosophers such as Ferré and Neville also cannot avoid this issue by treating time as a determination of God, at least so long as they remain theists in the Western tradition. Neither really is a theist in the Western tradition, so we doubt we have posed a problem for philosophers of their sort—we are inclined to think of Neville as a neo-Augustinian, saving eternity for God's mode of existing and taming time by explaining its plural modes, but not as a theist.

To bring the question into process language, the inexhaustibility thesis requires us to consider: what could possibly count as *evidence* for God's boredom or satisfaction or interest in the world, with regard to the totality of God's experience, or its limits? There is scripture of course, but scripture does clearly support the atemporal or eternalist view of God that classical theists try to defend. Eternity is Augustine's idea, not a scriptural one, except to the extent that Greek philosophy invaded Hebrew and Christian thought during the 300 years prior to the life of Jesus (and that was a significant invasion). The scriptures overwhelmingly depict a God active in history, changing "His" mind, subject to regret, hearing prayers, making decisions about things, waging a war with competing forces in some cases, commanding the slaughter of Midianites, and taking out vengeance against Onan himself, even making bets with "the enemy" on the likely responses of his favored servants to unmerited suffering. Only a few passages suggest a God existing in an eternal present as contrasted with many thousands that depict God in a temporal mode. We won't find out about whether the supposedly eternal God is curious or bored by the world from that sort of evidence. The *temporal* God definitely is *not* satisfied if scripture is the measure, but even less so if nature is the measure.[13]

How should we talk about whether God can grow bored with the universe? The reason ontological boredom is a crucial test is because it is suggestive of a being-toward-death in God, whereas if we considered ontological satisfaction or even curiosity, these would not deliver such existential structures. The latter two modes suggest a justified past and an adventurous future for God, while divine ontological boredom points toward divine finitude, specifically, the inability or unwillingness to accumulate new, deeper, and more intense experience. To claim it is *impossible* that God should grow bored with the universe invokes a kind of necessity that contradicts the idea of God's personhood and personality, and also the open character of the universe, along with God's "freedom" (whatever that may mean). Thus, we bring up the issue, i.e., God's boredom/exhaustibility, as a road to a more provocative topic, namely, the mortality of God (or divine being-toward-death).

There are, it seems to us, three possibilities:

(a) The Ever-growing. Imagine the universe keeps growing to a greater unity under contrast (to use Whitehead's phrase) within some intelligible (to God) limits: Why, assuming God is a personal being, should it be *a priori* impossible that God might outgrow, or grow tired of this universe *even at that*? It should be possible for God to outgrow even an increasingly interesting universe, and if it isn't possible for God to do that, we couldn't know it isn't. No empirical evidence for or against it settles the descriptive adequacy of the exhaustibility thesis, and thus, we may retain as a viable option that God *can* grow bored regardless of how interesting the universe is becoming.

(b) The Ever-dying. Imagine the universe eventually runs down and burns out (i.e., fails to grow), and this entropy limits God's capacity for interest in it: Why should it be seen as *a priori* impossible that this universe, as it runs down, might grow less interesting to God than it *now* is, regardless of whether God continues growing? (Note that Bergson takes this idea very seriously and says that the *élan vital* is finite, eventually using itself up as the universe sinks into inertness.) In this case, even a pessimistic view of the natural universe, coupled with optimism about God's capacity for continued experience, does not eliminate the prospect of ontological boredom, since other things might be boring to God even if God is not limited by the divine awareness of this universe. (See chapter 8, on cosmic epochs.) It wouldn't be impossible for God to grow bored even if God is not limited or determined by an entropic universe; if it *is* impossible, we couldn't *know* that it is. This way of imagining the universe also doesn't settle the question of divine exhaustibility. There is no *a priori* reason and/or evidence for asserting that the created universe determines God's capacity for ontological boredom.

(c) Homeostasis. Imagine the universe maintains an equilibrium (in cycles or something like it) between burning out and growing more complex and unified: Why is it *a priori* impossible that God should grow increasingly less interested in such an equilibrium as the general types of experiences God *has already had* come around again? In short, the theory of cosmic epochs does not rescue God from the peril of being exhausted. This view is close to deism. We might think of it as "the monotony problem." Let the created universe be as predictable or as unpredictable as it will, there is no reason, a priori, to conclude that God cannot grow bored with it, and no empirical evidence, for or against, can point us in a clear direction, so the exhaustibility thesis is still a problem. Those who want to say God's experience is *in*exhaustible (such as Hartshorne) are left with nothing but a possibility. There is no way of defining an inexhaustible God, or universe, in a way that is answerable to *human* experience, *even by analogy*.

Therefore, from the above three arguments, we cannot rule out the idea that God experiences being-toward-death—God's own death, that is.

Thus, thinkers who presume to treat God *as* process, or, on the contrary, as beyond determination, have no purchase on the question of possibility as it relates to the philosophical God. God is *not necessarily* our happy (or co-suffering) co-creator. Neither is God necessarily Being or Beyond Being or the Undetermined. None of these views can handle the exhaustibility problem or confirm the inexhaustibility thesis. Hartshorne's doctrine of dual transcendence comes as close to doing so as any position. We have refuted it. It is incoherent on the exhaustibility of the divine nature.

We discover here the reciprocal to Whitehead's perpetual perishing of subjective immediacy, which is that being-toward-death has the *effect* on the future that perpetual perishing has upon the past. If God is a temporal being, *with a future*, we cannot rule out the likelihood of God's possessing such an entropic structure. Hartshorne thinks he can rule this out, and he insists Whitehead should have agreed, but Hartshorne is working beyond anything logic, analogy, or experience can warrant. Apart from the three ways of thinking of the universe given above, the ever-growing, the ever-dying, and the homeostatic, what fourth possibility is there for the relation of God and the created universe, from a process perspective? Cosmic punctuated equilibrium is just a version of the first as modified by the third. It doesn't change the questions about exhaustibility. And finally, given the difficulty with determining God's level of interest in the universe, how can we know *a priori* that God never dies? What possible empirical or logical evidence can there be for or against this idea? The everlasting God of traditional theism and even Hartshornean neo-classical theism isn't *philosophically* viable.

We think process metaphysics ought to offer answers, however tentative, to these questions, and that the Abrahamic religions need answers also—not simply dogmatic assertions to the contrary. How might we construe the nature of God such that God *always* transcends the universe (whether by Hartshorne's dual transcendence or the eternalist thesis of Western theology), and *also* never grows bored with the universe (i.e., has a future)? No matter what we assert about God, ontological boredom remains a genuine possibility *for* God.

We think process metaphysics is challenged here in a way the Abrahamic traditions are not, since answers based upon God's *self-sufficiency* given in the Abrahamic traditions have been set aside by process philosophers, so as to maintain a meaningful, real relation between God and the universe. The idea of divine self-sufficiency, whether explicated as an unmoved mover, or as the Trinity, does "answer" these questions in a manner consistent with the conceptions of God that these traditions accept. These traditions have their own pitfalls, but process philosophers sacrifice perfect self-sufficiency in God for the idea that God is meaningfully enriched by the experiences of creatures and creation. Hence God, who is "the most indebted being in the universe" on Hartshorne's account, is also the most endangered (if we follow out the logic of that claim). The desire to attribute to God maximal affectivity cuts off the self-sufficiency strategy for answering the "boredom" question.

The process God is always conceived as able to surpass the universe and lure it in a given direction, and also able to surpass any given state of the divine self. How can the universe be guaranteed *never* to fail to bore, or disappoint its creator unless *it* has the same "power"? But if the universe is susceptible of the same infinite variety and contrast as its God, what makes *it* not God? We may say, with thinkers such as Neville, because the former is created while the latter is creator, that these are two sides of one symbol. But then there is clearly at least one crucial sense, i.e., as source of creativity, in which the universe is *not* capable of the same infinite sort of self-transcendence as God, and if it be lacking in this sense, what will prevent us from tracing the logic down to the conclusion that the universe may burn out or run in cycles which will bore God?

Thus, not only is Hartshorne's characteristic move imperiled, so is the move made by Neville, but only on the assumption that the transcendent God is a personal being. If we give up that idea, Neville can have his God, but his God isn't *God*, which is to say that Neville doesn't really disagree with Whitehead, properly interpreted. Perhaps such considerations provide the motive for Neville to dispense with the idea that God has a soul. It is the only consistent thing he could do, and traditional theists don't have this option. So Neville sidesteps the question of the exhaustibility of divine experience by making the divine experience a derivative determination of Ontological Act of Creation, while "God" is a version of existence that is less than "being itself." Neville cannot say that "being itself" is personal, but we don't think it gets him out of proving that *being itself* is inexhaustible. God by any other name (Oca) is no less a *philosophical* problem.

Returning to the process conception of God, and adapting an argument from Plato's *Phaedo*, shall we require God to give the universe a new push once in a while so "He" doesn't get bored? How long before God becomes bored of *that* (i.e., renewing the creative energy so the universe doesn't run down, or the novelty so that it doesn't just go around in circles)? It is easier to say that the universe has a self-creative aspect that *differs* from God's self-creative aspect in degree or in kind, but if so, it would seem that the universe must be *less* than God in respect of self-transcendence, or capacity for growth, for otherwise the universe just *is* God, and pan*en*theism gives way to pantheism for want of distinguishing criteria, as Corrington has argued. Process philosophers cannot allow the universe to be *more* than its God without risking the failure and death of God (and hence, a cessation of the accretion of value).

Is There a Solution?

The key to this problem lies in two considerations: the "evolutionary" consideration, and the "uniqueness" consideration, both of which ground the claim of unpredictability, within certain limits, for the universe, with the limitations understood in light of "contributionism" (we allow a modified version of Hartshorne's interpretation of the accretion of value). To avoid the idea of a bored deity, we need a deity capable of being surprised by the universe, and a universe with a creative power of its own, and we must give some determinate sense to what this sort of "surprise" is. First we must pose the problem of God's boredom in its most pointed terms, which is done from the standpoint of level (b) above, the ever-dying cosmos. We will use both logical/literal language and analogical language.

Hartshorne often argued that finite creatures cannot sustain indefinitely the addition of new experiences to their previous ones. After a certain saturation over time, the capacity of finite beings to take in the full value of these new experiences declines to the point that their variety and uniqueness are unavailable to the finite consciousness, which then grows bored in its monotonous existence, and eventually dies.[14] We have all seen this phenomenon in human (and even non-human) old age. Why should this analogy fail to describe the experience of the deity? Granted that no process theist believes it does, on what grounds can it be *excluded*? Granted that no adherent of the Abrahamic traditions believes God gets bored and dies, what makes the believers so confident?

Let us assume the analogy between divine and human experience is apt when we imagine God as "remembering" the past, but unlike human beings, doing so without loss, and containing, in addition, all the content of experience that was not even available to the consciousness of the experiencers at the time it occurred. This would mean that there is no negative prehension in God. As finite, non-divine experiencers mature, each new experience constitutes a smaller proportion of their total experience. Some people have said that this is why time seems to pass more quickly as we age—each new minute constitutes a progressively smaller proportion of our total experience. Why can we not say that the same is true of God, by analogy, especially if we hold that God is a temporal being (which Whitehead does not, of course)? Assuming God's experience is cumulative, why isn't it the case that each new moment constitutes an ever-smaller fraction of God's total experience? This is the "evolutionary" consideration, where evolution is viewed in light of the tendency of experience to accumulate and build upon previous experience, irreversibly.[15]

What would exclude a process God, of this non-Whiteheadian sort, from accumulating experience? If God's experience does accumulate, is it not the case that each new experience constitutes a progressively smaller fraction of the whole? Is God exempt from the idea that $1 + x > 1$? The process God is in fact conceived as being exempt from genuine increase, but not from evolving. Even Whitehead's infrequent mentions of the consequent nature of God clearly include *ordered change* and accumulation of experience. The absence of "increase" means only that the accretion of value does not add value to the divine experience. How a being can evolve with no accretion of value is a paradox we must consider.

Let us move then to the "uniqueness" consideration. Our sense of proportionality (how each new experience stands in relation to the whole of that finite creature's experience) depends, in the case of finite creatures, upon there having been a definite time at which a given being came *into* being (either the moment of conception, or any starting point so long as there *is* some durational epoch to which commencement belongs). For finite beings the estimation of the proportion of each new experience to *total* experience may be understood (even calculated) relative to the total collection of experiences that preceded it for that same being. In the case of the process God, since we cannot analogously assert such an *empirical* starting point within the created universe, we also cannot calculate, in principle, what proportion of God's total experience is represented in any given *unique* experience.

This uniqueness consideration may undermine the idea that God's experience of the universe is cumulative, or that God "evolves." But it does not. Addition is not increase unless it is addition to a finite set (and here we may include collections, but not constellations of possibilities, which suffer no alteration of any kind—see chapter 9). Adding one "individual" (as a potentiality) to a determinate constellation of possibilities falls short of any reasonable definition of "increase."[16] God might be conceived of as evolving both with and without "increase."

This result might seem to recommend the traditional solution of seeing God as atemporal. That is Whitehead's usual characterization of the deity— a point not lost on apologists of the perennial philosophy, whose enthusiasm for Whitehead would be most puzzling were the Claremont theologians right about Whitehead's God. The traditionalists show no such predilection for Hartshorne's God, which might be a clue to the wise. Consider the mess the process God has become, in our forgoing analysis. By refusing to be determined *as* finite beings are, while also being determined *by* finite beings, the process God becomes a paradox, either unnecessary due to "His" great changeability, or, due to "His" unchangeability; the process God disappears either way into the idea of process itself.

Auxier has argued elsewhere that the traditional view, if it is eternalist, also militates against conceiving God as a genuinely personal being, and hence we are locked on the horns of a dilemma: either we claim that God is eternal, making "His" personal mode of existence problematic (the traditional view), or we allow that God is temporal, i.e., God evolves, and face the issue of a lack of increase in God's experiences for lack of a starting point in God. In Whitehead's language, the primordial nature of God is cut off from creativity (a point Neville and others have made). If God has no "starting point," there *is* a sense in which each new experience bears *no* final or definite *proportionality* to God's total experience, *even for God*. We will eventually challenge the traditional view, but for now let us assume that process theists *might* come up with a way of handling the problems with divine evolution.

There is indeed a promising solution: This *absence* of proportionality allows the *uniqueness* of each experience to take center stage in our considerations of God's experience. No experience is *like* any other, at bottom, for God, we could say. God does not, therefore have "an actual world" in any sense consistent with Whitehead's ontological principle. This is Whitehead's view, in fact. When God experiences event x, there is no reason to think that its proportional relation to all other experiences God has is a *governing* feature of God's *experience* of x. This is because God does not *require* concepts to mediate a class of experiences, but rather, may have experiences concretely, whole, and unmediated. God, on this view, does not need generalizations, or analogies to "include" x. God experiences x *as x!*, and not *just* in its relation to similar experiences that are x-like. Nor does God just have "His" *version* of x. God has x-as-such-and-as-a-whole, with all of its relations, in its utter uniqueness. Here the analogy between human experience and divine experience can be legitimately limited, since concepts are needed only by beings who *cannot* take in x in its full uniqueness.

Thus, if we do not attribute conceptual thought to God, then God may experience x immediately *as x!*, even though God will also experience all the relations of x to other previous experiences, and will also include *how* x has responded to the contributions of all those other experiences as data for its own synthesis. Thus, where concepts and symbols mediate *for us* all experiences, *this* God of process does not *require* concepts and symbols to experience, include (and ground) x as the unique event it is. Yet, we do not thereby exclude the idea that God might employ concepts and symbols. We are not required to exclude God's act of thinking from our idea of God, we are only required to see it as a contingent feature of the divine, and to admit we do not know whether it is true that God *thinks*. We argued in the last chapter that to make it a necessity that God does *not* think dispenses not only with negative prehension in God, but also God's personality or "soul." Hartshorne cannot consistently maintain his idea of God in the face of this consideration. Whitehead can, but only by holding the God of philosophy at arm's length from the God of religion or theology.

The Self-creative Act

Here we come to appreciate the importance of the idea that in each actual occasion there must be a wholly *self*-creative phase or aspect unavailable to God in advance, in its specificity. The claim that in each new experience there is a *self*-creative moment that is *more* than can be discovered from the contributions to it of all *prior* experiences, is just another way of saying that each experience just *is* the experience it is—and no other—precisely because *that* is how God is affected by it and affects it. The mutual relation of affecting/affected requires conceptual mediation, and hence negative prehension, because *exclusion* of all that a unique entity *is not* is prior to its satisfaction as a unique contribution to its actual world.

The finite entity is felt in the mode of possibility *as well as* actuality by God, and some of the things it *might become* are *excluded* for the sake of what it *will be*. On this view God cannot anticipate with perfect accuracy the *self*-creative aspect of each or any event until it is completely concrete, i.e., satisfied. Asserting that each event is "unique" implies as much. There must be some valuative character in each actual occasion that was unforeseeable, even by God, and in virtue of which that occasion is irreducible to any other.

This argument is the basis of saying God negatively prehends finite wills until they are exercised concretely. The finite will, then, is grounded in the *self*-creative aspect of each actual occasion.[17]

The universe is forever open because each event is unique, and is individual in proportion to its self-creative potentiality. No two *events* have precisely the same potencies because each responds to a different *universe* as data for its own self-creative activity.[18] Obviously, complex societies of actual occasions have an open field of potencies in proportion to the intensity of their immediately preceding synthesis. Still, they might grow or die or equilibrate, but all relevant venues are open ones. The accretion of creative syntheses can either build on earlier activity by actualizing potencies that increase intensity, or by openness to the unification of a greater variety of data in the next act; or it can actualize potentialities that lead it toward an entropic decline, a neutralization of its own energies by diffusion; or it can oscillate with homeostatic effect.

The creative exercise of will, then, is the freedom to become more complex and deeper in actuality, while the entropic act is the letting loose of the creative tension generated by the disequilibrium of energies held together in complex interdependencies—the very diseqilibrium that compels the entity to act, hurls it toward the future. As far as we know, finite creatures (created beings?), eventually cannot hold the center, and the creative tension diminishes, or perishes. The accumulated experiences become too weighty and numerous for the creature to prehend meaningfully. Then conceptual feeling dies (meaning that it becomes negligible), and potentialities contract, since each new experience is an increasingly smaller proportion of the whole, and each past experience is prehended ever more abstractly as new experiences are had. This is a fair definition of death.[19]

The process God's case has to be seen as disanalogous in some ways. God's experience can evidently accumulate without need for a definite starting point *in proportion to which* each experience is a progressively smaller addition to the whole. Accumulation is not necessarily increase in God's case. What is missing is not the starting point *per se*, but the proportionality of new experiences *to* that point. The primordial nature of God is equivalent to possibility as such, considered in abstraction from all activity. Possibilities do not bear proportion to one another except in relation to a concrete or a posited actuality. So long as we are free to postulate a possibility *as* an actuality, we can always think of a different God, a different whole to which finite experiences belong, but which does not increase the totality of possibilities in the cosmic order, and does not commit itself to saying God either did or did not *create* the possibilities (a question we finite beings haven't the resources to answer, only to ask).

If this view is correct, then we can see that the universe is always novel and unpredictable to God, within limits, and that God will experience every event immediately in its uniqueness—after it happens. Not just anything can happen (i.e., the past qua past cannot suddenly start changing), but each event is just what it is in spite of the contribution of the past, in view of its *self*-creative act.

Does God Sing the Blues?

But what staves off the charge that God could either outgrow this continual novelty, or become uninterested in it for some reason? What holds a process God in relation to the evolving universe? Hartshorne might say that to assert a bored God is to speak in a way contrary to God's necessary nature or essence. But this goes only to level one in the first schema laid out above, i.e., God's *personhood*—applying only to language associated with the definition of God as governed by theology and logical language. In order to answer the charge completely we must engage levels two and three, especially the latter, the inexhaustibility thesis.

We can receive no final assurance of "ever renewable uniqueness in relation," always and with certainty, will satisfy divine aesthetic needs. God *might* grow bored with the universe unless we embrace a pre-established harmony regarding the rate of God's growth and the growth of the universe, and assert a *necessary* reciprocity to which we are not philosophically entitled by any evidence. This brings us back to "the evolutionary consideration."

Assuming that by "evolution" we mean something like "ordered, cumulative change," and assuming we note the asymmetry and apparent irreversibility of such changes, we have a ground for thinking that a contingent pre-established harmony may not be a hypothesis utterly without warrant in process philosophy. Indeed, Whitehead explicitly embraces it, to the annoyance (and incomprehension) of many of his followers. But by pre-established harmony we no longer mean, as did Leibniz, Malebranche, and Spinoza, a relation between mind and body; we mean that there is an intelligible relation (perhaps contingent, perhaps necessary—we will probably never know) between the pace of change in God's consequent nature, as a concrete being, and the development of the universe.

Note the asymmetrical character of evolutionary processes. Philosophically speaking, the process God cannot "undo" the past, for to act in such a manner would require an act in the present which takes as its data the past actualities to be "undone" (whether the actual God, if there is one, can "undo" the past we leave aside; our point is that we cannot make sense of the idea philosophically). Even if it were possible to remove from God's "memory" or concrete being the specific contributions made by an event, x, to all other events subsequent to it, to isolate x metaphysically such that it has no relations to anything real or actual, and thereby annihilate it metaphysically (we assume this is what "undo" means), the act by which these contributions are removed would remain a positive act which *does* contribute to all subsequent acts, and which has as a part of *its* content the very data which were to be "undone." Hence, the act is not really "undone," it is at most quarantined in a further act. Hence, we cannot understand God as "undoing" the past, and that alone is enough to make temporal processes asymmetrical, or "evolutionary" in the sense we mean. This also affirms Whitehead's idea that physical prehension is always positive prehension. In short, when your woman left you, you have the blues because she's gone, not because she never existed. The eternal sunshine of the spotless mind isn't available, especially to God.

One Damn Thing after Another

By definition, all asymmetrical processes must be cumulative in some way. To show this, assume there is at least one asymmetrical process that is *not* cumulative (including the accumulation of entropic losses). What then would make such a process asymmetrical and how could we know it? If the past of this process as a whole can simply disappear, and not accumulate in any way, then there is no more reason to call the process asymmetrical than symmetrical. If at least part of the past is always preserved, then the process is cumulative to exactly the degree it is asymmetrical ("impossible to undo"). There might be other ways of conceiving of asymmetry, but our point is not an ontological claim about how processes in the universe actually are, but only about what thinking of them clearly requires us to say. We have laid the basis for this observation in our account of the relation between physical and conceptual prehension.

As we said, the cumulative character of the past may be a problem for God in terms of continued capability to absorb new experiences without growing ontologically bored, but then we answered this with the observation that the immediacy with which God prehends the uniqueness of events, that relational feature of God *to* events, defies the proportion argument. Here we revisit this issue at a broader level. We assert, uncontroversially, that a cumulative, asymmetrical process, "an evolution," is also a relational phenomenon. Here we can appeal to the principle of panentheism, that all is *in* God although this "all" does not exhaust all that God *is*. But we think there is a better account, closer to Whitehead's actual view. By "God" we *mean* actuality of the postulated whole to which all of the "parts" belong, even if we leave open some of the features of the whole. God, so conceived, not only experiences immediately all discrete events in their uniqueness, but also experiences their relations to all other events (the "events" themselves just *are* configurations of these relations, plus the self-creative contribution of the actual entity—see the introduction to the book). These relations must then be seen as evolutionary in the sense of asymmetrical and cumulative. This move yields the following exceptionless generalization about the empirical universe: as far as we know, all evolutionary (i.e., asymmetrical cumulative) processes, taken as units, are developments within a larger whole. The many become one and are increased by one.

If we supposed there were an evolutionary development that is not a part of a larger whole, we supposed something that can be neither asymmetrical nor cumulative, since there is nothing compared to which it *is* either asymmetrical and cumulative or *not* asymmetrical and cumulative. We erroneously imagine here a God experiencing immediately a unique event x without also experiencing its relations to the other events that contribute to it and which it evaluates. That is clearly absurd, since it seems to suggest that either God fails to experience something that exists, or that God may not be aware of some portions of "His" own experience. If we can set these claims aside, it is clear that when God experiences evolutionary processes, God experiences them *as* evolving. It would also seem, therefore, that God experiences something like the past *as* over and done (actively contributing, but accomplished), the present *as* becoming, and the future *as* anticipated but not actual. The alternative is to say that the universe evolves, but God does not experience its actual evolution. That is the problem with the traditional view, saying along with Aquinas (and the church) that God has no real relation to the universe. This view deprives God of a crucial aspect of experience, and would correspondingly remove "Him" not only from a *full* understanding of "His" creation, but from an *adequate* understanding of anything.

Out, Out, Brief Candle!

Now we must ask: does it make any conceptual sense to assert that God's capacity to experience the universe could change or grow more quickly or less quickly than the universe evolves? We can get past the idea God might grow more quickly than the universe by noticing that the data which affect God just *are* the universe, as far as we are concerned, and it is more reasonable to suppose growth in God in proportion with these data than out of proportion with them. The theory of cosmic epochs in Whitehead is intended to stake out this territory for our metaphysical thinking, but without ultimately closing off the alternatives we can imagine. It might be possible to learn something about the relation between our cosmic epoch and others, by way of a study of pure possibility, but that study would be of dubious value. In our cosmic epoch, mortal creatures meet their final ends in part because the process of the universe increasingly exhausts their finite capacity to absorb new experiences proportionally to the occurrence and accumulation of the value of events.

Still, one cannot be certain whether the process of God's growth in capacity for experience may be disanalogous to the destiny of such capacities in other finite beings. Assuming (for the moment) the universe never outruns God's capacity for experience, that God does not "die" in that sense, we cannot finally rule out that God's capacity for experience might not outrun the process of the universe, as the universe runs down (entropy), and God might become ontologically bored with the dying universe. While we cannot rule this out, we can say we have no basis for thinking it is true, and *some* basis for thinking it is false, given that all we know about growth indicates proportionality, and nothing we can ever know about growth could count as evidence for disproportionality between God's development and the development of the universe (all our evidence comes *from* the universe). Where there can be no contrary evidence in principle, and there *can* be positive evidence, it seems fair to believe what the positive evidence suggests. Hence, it is reasonable to conclude God does not outgrow the universe, or that if "He" does, we could never know it.

The idea that God might evolve or grow *less* quickly than the universe is harder to dismiss; the case is analogous to a situation of divine Alzheimer's disease. And here we begin to come to the literal "crux" of the issue. Were it to happen that God grows less able to take in the universe as *it* grows, the universe would become in some sense "more than God." So why not affirm this idea? It is clear that this is analogous to the idea that we can produce or create things greater than ourselves (e.g., one might have children who become saints), and thus, God might give existence to a universe that outlives "Him," both physically and in the attainment of value. But this idea only makes sense when one considers "wholes" and "parts" and "greater than" in abstraction. This is not the claim that concerns us. God would be *at least* objectively immortal in the universe "He" created, and then we would have to find evidence of God's death or decline or departure in terms of subjectively immediate experience of the concrete universe. That won't happen. What could count as evidence that God is not experiencing the universe now, but did at some point in the past? Or that God is experiencing the universe *in a diminished way* now compared to some point in the past, or that God will have a diminished experience the universe at some point in the future, compared with now or the past?

If we invoke evil as evidence that God is "slipping" or losing a grip on the vivid experience of the universe, we risk exaggerating the human condition and its importance in the universe, and of myopia in thinking our own times are worse than other times, when in truth we have no direct, concrete experience of those other times. What is more unbearable? A universe in which there is much freedom being actively misused, or the void? The idea that the universe might become greater than God is not really the issue to worry about.

The metaphysical problem is that insofar as all finite beings work with the available data in their self-creating, and insofar as they *are* self-creating, then God by definition did not "create" them wholly. The self-creative moment of an actual entity in its actual world does not imply its *total* independence from other occasions, only its uniqueness. No occasion could achieve its unique synthesis without a context of response set by the data prehended in the response phase of its concrescence. If the universe is seen as an ordered society of occasions, then the only encompassing context of response within which this society can create itself is God's concrete being. If God's concrete being is ever less comprehensive than the universe, then the universe would have to become more and more self-creative in proportion as God's contribution dwindles in importance, and as the previous self-creativity comes to play an increasingly greater role as data in further response.

This situation, if it can be actual, would lead ultimately to the withering away of God, although not to God's final demise, since God would always play *some* role and make *some* contribution to the evolving universe. But as the self-creativity of the universe became more and more important to its further growth, it would have an ever-decreasing energetic basis for this creativity, and thus either would operate increasingly *ex nihilo*, or fall increasingly into monotonous repetition of what it has done before. In the former case it is fair to say that the universe *becomes God* in every meaningful sense, to the extent it can sustain and intensify its own *ex nihilo* creativity. In the latter case, Bergson wins the day—the universe burns out; the *élan vital* exhausts itself and becomes inert. For repetition would become constant motion, and constant motion (no acceleration in the sense physicists use the term) involves no change of state. One state of the universe eventually would not be distinguishable from the future states. Ontological monotony pushed to the extreme is non-being, insofar as we can make positive sense of the idea of non-being. While this may be the truth about the universe, it cannot be consistent with theism—unless we conceive of God as mortal in *all* respects.

And this is our point: The process God ought to be conceived as mortal in at least *some* respects. But we might conceive of this mortality as tied up with the eventual destiny of the universe. What happens in the universe contributes to or takes away from God's experiential richness and therefore what we might as well called God's "well-being." Whatever damages this well-being is something like God's "mortality," and whatever contributes to its greater growth is a part of God's "life." One might even say that God could "live" longer if we act so as to maximize the energetic ground of existence—act in such a way that increased unity in variety, while variety in unity comes to be actualized. If there is a "categoreal" imperative, this is it.

Life as We Find It

We cannot be certain what activities invigorate the divine life, but we think it is not crazy to suggest that the continuance and accretion of value in the divine life depends in some minute way upon what *we* contribute to it, recalling that each event is unique and can be the only source of *just that value* had by it and it alone. We do have the analogy of which activities invigorate and extend our own lives, and which do not. If we are to affirm life, including the value of the divine life, we ought to think seriously about how to contribute to preserving life—to live, to well, to live better. Thus, principles of health, etiquette, aesthetics, and most of all moral ideals and purposes, become energistic metaphysical norms. After all, that is the function of reason. We would have to come to a conception of peace, for instance, that was invigorating and creative, not just placid. One is reminded of Martin Luther King's conception of peace not as the *absence* of conflict but as the *presence* of justice.

The other pole that comes with asserting that there *is* a "living" God must reckon with the *implied mortality* of God. A living being, if the idea of its "life" is meaningful, must be thought of as in some sense mortal; we must allow that its life can be meaningful only to the extent that its life *may* succumb to death. A perfectly secured immortal existence is not "life." It falls short of producing a generic contrast. Without proposing a full Roycean *apologia*, this insight about God's "life" leads us in the direction of one of the Abrahamic traditions, namely Christianity. Unlike Judaism and Islam, Christianity builds in, unconsciously, a notion that God can die. Obviously the idea is pilfered from older traditions, but its power is concealed in its creative relationship with the idea of a God with a will for the world that grounds ethical action. We do not think that the joining of the Abrahamic God with the dying and resurrecting gods of the ancient Near East (Osiris, Dionysus, etc.) renders Christianity superior, and we think that Jews and Muslims are right to suspect that Christians are not, strictly speaking, monotheists. Neither is Whitehead, *strictly* speaking.

The divinization of Jesus in the pattern of Osiris and Dionysus argues against treating Christianity as a kind of monotheism, as Jews and Muslims understand it, and adding in the Holy Spirit and the Church as the Body of Christ complicates the story even more. We believe Judaism and Islam have a far better chance of coming to a self-consistent *account* of God, but that they may well miss in so doing a crucial lesson contained in the middle tradition: *that even God can die*. The traditional view of Christianity also offers a way of seeing, symbolically at least, a way to grasp God's living mode of existence neither as wholly finite and mortal, nor as, in its entirety, independently infinite (and to that extent impersonal and meaningless). Of course, Islam and Judaism have always held that there was a relation of the world to God, but explaining the relation of God to the world has been trickier. Let us follow the Christian path between the horns, although we don't know any orthodox Christians who are likely to approve of what comes next.

Indeed, this philosophical view of God's temporality and relatedness to the world, in the mode of divine mortality, opens up an interesting approach to theology and specifically Christology and soteriology, since it seems we might take the message of the incarnation and crucifixion to be something like God's saying "if you people think I can't die, let me show you how wrong you are . . ." This is a paraphrase of "forgive them for they know not what they do."[20] And indeed, assuming Jesus is fully human, we may point out that he must die as *we* die, i.e., not knowing for certain whether he will rise again. If Jesus dies in full knowledge of his coming resurrection, then he does not die *as we do*, in ignorance of what, if anything, comes after. If Jesus does not die *as we do*, then he is not one of us. If Jesus is not one of us, the blood atonement fails, since the entire point was to show us that it is possible to exist finitely in human form and yet be perfected through a perfect high priest who takes up His place in the celestial Holy of Holies (see Hebrews, chs. 8–10). The atonement does not depend upon the resurrection but upon the finite death of Jesus *qua* human. So it is crucial to Christian doctrine that Jesus be fully human, even to the point of being *fully* mortal. We have not run across many Christians who discuss this point.

But for Christianity, Jesus *is* God also, in some sense of the word "is." To the extent that Jesus is one with the Father (as he asserts), his death is also God's death, in some sense of "is." For Christians to deny that God is mortal, they must therefore deny reality of the cross. Hence, for traditional Christians, an idea of the mortality of God is built in to their conception of God, even if it is not talked about. Of course God is immortal in some respects, for Christians, but not in all respects, which is our point. Jesus is different enough after the resurrection that no one recognizes him, and we may see symbolized in this that it is not easy to recognize or grasp God's immortal aspects except *through* the mortal ones. Hence, we shall have to grasp the immortality of the process God by means of "His" mortality. Let us henceforth refer to the mortality of God as "His"[21] "soul."

The Soul of the Matter

We recognize the inversion implied by our move, since we are accustomed to think of our own souls as our *immortal* part, but we think we can get away with that inversion, because we include the idea that our souls are *finite* even while they may attain beatitude or eternal survival, or a party at some celestial Key West. At the very least, our souls have a point of origin and are to that extent finite. In the case of a divine soul, the implication of finitude must also hold in some sense, and with finitude, mortality. If God is a personal being, then God exists finitely and prehends negatively the wills of others and the events that were possible but never became actual. These modes of finitude are required for our saying that God *has* a soul, but not sufficient for saying God exists as a personal being. Personal beings have cumulative experience, which is to say they evolve. Evolving beings reach a point of saturation and can no longer make use of new experiences. They die. If this is not the character of God's mortality, and if God is not identical to the universe itself (and mind you, we haven't ruled out either of those), then we are obliged to conceive of God as a living being whose development parallels the unfolding of the universe. Saying that a being is "living" can only be meaningful where the possibility of death is real. Otherwise "God lives" is just a strained metaphor.

Now we may add that if God hasn't got a *mortal* soul, then God lacks a soul altogether, since divine mortality is the way through the dilemma Christianity has given us about how God can become human and *remain* God. Here we may answer the inexhaustibility thesis. God's personal mode of existence is mortal and exhaustible, as Jesus' existence was, but there is "the resurrection"—a certain inexhaustibility in God and the universe that is difficult to recognize from a mortal perspective. Yet, by analyzing mortality we come to understand the immortal, to the extent that we *can* understand it. The task is similar to grasping eternal objects *qua* eternal, which is no easier. If we attempt to grasp finite, mortal existence as a contraction and limiting of infinite, immortal existence, we never learn the heart of our own finitude; we rather explain it away as a defective instance of life. That is not only unempirical but also unwise. Our path to immortality, in living as in thinking clearly *about* living, lies in seeing what mortality suggests about the immortal, not the converse. As Whitehead points out:

> We cannot understand the flux which constitutes our human experience unless we realize that it is raised above the futility of infinitude by various successive types of modes of emphasis which generate the active energy of a finite assemblage. The superstitious awe of infinitude has been the bane of philosophy. The infinite has no properties. All value is the gift of finitude, which is the necessary condition for activity.[22]

We do not think we are likely ever to have an adequate understanding of any of the three levels (personhood, personality, and inexhaustibility) through which we speak of the process God, but we *do* think it is clear that the ground gets shakier as we ascend levels, and that our basis for supposing, for example, the complete inexhaustibility of God is far, far weaker than our basis for supposing that God exists (in a personal mode), and that God has a real relation to the world (actual personality). As articles of faith, the first is clearer and more assured than the second, and the second is more assured than the idea that God has an infinite and inexhaustible future.

That is why the God of philosophy is not of very much use to religious believers, although it is pretty useful to theologians. All this is by way of saying, with Nietzsche's madman, "I have come too soon." If God has a "soul," in the sense of a finite mode of existing that is tied to life, we have little reason to suppose it is immortal, and some analogical grounds for supposing it is mortal. The direction of God's activity, for process philosophers, is from whole to part (providing the creative context for creative activity by finite beings, which could be called incarnation or *catabasis*). The activity of finite beings is from part to whole, contribution of a satisfaction to the many, which could be called self-transcendence, contributionism, or even *anabasis*. To search for the transcendent God is neither necessary nor wise for philosophy. The God immanent in every act is the one we can know something about empirically.

But now, simply assume God's soul *is* mortal, that mortality is implied in conceiving of God as living and personal, and that *catabasis* brings with it some cost to God, and some risk of failure (e.g., Jesus, or Buddha, or Muhammed falls out of a tree and breaks his neck at eleven years old).

Now pretend, if you will, to be a Christian for a moment. It becomes even more important that we act so as to increase the vigor and life of God to the greatest extent we can, since there can be no value (at least so far as we can understand or imagine) without God—which is to say, the finite parts point to a whole regardless of whether they *constitute* that whole. So it is not only true that God helps those who help themselves, it is also true that those who help themselves help God (if "help" is understood as a norm along the lines of greater unity under contrast).

To put this point in traditional religious language, the mortal God, then, is in some sense Christ crucified, and the resurrection *not* assured, or at least not wholly comprehensible, since it depends on the free and self-creative acts of finite beings, like Jesus, to assist. The only thing assured is that God, once *having been* always *will have been*, and so can never be nothing at all. Contra Hartshorne, that is the most we can really know about the necessity of the divine life, from a process perspective. Yes, this looks like incarnational theology, but if we strip away the theological language, it is nothing other than the structure of prehension as we have explained it earlier in the book.

Finally, a connection between God's immanence in the universe and transcendence of it was thoroughly prefigured in the foregoing discussion. It is now plain that the mortality of God can be associated with modes of immanence and the immortality with modes of transcendence, with the latter subject being highly speculative on our part, and requiring an extensive analysis of the inexhaustibility thesis. Frankly we aren't interested in carrying that task further. We cheerfully leave it to the next generation of process theologians, if they care to engage it.

We may construe God's immanence in the universe in numerous ways apart from the incarnation of God in Christ, of course, and indeed, a philosopher has a duty to do so. To name only a few, God may be immanent in the sage, the Enlightened Being, the Prophet, the Lawgiver, the Bodhisattva, or even in the regularity of Nature—the last of which was Whitehead's particular specialty. All of these modes of immanence portend mortality in different ways. The example of Christ is helpful in its graphic depiction of God's mortality, illustrating the point to finite beings who cannot get the idea otherwise. Optimally, Christians learn to grasp God's divinity through Jesus' humanity, and once they have, the immanence of God in other forms—the Prophet, the Lawgiver, the Sage, the Enlightened Being, the Bodhisattva, and Nature—become no great stretch. To *begin* with the divinity and immortality of the Risen Christ, and *then* to attempt to grasp his humanity (or our own) upon that basis, is to take the incomprehensible as a road to the ordinary. The better path, we contend, leads the other direction: from the ordinary to the extra-ordinary, from the mortal to the immortal, from the immanent to the transcendent, from the concrete to the abstract, or, as Aristotle was wont to say, from what is true for us to what is true unconditionally.

Notes

Introduction

1 It is not entirely missing, though. See Bas Van Fraassen's charming tale of "The Tower and the Shadow," on pages 132–134 of *The Scientific Image* (Oxford: Clarendon Paperbacks, 1980). Van Fraassen, of course, is not a Whiteheadian.
2 Iris Murdoch, *Metaphysics as a Guide to Morals*, reprint ed. (New York: Penguin Books, 1994).
3 Our uses of narrative techniques and storytelling forms here, while relatively small, are nonetheless essential. We are, for that matter, scarcely unique in observing the primacy of narrative in "making sense," especially with regard to centering the concept of the self. In this regard, Olav Bryant Smith does a particularly solid job of drawing attention to the works of Kant, Heidegger, Whitehead, and Ricouer in giving narrative a key role in understanding. We find it especially satisfying to see such a careful work as Smith's explicitly linking Whitehead and Ricouer, and consider much of our own use of narrative to be validated by Smith's meticulous scholarship. We would situate Smith's understanding to that of Murdoch's mentioned above. For her part, Murdoch observes that, "Whitehead said that western philosophy was all footnotes to Plato. The truth of this epigram becomes especially evident in the second half of the twentieth century as we realize how many of our current problems Plato was aware of." [Murdoch, 3255 Kindle edition]

Smith takes Plato and Plato's method of philosophizing for granted, and so goes straight to the issue of narrative intelligibility via what he calls "emplotment," building on an idea first articulated many years before by Hayden White. "Emplotment," Smith says, "is the method of storying, i.e., creating a story." (5) This statement is explicitly made in the context of linking Whitehead and the "missing element—narrative" with Ricouer's work. We agree with Smith when he says that, "the creation of narrative through emplotment reconstitutes elements of experience in a novel manner to create novel worlds. A narrative synthesizes distinct elements of experience . . . into a new unity, i.e., a story." (143) Smith's primary focus is, as noted, upon "the self." But in Whiteheadian terms (that will be explained in greater detail as we proceed) even an electron is a "self" that has "experience." Meanwhile, those higher phases of experience that qualify as human, while profoundly interesting, are really just specifications of these more general cosmological principles. Narrative takes us into those principles as surely—and, sometimes, far more surely—as empirical adequacy and logical coherence.
4 See Gary L. Herstein, "Toward a New Metaphysics of Identity," MA thesis (Chicago: DePaul University, 1992), 63–72.
5 See chapter 4.

6 See the videos at: http://www.youtube.com/watch?v=MfuFSHUeExc and http://www.youtube.com/watch?v=-Bv6KfnuepA&list=RDMfuFSHUeExc, accessed February 7, 2014.
7 Whitehead uses the word "probable" in a sense that is quite independent of his concept of "possibility" (eternal objects), and the latter will come in for much discussion in this book. For Whitehead's theory of probability, see *Process and Reality*, corrected ed. by David Ray Griffin and Donald W. Sherburne (New York: Free Press, 1929 [1978]), 199–207; one may find his views prefigured and worked out in greater detail by C.S. Peirce and by his followers. Peirce's most famous articulation of the issue is in "Deduction, Induction, and Hypothesis," which first appeared in his *Popular Science Monthly* articles of 1877–1878. A revised and improved version was to be included in a book entitled *Illustrations of the Logic of Science*, now out, ed. Cornelis de Waal (Chicago: Open Court, 2013), ch. 6. It is this version that we recommend.
8 For the technical discussion of "event," see Alfred North Whitehead, *Science and the Modern World* (New York: Free Press, 1967 [1925]), 72–73, 119–124. We want to stress, however, that with this and every technical term in Whitehead, the context of inquiry dictates the meaning of the term. See Auxier's discussion of this point in the Fordham edition of *Religion in the Making*, cited below. In the case of the term "event," Whitehead uses it most often when he is inquiring into the way science studies nature.
9 This distinction might be the essence of our departure from the impressive interpretation of Whitehead given by F. Bradford Wallack in *The Epochal Nature of Process in Whitehead's Metaphysics* (Albany, NY: SUNY Press, 1980). While Wallack never claims to be giving a complete account of Whitehead's metaphysics, she does attempt to reconcile the theory of the actual entity with physics as it was practiced up to 1980, particularly insisting that Whitehead's concept of the actual entity was inspired by the quantum revolution in physics in its early days (and we will show hereafter that this claim is historically mistaken), and basically collapsing the idea of the "event" into the idea of the "epochal occasion." She asserts so many things that are spot on in the course of her book that it is painful for us to dissent so completely on this point. Physics and "natural philosophy" study *events*, and if those events are transformed into actual entities, it is for the sake of having a unit of explanation, not because Whitehead believes actual entities are part of concrete experience—and he does not assert that they are. So, for example, where Wallack says that "the actual entity is *any concrete existent whatsoever*" (7), she would have done better to say that the actual entity is the philosophical idea capable of explaining any concrete existent whatsoever. She cites, as the authority for her assertion *PR* 18, but without including Whitehead's crucial qualifiers at the beginning of the paragraph that he singles out four "generic notions" because "they involve some divergence from antecedent philosophical thought" (and here he lists the actual entity along with prehension, nexus, and the ontological principle); he re-identifies these as philosophical "notions" before listing the three Wallack cites (omitting the ontological principle), and makes it very clear that these are notions framed exclusively for philosophical discussion that avoids misplaced concreteness. One cannot remove the concept of the actual entity from the philosophical purposes for which it was framed and employ it to do the work of the existents that are studied by physics or any other natural science (i.e., events). See also Jorge Nobo, *Whitehead's Metaphysics of Extension and Solidarity* (Albany, NY: SUNY Press, 1986), 414–415. The criticism we offer here comes down to saying that reconciling Whitehead's theory of extension with physics, even when it is done well, is not the same thing as seeking an adequate, applicable and logically rigorous philosophical cosmology. See Wallack, *The Epochal Nature of Process in Whitehead's Metaphysics*, 295; and Chris Van Haeften, "Atomicity and Extension," *Process*

Studies, 32:Supplement 3 (2003), esp. 12–14, and "Extension and Epoch: Continuity and Discontinuity in the Philosophy of A.N. Whitehead," *Transactions of the Charles S. Peirce Society*, 37:1 (2001), 59–79.
10 See especially Whitehead, *Adventures of Ides* (New York: Free Press, 1967), the opening two paragraphs of chapter 15, 220.
11 See for example, Keith Robinson, ed., *Deleuze, Whitehead, Bergson: Rhizomatic Connections*, (New York: Palgrave Macmillan, 2009); and Keith Robinson, *Deleuze, Whitehead, Bergson* (New York: Palgrave Macmillan, 2008).
12 Steven Shaviro, Without Critera: Kant, Whitehead, Deleuze, and Aesthetics (Cambridge, MA: MIT Press, 2009).
13 For Harman, his blog is found at: http://doctorzamalek2.wordpress.com/. Also see Matthew David Segall, *The Physics of the World Soul: The Relevance of Whitehead's Philosophy of Organism to Scientific Cosmology*, and *On the Matter of Life: Towards and Integral Biology of Economics* (both eBooks by lulu.com).
14 Examples would be Dennis Soelch, Aljosche Berve, and several other young contributors to *Beyond Superlatives: Regenerating Whitehead's Philosophy of Experience*, eds. R. Faber, J. R. Hustwit, and H. Phelps (Cambridge: Cambridge Scholars Press, 2014).
15 Particularly of note is Keller's newest book, *Cloud of the Impossible: Negative Theology and Planetary Entanglement* (New York: Columbia University Press, 2015).
16 See the account at http://en.wikipedia.org/wiki/Blind_men_and_an_elephant, accessed February 11, 2014.
17 See Victor Lowe, *Alfred North Whitehead: The Man and His Work*, ed. J. B. Schneewind, vol. 2 (Baltimore: Johns Hopkins University Press, 1990), 13–14. See Auxier's critique of Lowe's methods and understanding of Whitehead's development in "Influence as Confluence: Bergson and Whitehead," in *Process Studies*, in the special focus section on "Bergson and Whitehead," 28:3–4 (Fall–Winter 1999), 267; 301–338; 339–345.
18 See Isabelle Stengers, *Thinking with Whitehead: A Free and Wild Creation of Concepts*, trans. Michael Chase (Cambridge, MA: Harvard University Press, 2011), 4. See also the thorough treatment of these problems in Ronny Desmet, "A Refutation of Russell's Stereotype," in *Whitehead: The Algebra of Metaphysics*, eds. Ronny Desmet and Michel Weber (Leuven de Neuve, BE: Les editions Chromatika, 2010), 127–209.
19 See Lowe, *Alfred North Whitehead*, 12–13, 15–16.
20 Our current book project, *The Continuum of Possibility*, takes this formalization beyond a sketch.
21 See Stengers, *Thinking with Whitehead*, 118–119.
22 Most of Stenger's work has been undertaken with Whitehead in the background or altogether absent. She is not a "Whitehead scholar," but is rather a thinker whose achievement *depends* upon Whitehead's ideas in various ways. The assessments of her *Thinking with Whitehead* in the Focus section edited and introduced by Keith Robinson, *Process Studies*, 37:2 (Fall–Winter 2008), 74–189, with Stengers' response, is a good way to get a sense of her place in Whiteheadian philosophy.
23 William Christian, *An Introduction to Whitehead's Metaphysics* (New Haven: Yale University Press, 1959).
24 Ivor Leclerc, *Whitehead's Metaphysics: An Introductory Exposition* (Atlantic Highlands, NJ: Humanities Press, 1958).
25 Nobo's *Whitehead's Metaphysics of Extension and Solidarity*.
26 John Lango, *Whitehead's Ontology* (Albany, NY: SUNY Press, 1972). One thing we especially appreciate about Lango's work is that he seems to recognize just how far off the mark most Whitehead interpretation is. He says: "Although there is virtually no awareness among interpreters of Whitehead's metaphysics

that actual entities are known only through metaphysical speculation, it is generally understood that ordinary human perception does not convey knowledge of an individual actual entity." (5–6) We shall have reason to take issue with what Lango means by "speculation" in our notes below, but we agree that very few Whitehead interpreters grasp that his method in metaphysics is hypothetical, through and through, and Lango does get this right. He is also right about the half-conscious character of most of Whitehead's interpreters. If they know not to look for actual entities to appear in their field of vision, it isn't because they know what Whitehead means by "perception," which, frankly, is pretty far from the usual meaning (as Lango also points out).

27 Wallack, *The Epochal Nature of Process*. Wallack's book has not received the attention it deserves. It was dutifully reviewed in *Process Studies* and in the *Transactions of the Charles S. Peirce Society* when it appeared, and Nobo makes respectful reference to it, along with Van Haeften and a few others among the best interpreters of Whitehead, but it seems to have been unduly neglected by most. Stephen David Ross gives Wallack adequate notice in *Perspective in Whitehead's Metaphysics* and George Lucas at least mentions her in *The Rehabilitation of Whitehead*. But among those who ought most to be concerned with Wallack argued, such as Murray Code, Tim Eastman, Michael Epperson, Richard Weidig, and even such humanists as Stephen Franklin, she is never mentioned. Lewis Ford never cites the book in any place we can find (a thorough but perhaps not wholly exhaustive search), and since Wallack calls for a "revolution" in Whitehead interpretation that would undermine Ford's reading, and since it is impossible that he did not know about the book (he was editor of *Process Studies* when the review was published), it is peculiar that he does not mention it, at least to argue against it. Its thesis is more than relevant to many claims Ford makes, and Wallack explicitly criticizes Ford in the course of her argument (119, 134, 197, 257), and the criticisms taken together constitute a wholesale repudiation of his entire approach, as it had been articulated to that point in time. Unlike Nobo, with whom Ford was obliged to carry out numerous public disputations, Wallack was not well-placed in academia, was a woman, did not have a Ph.D., and had the bad luck to read Whitehead honestly against the grain for a generation that did not want to entertain the idea that the "actual entity" was anything apart from a little puff of atomic existence.

28 Judith A. Jones, *Intensity: An Essay in Whiteheadian Ontology* (Nashville: Vanderbilt Press, 1998). Our current book project engages Jones more thoroughly.

29 Victor Lowe, *Understanding Whitehead* (Baltimore: Johns Hopkins Press, 1962).

30 Auxier undertook an extensive refutation of Lowe's assertions about the Bergson-Whitehead relationship in an earlier independent article. "Influence as Confluence," 267, 301–338, 339–345.

31 See Donald W. Sherburne, *A Key to Whitehead's Process and Reality* (Chicago: University of Chicago Press, 1966); Elizabeth Kraus, *The Metaphysics of Experience: A Companion to Whitehead's Process and Reality* (New York: Fordham University Press, 2nd ed. 1998); and John Cobb, Jr., *Whitehead Word Book: A Glossary with Analytical Index to Process and Reality* (Claremont, CA: P&F Press, 2008), available on-line at: http://processandfaith.org/sites/default/files/pdfs/WordBookWeb.pdf, accessed February 11, 2014.

32 See Richard Weidig, *The Creative Advance: From History to Mystery* (Berkeley, CA: Berkeley West, 1987); and Ralph Pred, *Onflow: Dynamics in Consciousness and Experience* (Cambridge: MIT Press, 2005).

33 Michael Epperson, *Quantum Mechanics and the Philosophy of Alfred North Whitehead* (New York: Fordham University Press, 2004). We have mentioned earlier that Epperson neglects the book that is perhaps closest to his own project and, to our thinking, comes far closer to addressing it constructively, which is

Wallack's *The Epochal Nature of Process*. Perhaps there is still time for a valuable corrective to Epperson's ideas available here, and we note that since both Wallack and Epperson refer often and affirmingly to Henry Stapp, that might be a fruitful place for a dialogue to begin.
34 Michael Epperson and Elias Zafiris, *Foundations of Relational Realism* (Lanham, MD: Lexington Books, 2013).
35 Ibid., 21.
36 See Ibid., 25–27, in particular.
37 See Epperson and Zafiris, *Foundations of Relational Realism*, 211.
38 Granville C. Henry and Robert J. Valenza offer an interesting example of an algebraic tool (graph theory) applied to some Whiteheadian ideas in "The Principle of Affinity in Whitehead's Metaphysics," *Process Studies*, 23:1 (Spring 1994), 30–49. See also Granville Henry, *Forms of Concrescence: Alfred North Whitehead's Philosophy and Computer Programming Structures* (Canbury, NJ: Associate University Presses, 1993).
39 Murray Code, *Order and Organism: Steps to a Whiteheadian Philosophy of Mathematics and the Natural Sciences* (Albany, NY: SUNY Press, 1985).
40 See Timothy E. Eastman and Hank Keeton, *Physics and Whitehead: Quantum, Process, and Experience* (Albany, NY: SUNY Press, 2004).

Chapter 1

1 Alfred North Whitehead, *The Concept of Nature* (Cambridge: Cambridge University Press, 1920), p. v.
2 See the edition of Whitehead's *Religion in the Making* that contains Auxier's comprehensive glossary (New York: Fordham University Press, 1996), edited and introduced by Jude Jones. This example from p. 182. The glossary exhaustively cross-references and defines all technical terminology and fills pp. 161–251 of this edition.
3 Ibid., 128.
4 Ibid., 93.
5 Ibid., 100.
6 An extreme example of this maddening habit comes in *The Concept of Nature* when Whitehead explains speculative as distinct from logical "demonstrations," and gives one example of how a lecturer (in biology presumably) "demonstrates by the aid of a frog and a microscope the circulation of the blood for an elementary class of medical students. I will call such demonstration 'speculative' demonstration, remembering Hamlet's use of the word 'speculation' when he says, 'There is no speculation in those eyes.'" (6) Not only does Whitehead define speculation by telling us when it is absent, but relies on our immediate recognition of the context—an accusing ghost that knows who killed him *does not* speculate. But even more annoyingly, Whitehead misremembers who said this— it was Macbeth screaming at Banquo's accusing ghost in Act III, Scene IV, but at least he did accurately recall that this was said of an angry ghost and written by Shakespeare!
7 It is common for Whitehead to comment at various stages on the relation of the current work to previous works, and these comments are nearly always "supplementary" in nature—i.e., if a reader wants more information about this or that, see an earlier work. Sometimes Whitehead does make more substantive comments, such as the following in *The Concept of Nature*:

> This volume on the 'Concept of Nature' forms a companion book to my previous work *An Enquiry Concerning the Principles of Natural Knowledge*. Either book can be read independently, but they supplement each other. In

part the present book supplies points of view which were omitted from its predecessor; in part it traverses the same ground with an alternative exposition. . . . On the other hand, important points of the previous work have been omitted where I had nothing fresh to say about them. . . . . I am not conscious that I have in any way altered my views. Some developments have been made."

(vi–vii)

This statement applies to the relationships among all of Whitehead's works, and such statements are frequently repeated through the corpus.

8 This tendency is prevalent among what is commonly known as the "genetic" interpretation of Whitehead's development.
9 Whitehead says that "the very purpose of philosophy is to delve below the apparent clarity of common speech." *Adventures of Ideas* (New York: Macmillan, 1967 [1933]), 222.
10 This point is made nicely by Lango in *Whitehead's Ontology*, 5, 8–11.
11 We take up this topic of the role of predication in some detail in our forthcoming book, *The Continuum of Possibility*.
12 A very astute account on the purposes and methods in *Enquiry into the Principles of Natural Knowledge* and *The Concept of Nature* is found in Robert D. Mack, *The Appeal to Immediate Experience* (New York: King's Crown Press, 1945), 27–32. Mack anticipates the account of naturalism we will offer in chapters 10 and 11. However, Mack (as with so many others) does not appreciate how important is the difference between the kinds of inquiry natural philosophy requires in the triptych and cosmology or metaphysics requires in *Process and Reality*. Mack reads the shift in topic of inquiry as a shift in viewpoint, which is not correct, but also as a shift in emphasis, which is correct.
13 The choice here has philosophical importance, since "to express" allows externalization without requiring that what is expressed actually evacuates the entity, whereas to egress denotes leaving the entity empty. In the case of Whiteheadian expression, Elvis *never* leaves the building, the building only expresses its Elvis value. We will add and rehabilitate the word "egress in our next study, but see the opening paragraphs of Poe's "The Masque of the Red Death" for a poignant use of ingress and egress.
14 See Whitehead, *Process and Reality*, 24. See Category of Explanation xvi, and 29–30.
15 See Philip J. Davis and Reuben Hersh, *The Mathematical Experience* (Boston: Houghton Mifflin, 1981), 136–151.
16 It is good to remember that many of Whitehead's books were based on various series of lectures he presented, and the amount of effort he devoted adapting those lectures to the format of reading was variable to say the least. It is difficult to believe, for example, that *Process and Reality*, the book, was the Gifford Lectures. How many people sat through that? What on earth did they think Whitehead was talking about? Heaven only knows.
17 Whitehead, *Adventures of Ideas*, 220.
18 See Paul Arthur Schilpp, ed., *The Philosophy of Alfred North Whitehead* (LaSalle, IL: Open Court Press, 1951 [1941]); and Wilbur Urban, "Cassirer's Philosophy of Language," in *The Philosophy of Ernst Cassirer* (LaSalle, IL: Open Court Press, 1958 [1949]), 437–438.
19 Urban, "Cassirer's Philosophy of Language," It is interesting that Urban was also referred to by Langer as setting her problem up in the preface to the first edition of *Philosophy in a New Key*.
20 Whitehead, *Adventures of Ideas*, 235.
21 Whitehead, *Process and Reality*, 4.

22 Whitehead, *Process and Reality*, 4. As a guide to the role of metaphor in metaphysics, see Murdoch, *Metaphysics as a Guide to Morals*.
23 Whitehead, *Process and Reality*, 4.
24 For a fuller account of what is meant by "radical empiricism" here and Whitehead's relationship to it, see chapter 2.
25 An exception to this is the various articles published by James Bradley in *Process Studies*. See the bibliography.
26 This is our shorthand designation for the three major works in natural philosophy Whitehead published between 1919 and 1922, namely, *An Enquiry into the Principles of Natural Knowledge* (1919), *The Concept of Nature* (1920), and *The Principle of Relativity* (1922). For an account of the relevance of these works to Whitehead's later work, see Gary L. Herstein, *Whitehead and the Measurement Problem of Cosmology* (Frankfurt am Main: Ontos Verlag, 2006), esp. 11 ff.
27 Ford has been talking about Whitehead's alleged discovery of "temporal atomicity" for many years, and others besides us have challenged his thesis, but his most recent defense of the idea is "The Indispensability of Temporal Atomism," *Process Studies* 38:2 (Fall–Winter, 2009), 279–303. As with all of Ford's defenses, this one depends on his own imaginative reconstruction of the development of Whitehead's metaphysics.
28 Many thinkers have misunderstood Whitehead's use of the term "organism." For example, in Norirtoshi Aramaki: "A Critique of Whitehead in Light of the Buddhist Distinction of the Two-Truth Doctrine," *Process Studies*, 36:2 (Fall–Winter, 2007), 294–305, Professor Aramaki treats "organism" as literal and ontological, rather than logical and explicative, and criticizes Whitehead for, in effect, not knowing enough biology. (See, for example, Aramaki, 296–297.)
29 Contra Ford and Stengers, Isabella Palin argues for a continuity of themes and methodologies in Whitehead. See Palin, "On Whitehead's Recurrent Themes and Consistent Style," *Process Studies*, 37:2 (Fall–Winter, 2008), 78–97.
30 Whitehead, *Adventures of Ideas*, 221.
31 This point will be treated in detail in chapter 7.
32 John Lango complains about this when he realizes that in order to confirm a formal definition of the "actual entity," he must be able to produce at least one example of the thing itself. As he says:

> . . . the fundamental entities in Whitehead's ontology—the actual entities—are purely hypothetical, postulated by his metaphysics, but not known in any other way. We cannot observe an actual entity with our senses; we cannot infer an actual entity through introspection. We must therefore obtain an understanding of actual entities through speculation.
>
> (*Whitehead's Ontology*, 5)

This is correct as stated, but it shows how many people expect Whitehead's ontology to be about stuff we might find in the world if we could examine the world with enough subtlety. Obviously Lango knows better (6). Yet, he is in a sense wrong about whether we can observe the actual entity. If the ontology succeeds on Whitehead's own criteria, it would be closer to the truth to say that there is no instance of experience in which the actual entity fails to be exemplified.
33 Whitehead, *Religion in the Making*, Preface (unnumbered page). Unfortunately, the Fordham University Press edition of this book (1996) accidentally omitted the Preface (even though it appears in the Table of Contents), so one must consult an earlier edition for these extracts.
34 Whitehead, *Process and Reality*, xi. This may strike some readers as an odd characterization of the book, but if we understand that Whitehead thinks philosophy took a wrong turn with Kant (see the Introduction to this book), due to Kant's

over-generalizing of presentational space, and if we recall that prior to Kant, philosophers did not make a clear distinction between psychology and epistemology, it is easier to understand why Whitehead would seek to rehabilitate Enlightenment epistemology. For Whitehead, the actual and the possible resolutely remain on equal footing—see the categoreal scheme, where "actual entities and eternal objects stand out with a certain extreme finality." (*Process and Reality*, 22). For a relevant and convincing account of the tendency in the Enlightenment to mix psychology with epistemology, see Ernst Cassirer, *The Philosophy of the Enlightenment*, trans. C. A. Koelln and J. P. Pettegrove (Princeton: Princeton University Press, 1951), ch. 3. Other interpreters have noticed but misconstrued the importance of this sentence from Whitehead's Preface to *Process and Reality* about "Modern" thought. Donald Sherburne thinks it does not matter "whether one thinks that this particular phase of thought ended with Hume, with Santayana, or with Sellars, Quine, and Davidson." Sherburne believes that "the main point is that Whitehead sees himself as writing at the end, not of philosophy, but, as he says, of a *phase* of philosophy." (Donald W. Sherburne, "The Process Perspective as Context for Educational Evaluation," *Process Studies*, 20:2 (Summer 1991), 80. The word "recurrence" means that something is coming back, not that something is over. It makes a very great difference that Whitehead marks the end of the phase he mentions with Hume because, as he explains, Kant changed the questions after Hume and set philosophy on the wrong course.

35 The further elucidation and explanation of the unusual use of the term "theory" is later offered in Whitehead's account of propositions in *Process and Reality*:

> [W]ith the growth of intensity in the mental pole, evidenced by the flash of novelty in appetition, the appetition takes the form of a 'propositional prehension'. These prehensions will be studied more particularly in Part III. They are the prehensions of 'theories'. It is evident, however, that the primary function of theories is as a lure for feeling, thereby providing immediacy of enjoyment and purpose."
>
> (184)

Whitehead is thinking about theories along lines similar to Peirce, in the latter's theory of abduction (which he also calls "hypothesis").

36 Whitehead, *Adventures of Ideas*, 103. See also the "Epilogue" to Part I on p. 100.
37 Ibid., 284.
38 Ibid., 284.
39 See Whitehead, *Process and Reality*, 3–4.
40 We aim here to present Whitehead's view of "adequacy" accurately, but we are aware that this strong claim about adequacy could be questioned. The inquiry into something *sui generis* might yield results that are unrelated to the results of other inquiries even for the same inquirer. We have mentioned this problem in our discussion of the ontology of events in the Introduction.
41 See chapters 7 and 8 for a very detailed account of this structure.
42 See Whitehead, *Process and Reality*, 19.
43 See ibid., 235.
44 Almost all Whitehead scholars and commentators confuse these different levels of generality. The examples would be limitless, but we take one from John Lango, largely because he is an astute reader of Whitehead. But he says:

> Causal perceptions of objects external to the perceiving human mind may indeed be vague, but surely the human mind, a linear series of actual entities (*PR* 163–167) can be conscious of a single "occasion" (i.e., finite actual entity) of its own self.
>
> (Lango, *Whitehead's Ontology*, 7)

The passages cited are from the theory of concrescence. They have nothing directly to do with the theory of perception; further, the only place that Whitehead says anything remotely similar to what Lango claims here is embedded in his discussion of propositions (and hence, the theory of judgment, PR 198—he is talking about how 1+1=2 fails in metaphysical generality), and has no direct application to the problem of perception, which is far more abstract. If by "human mind," Lango means something like "personal route," he needs to revisit what a personal route is for Whitehead (again, see PR 198), which involves nothing so complex as human experience, let alone a human *mind*. In this one sentence, Lango conflates nearly every level of generality Whitehead distinguishes. The problem is a common one.

45 This topic will be taken up in detail later in the book, but the relationship between definiteness and determinateness is an important and difficult area of interpretation. Leemon McHenry made the interesting suggestion that the difference between the categories of explanation and categories of obligation in the Categoreal Scheme of *Process and Reality* is that explanations tend toward definiteness (showing how occasions concresce), while the obligations show the law-like tendencies among entities, revealing their determinateness. We may be reading more in to McHenry's article than he intended, but this interpretation of definiteness and determinateness in Whitehead agrees with ours in this book. See McHenry, "The Axiomatic Matrix of Whitehead's Process and Reality," *Process Studies*, 15:3 (Fall 1986), 172–180, esp. 178–179. Also valuable in this respect is the way Whitehead explains these terms in the 1927 essay "Time," collected in *The Interpretation of Science: Selected Essays*, ed. A. H. Johnson (Indianapolis: Bobbs-Merrill, 1961), 240–247. In speaking of "categories" of supersession, prehension, and incompleteness, in that essay, the interplay of which gives rise to "the concept of time," Whitehead argues that the elements at issue are of two kinds, "other occasions" than the actual entity in its actual world, and "universals—or, as I prefer to call them, the eternal objects; these eternal objects are the *media* of actuality, whereby the *how* of each actual occasion **is determinate**. Because **the other occasions are each in a definite way** required for the organization of any one occasion A, the world is called a system. The **definite** way in which A includes other occasions in its concretion is here called 'Prehension'." (241, bold ours, italics Whitehead's) We will take some trouble to explain in a later note pertaining to Jorge Nobo's interpretation of Whitehead, that the way the term "Prehension" as used here is different from its standing in *Process and Reality*, and since the *concept* of time is Whitehead's real concern in the 1927 essay, the model for understanding this inquiry is that provided by *The Concept of Nature*. We are looking for a *concept*. Concepts are the basic organizational structures of natural knowledge, or knowledge of nature. Time is treated within that framework in this 1927 essay. Whitehead presupposes, in effect, the same thing he says in *The Concept of Nature*, which is:

> Our knowledge of nature is an experience of activity (or passage). The things *previously observed* are active entities, the 'events.' They are chunks in the life of nature. [So much for the supposed "later discovery" of temporal atomism in Lewis Ford's "compositional analysis"] These events have to each other relations which *in our knowledge* differentiate themselves into space-relations and time relations. But this differentiation between space and time, though inherent in nature, is comparatively superficial; and space and time are each partial expressions of one fundamental relation between events which is neither spatial nor temporal. This relation I call "extension."
>
> (*Concept of Nature*, 185, our emphasis)

To get a handle on the *concept* of time as related to natural knowledge is to frame an inquiry that shows how time expresses extension *in our knowledge*. The category of supersession provides the main idea in this essay. It is mainly *physical* prehension and *physical* time, understood in abstraction of the mental pole, and thus not wholly concrete (see 240–241) that contribute to the *concept* of time, insofar as that concept contributes to our knowledge. The mental pole is discussed in the 1927 essay only insofar as it enters into physical existence. The mental pole, from this point of view, is, as he says, "merely analytic." (*The Interpretation of Science*, 242) Still, the habit of using the term "determinate" and determinate order to describe the way in which eternal objects are present in actual entities, and the term "definite" to describe the presence of "other actual occasions" in an actual entity is consistent in Whitehead's inquiries in this 1927 essay, and in *Process and Reality*, and in *Science and the Modern World*. These inquiries do overlap, and the inquiry into "Time" is consistent with the other longer inquiries. The distinction between determinate order and definiteness will do a good deal of work in our book, subsequently, so we will return to it with further arguments and explanation. For now it is sufficient to note its importance and provide some initial exposition from that 1927 essay.

Chapter 2

1 See Whitehead, *Process and Reality*, xii.
2 We are not the first to notice that Whitehead is a radical empiricist. For example, Victor Lowe states: "Whitehead was probably acquainted with James's *Psychology* and perhaps heard much of the ingenuity of the concept of the specious present from McTaggart and others. From this point of view it is clear that his early empiricism was more radical than atomistic." *Alfred North Whitehead: The Man and His Work*, ed. J. B. Schneewind, vol. 2 (Baltimore: Johns Hopkins University Press, 1990), 105. Lowe takes for granted that Whitehead evolved away from this view toward an atomistic view, which we are here disputing. Whitehead's familiarity with James increased as his writing went forward, such that James became a regular touchstone and reference. See also Robert M. Palter, *Whitehead's Philosophy of Science* (Chicago: University of Chicago Press, 1960), 24–25; and Nancy Frankenberry, *Religion and Radical Empiricism* (Albany, NY: SUNY Press, 1987). There are many other commentators who have said similar things, and not merely about Whitehead's early writings. Undoubtedly, informed readers will think of the book by Craig Eisendrath, *The Unifying Moment: The Psychological Philosophy of William James and Alfred North Whitehead* (Cambridge, MA: Harvard University Press, 1971), in this connection. But Eisendrath is hostile to radical empiricism (see 67–68) because he is so taken with the subjectivist principle (and none could deny that Whitehead has an amazingly powerful defense of it). Radical empiricism works otherwise. Our subjective experiences are *examples* of the structure of possibility (eternal objects) in our immediate epoch, and beyond this we cannot generalize their peculiarities in desire (appetition), consciousness, knowing, judging, or any other category that most people would interpret in a psychological way. We explain Whitehead's reconciliation of his radical empiricism and realism under the heading of what we call, later, his "radical realism." We believe this to be Whitehead's actual view.
3 See Auxier, "Influence as Confluence," in the special focus section on "Bergson and Whitehead," *Process Studies* 28:3–4 (Fall–Winter 1999), 267; 301–338; 339–345. See also Auxier, *Time, Will, and Purpose: Living Ideas from the Philosophy of Josiah Royce* (Chicago: Open Court, 2013), chs. 4 and 7.
4 Whitehead, *Adventures of Ideas*, 222 (our emphasis).
5 See ibid., 222–223.

6 See ibid., 223.
7 The problem with philosophy trying to overreach the limits of logic to make it a basis for ontology is as old as Plato's appeal to geometry, although some philosophers, such as Descartes, were always suspicious of the syllogism as a basis for ontology, preferring mathematics. The problem of the unholy trinity is the same as the one Kant warned against. Radical empiricists insist upon the irreducible relevance of the real, but do not insist upon a single ontology of the real, even ideally.
8 The relationship between Whitehead and the tradition of "substance metaphysics" is subtle. In no way do we wish to oversimplify it. We speak almost more of the *effect* of the tradition on professional philosophical discourse of the last two centuries than of substance metaphysics itself. For a subtle story about how Whitehead relates to this tradition, and one in line with our interpretation, see Reto Luzius Fetz, *Whitehead: Prozessdenken und Substanzmetaphysik* (Freiburg: Verlag Karl Alber, 1981), Section 3. We are appreciative of the translation of the relevant parts of this outstanding book by James W. Felt, in *Process Studies*, including: "In Critique of Whitehead." *Process Studies*, 20:1 (Spring 1991), 1–9; and most importantly, "Aristotelian and Whiteheadian Conceptions of Actuality: I," *Process Studies*, 19:1 (Spring 1990), 15–27; and "Aristotelian and Whiteheadian Conceptions of Actuality: II," *Process Studies*, 19:3 (Fall 1990), 145–155. Also, Auxier has addressed this issue in some detail in *Time, Will, and Purpose*, ch. 2.
9 Whitehead is fairly impatient with scientists who persist in treating "law" as if it were a guarantor of necessary knowledge. He wryly notes: "Nature is patient of interpretation in terms of Laws which happen to interest us," *Adventures of Ideas*, 136. See also the discussion in Whitehead, *The Function of Reason* (Boston: Beacon Press, 1958 [1929]), 48–53. We will take up this question in greater detail in chapter 10 in the sections dedicated to how "bad metaphysics" does not make for "good science."
10 A good overarching critique of the failed project of analytic philosophy is Nicholas Capaldi, *The Enlightenment Project in the Analytic Conversation* (Dordrecht: Kluwer Academic Publishers, 1998). For an accessible account of the end of necessitarian science, see Ilya Prigogine and Isabelle Stengers, *Order Out of Chaos* (New York: Bantam, 1984). We shall have much more to say about this issue in the course of this book, and our next book deals with these issues in still greater detail.
11 The classic text, G. E. Hughes and M. J. Cresswell, *An Introduction to Modal Logic* (London: Methuen & Co., 1968), simply starts out by saying that logical necessity is a "topic which bristles with philosophical difficulties," but then proceeds to make necessity the most basic modal notion, followed by impossibility, contingency, and possibility, which they do not even attempt seriously to define, saying rather that "the sense of these expressions should be sufficiently clear from what we have said in the case of necessity." (22–23). These other notions are not sufficiently clear from what has been said of necessity, and to define any of these in terms of necessity, even much subtler accounts of necessity (such as Kripke frames), is simply archaic. All these accounts take for granted the basic gesture of extensional logic, where "extensional" here must be carefully distinguished from Whitehead's use of that term. Whitehead's logic proceeds in the opposite direction, "intensionally," where grouping precedes selection, and selection has no bottom "element." See Whitehead, *A Treatise on Universal Algebra*, 35–38, for an example of intensional elements.
12 Our detailed arguments concerning possibility will be taken up in chapters 8 and 9.
13 In *Process and Reality*, Whitehead makes a point of noting how realistic philosophies have been vulnerable to positing vacuous actualities (see 29).

14 The distinction between living and inert ideas is explained by Whitehead in *The Aims of Education and Other Essays* (New York: Free Press, 1929), 1–2.
15 See Whitehead, *Science and the Modern World*, 159–160. Here he makes clear that what he means by an "eternal object" is a possibility *for* an actuality, but there are varied modes of ingression, and while an eternal object "cannot be divorced from its reference to other eternal objects, and from its reference to actuality generally . . . it is disconnected from its actual modes of ingression into definite actual occasions." (159) This "disconnection" is a principle of discontinuity between actual and possible. The relation between possible and actual is pluralized and made contingent, while necessity is reinstated (in multiple modalities) among possibilities *independent* of their relation to particular actualities. We formalize this dependence/independence relation in chapter 9. Whitehead restates this same basic point in *The Function of Reason* (Boston: Beacon Press, 1958 [1929]), 9, with explicit reference to possibility. We will spend significant time on this difficult relationship in chapters 8–11.
16 See Whitehead, *The Function of Reason*, 78–79.
17 See ibid., 71. For an earlier affirmation of this viewpoint, see *The Principle of Relativity* (Cambridge: Cambridge University Press, 1922), 63.
18 William James, *Writings of William James*, ed. John J. McDermott (Chicago: University of Chicago Press, 1992), 136.
19 We find Whitehead's statement of his radical empiricism in a number of places (including the passage with which we began this chapter), but an especially important one is the following: "It must be remembered that just as the relations modify the natures of the *relata*, so the *relata* modify the nature of relation. The relationship is not a universal. It is a concrete fact with the same concreteness as the *relata*," *Adventures of Ideas*, 157. See also the same work, where he extols us to "interrogate experience for evidence of the interconnectedness of things." (220) One of the reasons these are important passages is that they were written in 1933, late in his career. See also *The Principle of Relativity*, 49, for the observed character of concrete relations and *relata*, and also the reinforcement of his radical empiricism; cf., ibid., 73–74. We will point out numerous statements of Whitehead's radical empiricism in the course of the book. The case does not rest solely with this chapter or section. If one desires a more precise description of "particular experience," we would offer Husserl's account (which descended from James): "The living S that is still fresh, still exercising an affection, coincides with the property that has been drawn from it. But then we do not have a unity of knowledge, then the S is not characterized as the substrate of a determination for the ego, and the determination itself is not characterized as a determination. Should this be the case, then the identification must be one that is actively carried out, it must be an act running through the thematic unity of both terms, an act that we can describe the following way: The S as theme initially undergoes a general examination that is lacking any determination. An affecting moment α, which is passively 'enclosed in S,' now penetrates to the active ego. But this ego is abidingly interested in S; as such it 'concentrates' its interest that is, its S-interest in α. The fullness of givenness of the S is enriched in the grasping; but this takes place because it itself is given to consciousness as S only in its particularity. The concentration on the particularity therefore fulfills and enriches the interest in S." Husserl, Edmund, Husserl, Edmund, *Analysis Concerning Passive and Active Syntheses*, trans. Anthony J. Steinbock (Dordrecht: Kluwer, 2001), 295. We thank Matthew Donnelly for bringing our attention to this passage.
20 See Whitehead, *Adventures of Ideas*, 220 and following.
21 Whitehead, *Process and Reality*, 61.

22 Ibid., 35. It is advisable to read the entire passage with which this famous phrase is contextualized. It is clearly not a claim about concrete fact, but a claim about how we can use extensive abstraction to *think* about the problem of physical space and time (recalling that the entire concept of *physical* space and time is a product of the extensive method, not the experience with which it begins, concrete fact).
23 Confusion concerning temporality in Whitehead's philosophy is not hard to find in the literature. See for example Edgar Towne's essay, "All Causality Occurs in a Present: G.H. Mead's Proposal to Process Philosophy," *Process Studies* 39:1 (Spring-Summer 2010), 87–105. The first sentence of his abstract and the first sentence of his main text should be compared. The abstract reads: "G. H. Mead and A. N. Whitehead agree that all causation occurs in *a* present. . . ." As stated, this is true enough. But then Towne asserts: "Alfred North Whitehead . . . and George Herbert Mead . . . agree that all actual causation occurs in *the* present. . . ." (87, our emphasis.) The difference between the indefinite article (in the abstract) and the definite article (in the main text) is vital. Unfortunately, this error undermines Towne's argument, since the word "present" has two different, and perhaps irreconcilable, meanings throughout.
24 Such injuries can and do occur, although the consequences of them have been slightly different than the Nolans present in their story. The British musicologist and conductor Clive Wearing is a famous example. See http://en.wikipedia.org/wiki/Clive_Wearing, accessed September 26, 2012.
25 Vonnegut makes it very clear that he is only playing around with time, since he keeps repeating that he is writing about the timequake in 1996, which is not possible on his own schema, since no one would have known in (the first) 1996 the timequake would occur in 2001, and no one would have been free to write about it during the rerun. His constant repetition of the date of its writing is a challenge to the reader either to think of him as a Tralfamadorian, or a joker.
26 In a way, Dedekind was not only the Professor Marvel who peddles irrational numbers, but he is also the "man behind the curtain" when it comes to the self-awareness of our invention of numerical continuity—pull back the curtain and we find just the human will. As Leopold Kronecker famously said, "God made the natural numbers; all else is the work of man." See Richard Dedekind, *Essays on the Theory of Numbers: Continuity and Irrational Numbers*, auth. trans. Wooster Woodruff Beman (New York: Dover, 1963 [1901]), 8–19.
27 See the "Duality" and "Logic" sections of the article on CMOS (complementary metal-oxide semiconductors) here: http://en.wikipedia.org/wiki/CMOS, accessed February 12, 2014. For CMOS systems, 0 and 1 are insufficient; there has to be a value for "inactive" as well.
28 This is far from being the last readers will see of Zeno's paradoxes, which play a surprising and formidable role in our account of the theory of extension. See chapter 4.
29 See Whitehead, *Process and Reality*, 55, for clear confirmation on the issue of particular experience, and the same point is found in many other places.
30 Parts III and IV of *Process and Reality* model continuity (i.e., overcome discreteness) in different respects. The theory of prehension implies, finally, that the entire universe is prehended from each perspective in it. In Part IV, the whole of the universe particularizes itself in modes of connection for every actual entity—increases the world by one. That is an event. Our next book develops this point in great detail.
31 For example, eternal object (i.e., possibility) x for Entity 1 and x for Entity 2, while discontinuous on the basis of the genetic analysis of the prehensions of either 1 or 2, is still structurally x. Recall where Whitehead says he cannot

demonstrate that *x* is *x* in some final sense, but he can think of no empirical hypothesis that is advanced by its denial. See Whitehead, *Principle of Relativity*, 69. He admits that the evidence for a hypothesis of *x*'s identity here is "slight," but it is the simplest hypothesis. It does not matter whether the *x* is an eternal object for each entity in question, or a physical object in the actual worlds of entities 1 and 2, since entities 1 and 2 occupy different actual worlds. The point is that we will treat *x* as intersecting with *x* in the possible worlds of any and all actual entities we compare. This is a minimal logical continuity, hardly the assertion of "temporal atomicity"—the evidence for which is not only slight but wholly absent. We will treat this issue in greater detail in chapters 7–9.

32 See McConnell, *Bowne*, 277–278.

Chapter 3

1. Alfred North Whitehead, *The Aims of Education and Other Essays* (New York: Free Press, 1967), 77–89.
2. Alfred North Whitehead, *The Interpretation of Science* (Indianapolis: Bobbs-Merrill Co., 1961), 187–203. For the record, we explicitly avoid drawing on Code's seminal work, *Order and Organism* (Albany, NY: SUNY Press, 1985), despite its at least superficial relevance to our project here. However, Code himself has expressed serious reservations about his work to one of us (Herstein) in conversation.
3. Whitehead, *The Principle of Relativity*, 4.
4. Alfred North Whitehead, *Modes of Thought* (New York: The Free Press, 1968 [1938]), 174.
5. Dewey, *Democracy and Education*, (New York: Palgrave Macmillan, [1916]), 328.
6. Whitehead, *Aims of Education*, 115.
7. Ibid., 53.
8. See Alfred North Whitehead, "Indication, Classes, Numbers, Validation," in *Essays in Science and Philosophy* (New York: Philosophical Library, 1948 [1937]), 240. He knew *Principia Mathematica* failed in its central aim.
9. See Whitehead, *Aims of Education*, 80.
10. Ibid., 77.
11. Ibid., 78.
12. Ibid., 79. Particular attention should be paid to Whitehead's use of the word "logical" here. We will say more on this momentarily.
13. Ibid., 80 (our emphasis).
14. See ibid., 81.
15. Ibid., 79.
16. Ibid., 84.
17. See, for example, Morris Kline, *Mathematical Thought from Ancient to Modern Times*, 3 vols (Oxford: Oxford University Press, 1990); and Edna Kramer, *The Nature and Growth of Modern Mathematics* (Princeton: Princeton University Press, 1983).
18. See Whitehead, *Aims of Education*, 80.
19. Alfred North Whitehead, *An Introduction to Mathematics* (Oxford University Press, Oxford, 1958 [1911]), 48.
20. Whitehead, *Aims of Education*, 84.
21. Examples abound, but misinformation about climate science, healthcare cost and effectiveness, economic and educational studies, and many others, undermine the likelihood that policy decision will be made intelligently.
22. See Whitehead, *An Introduction to Mathematics*, 183 ff.
23. In mathematics, the terms "measure" and "measurement" have very distinct and precise meanings. For the purposes of this book, however, we will be using

the terms interchangeably, corresponding to what mathematicians mean by measurement.
24 See Gary L. Herstein, *Whitehead and the Measurement Problem of Cosmology* (Frankfurt am Main: Ontos Verlag, 2006); and Whitehead, *Principle of Relativity*, ch. 3.
25 David H. Krantz, R. Duncan Luce, Patrick Suppes, and Amos Tversky, *Foundations of Measurement*, 3 vols. (Minneola, NY: Dover Publications, 2007 [1971]).
26 Whitehead, *Aims of Education*, 84.
27 For further details, see Herstein, "Toward a New Metaphysics of Identity," section 2, 39–58.
28 Whitehead, *The Interpretation of Science*, 187.
29 Ibid., 194, our emphasis.
30 James Bradley nicely summarizes this advantage of favoring the algebraic view. He says:

> . . . since the time of [J.H.] Lambert and Kant (cf. Cassirer *Substance and Function*), the concept of the function offers a model of analysis that avoids the abstractness of the concept of genus, which, considered in itself, neglects all specific difference. Here, for Whitehead, the particular significance of the algebraic-logical function resides in the notion of 'any' or the variable (see *Essays in Science and Philosophy*, 104; cf. *Modes of Thought*, 106–107). . . . the function is able to 'express the concurrence of mathematical-formal principles with accidental factors,' which Whitehead also describes as 'the suffusion of the connective by the things connected' (*Essays in Science and Philosophy*, 128). More explicitly: the function represents the rule of connection which, in virtue of the different values the variable can take, contains within itself all the particulars for which it holds. This is clearly intended to be the character of the categoreal scheme.
>
> (PR 20)

This is from Bradley, "Transcendentalism and Speculative Realism," *Process Studies*, 23:3–4 (Fall–Winter 1994), 166–167.
31 Whitehead, *The Interpretation of Science*, 194.
32 Ibid., 196.
33 Ibid., 197.
34 Ibid., 196.
35 Ibid.
36 Ibid., 197.
37 Ibid.
38 Whitehead, *Aims of Education*, 85 (our emphasis).
39 Ibid., 86.
40 Ibid. This may not sound like the trigonometry you learned in school. On the other hand, it illustrates the problem with mathematical pedagogy pointed out above. Your teacher also did not tell you that a square route was a formula for making any number into a geometrical square, or that a line is a moving point, while a plane is a moving line. You were not told that concepts of "shortness" and least distance change whenever the moving point is on a curved surface. In sum, mathematical pedagogy is carried out by people who typically haven't been trained to think about mathematics in these deeply relational and algebraic terms.
41 Whitehead, *Enquiry into the Principles of Natural Knowledge*, 197.
42 Ibid., 198.
43 John Dewey, *Logic the Theory of Inquiry, the Later Works of John Dewey*, ed. Jo Ann Boydston, vol. 12 (Carbondale, IL: Southern Illinois University Press, 1986 [1938]).
44 See Jaakko Hintikka, "Is Logic the Key to Good Reasoning?" *Argumentation*, 15 (2001), 35–57.

45 See Michael Friedman, *The Parting of the Ways* (LaSalle, IL: Open Court, 2004). See also Nicholas Capaldi, *The Enlightenment Project in the Analytic Conversation*, for histories of the analytical movement that do not presuppose its immediate superiority.
46 On the importance of formal logic, see Whitehead, *Modes of Thought*, 106–107.
47 See Hintikka, "Is Logic the Key to Good Reasoning?".
48 The classic introductory work in model theory remains C. C. Chang and H. Jerome Keisler, *Model Theory*, 3rd ed. (Amsterdam: North-Holland Press, Amsterdam, 1990); also (New York: Dover Publications, 2012). The application of constructive game methods to model theory is treated in detail in Wilfrid Hodges, *Building Models by Games* (Cambridge: Cambridge University Press, 1985). A more basic introduction that emphasizes the game technique is Kees Doets, *Basic Model Theory* (Stanford, CA: CSLI Publications, 1996). For the connections between model theory and modal logic, which includes an overview of Kripke frames, see Valentin Goranko and Martin Otto, *Model Theory of Modal Logic*, http://www.mathematik.tu-darmstadt.de/~otto/papers/mlhb.pdf, accessed February 3, 2014.
49 See, for example, Wilfrid Hodges, *Building Models by Games* (New York: Dover Publications, 2006).
50 Jaakko Hintikka, "Is Logic the Key to All Good Reasoning," *Argumentation*, 15 (2001), 45.
51 Sara Negri and Jan von Plato, *Structural Proof Theory* (Cambridge: Cambridge University Press, 2001). In particular, the section entitled "Structural proof theory" lays out the context and purpose the approach, Kindle edition, locations 60–106.
52 An early and comprehensive survey of the prospects was provided by Henry in *Forms of Concrescence*. Obviously computing has developed since that time, and Whitehead's relevance to it has increased. Henry makes many of the points about Russell and Whitehead we have made (see esp. 121–122).
53 Paul Bogaard makes the point that Lewis Ford, "enthused by [Ivor] Leclerc's reliance upon the Aristotelian distinction between actual and potential" then advocated treating molecules as actual occasions "of a larger scope." But as Bogaard rightly observes, "it remains a tug-of-war between the desire to preserve the actuality of the constituents and to acknowledge the actuality of the complex whole they comprise. Ford's suggestion that we simply reapply the full Whiteheadian analysis at the level of the molecule . . . brings all the sophistication of Whitehead's treatment to bear on these opposing requirements, but at the cost of locating its full significance at the molecular level." This is just the sort of forced dichotomy—that the "full significance" of Whitehead's treatment must be found at one particular level of the physical world—we argue against in this book. Bogaard continues, saying that "not even Ford, I trust, wants to apply the 'actual occasion' analysis at every level." We *do* want to apply that analysis at every level, at least insofar as explanation is what we aim at. We shall expend great effort showing how possibility, actuality, and potentiality operate in Whitehead's philosophy (see especially chapter 8, 9, and 11). See Bogaard, "Whitehead and the Survival of 'Subordinate Societies'," *Process Studies* 21:4 (Winter 1992), 220–221.
54 See for example, Anthony G. Cohn and Achille C. Varzi, "Mereotopological Connection," *Journal of Philosophical Logic*, 32 (2003), 357–390; Stephen Blake, *A. N. Whitehead's Geometric Algebra*, http://www.stebla.pwp.blueyonder.co.uk/Whitehead.html, 2005, accessed February 3, 2014; Ian Pratt and Dominik Schoop, "Expressivity in Polygonal, Plane Mereotopology," *Journal of Symbolic Logic*, 65:2 (June 2000), 822–838; D. A. Randell and A. G. Cohn, "A Spatial Logic Based on Regions and Connections," in *Principles of Knowledge Representation*

and Reasoning, eds. B. Nebel, C. Rich and W. Swartout (Proceedings of the Third International Conference [Los Altos, CA: Morgan Kaufmann, 1992]), 165–176.
55 Herstein discusses the relation between Whiteheadian narrative and *Finnegans Wake* at greater length in his *Internet Encyclopedia of Philosophy* article on Whitehead: http://www.iep.utm.edu/w/whitehed.htm, accessed January 18, 2017.
56 Paul Halmos, *Algebraic Logic* (New York: Chelsea Publishing Co., 1962), 17 (our emphasis).
57 Interest in algebraic logic evaporated with the work of Frege, Dedekind, and Peano, and did not come back into favor until the work of Tarski in the late 1930's and after. Even then, Halmos' book was still rather apologetic in his defense of the algebraic approach in 1962.
58 Note the basic definition of algebra only emphasizes structure-preserving transformations. A "group" is a more specific type of transformation since it can be undone or reversed. See Hermann Weyl, *Symmetry* (Princeton: Princeton University Press, 1952).
59 See Herstein, *Whitehead and the Measurement Problem of Cosmology*, especially ch. 3 and ch. 6, sec. 3.
60 As noted, one of the best routes of entrance into any area of mathematical study is through the history of that subject. This is certainly true of abstract algebra, and its contemporary expression in category theory. The best such history to date is Leo Corry, *Modern Algebra and the Rise of Mathematical Structures* (Basel: Birkhauser Verlag, 2004). The least technical introduction to category theory currently available is the excellent volume by F. William Lawvere and Stephen Hoel Schanuel, *Conceptual Mathematics: A First Introduction to Categories* (Cambridge: Cambridge University Press, 1997). The classic mathematical treatise on the subject of category theory is Saunders MacLane, *Categories for the Working Mathematician* (New York: Springer Verlag, 1971). Notable for its explicit developments of the connections among category theory, logic, and generalized spatial reasoning is Saunders MacLane and Leke Moerdijk, *Sheaves in Geometry and Logic: A First Introduction to Topos Theory* (New York: Springer-Verlag, 1992). Also worth mentioning here are Michael Barr and Charles Wells, *Toposes, Triples and Theories* (New York: Springer-Verlag, 1985); and Andrea Asperti and Giuseppe Longo, *Categories, Types and Structures* (Cambridge, MA: MIT Press, 1991).
61 For a basic introduction, see John L. Kelley, *General Topology*, 2nd ed. (New York: Springer-Verlag, 1985). In addition, Steven Vickers has demonstrated the substantive relationship between logic and topology; see his *Topology via Logic* (Cambridge: Cambridge University Press, 1989).
62 See Leo Corry, *Modern Algebra and the Rise of Mathematical Structures*, 2nd ed. (Basel: Birkhauser Verlag, 2004), esp. Introduction.
63 See Robert Goldblatt, *Topoi: The Categorial Analysis of Logic* (New York: Dover Publications, 1984); and J. L. Lambek and P. J. Scott, *Introduction to Higher Order Categorical Logic* (Cambridge: Cambridge University Press, 1986); and J. L. Bell, *Toposes and Local Set Theories* (New York: Dover Publications, 2008).
64 The name deliberately invoking the connection to topology.
65 See Goldblatt, *Topoi*, xi.
66 We want to make clear that we are not endorsing (indeed, are not even interested in) Russell's metaphysical theory of types, and we endorse Whitehead's decision to dump that part. If we retain the term "type," we hope it will not mislead anyone. See Whitehead's use of the term in "Indication, Classes, Number and Validation" (1937), in *Essays in Science and Philosophy*, esp. 229.
67 See J. L. Bell and A. B. Slomson, *Models and Ultraproducts* (New York: Dover Publications, 2006), 72.

68 Michael Dummett, *The Logical Basis of Metaphysics* (Cambridge, MA: Harvard University Press, 1991).
69 Quite aside from the fact that he devoted an entire major division of *Process and Reality* to each, one can see this in such things as the interplay between atomicity/actuality and continuity/potentiality (for example *Process and Reality* 61, 62, 67, 69, 71), and internal vs. external forms of relatedness (287–288).
70 See Whitehead, *Modes of Thought*, 174: "Poetry allies itself to metre, philosophy to mathematic pattern."
71 See for example, Whitehead, *Process and Reality*, 289. We dedicate this paragraph to Don Sebastian and Don Pedro.
72 See Whitehead, *Process and Reality*, 287–288.
73 Whitehead's "solidarity thesis" at *Process and Reality*, 40, is here intended. The full importance of conceiving the whole as "the universe as a solidarity" is nicely brought out by Nobo in *Whitehead's Metaphysics of Extension and Solidarity*.
74 See Whitehead, *Interpretation of Science*, 192–193.
75 See John von Neumann, *Continuous Geometry* (Princeton: Princeton University Press, 1998 [1936]).
76 As evidenced in Whitehead's UA, and his *The Axioms of Projective Geometry* (Cambridge: Cambridge University Press, 1906 and 1907), and his entry "Mathematics," in the *Encyclopedia Britannica*, 11th ed., vol. 17 (Cambridge: Cambridge University Press, 1911), 878–883.
77 See Herstein, *Whitehead and the Measurement Problem of Cosmology*, ch. 4, esp. pp. 130–138.
78 Although the logician and computer scientist Giangiacomo Gerla has commented on these analogies in a variety of places. See Giangiacomo Gerla, "Pointless Geometries," *Handbook of Incidence Geometry*, ed. F. Buekenhout (Amsterdam: Elsevier Science, 1995), ch. 18, 1015–1031.

Chapter 4

1 See Whitehead, *Process and Reality*, 294.
2 The reasons for the scare quotes will become evident soon enough. See R. Kirby Godsey, "Relation and Substance in Whitehead's Metaphysics," *Tulane Studies in Philosophy*, Studies in Process Philosophy II, 24 (New Orleans: Tulane University, 1975), 12–22. If extension plays a lesser role in *Process and Reality* than in the Triptych, as Godsey claims, why, then, does Part IV exist? In addition, what Godsey asserts as Whitehead's repudiation of extension as a fundamental concept in *Process and Reality*, (Godsey, 21) is, in fact, only Whitehead's shift from a primary logic of mereology to one of mereotopology.
3 See, for example, Whitehead, *Enquiry into the Principles of Natural Knowledge*, 167, 189.
4 See Whitehead, *The Concept of Nature*, 162.
5 Ibid., 5.
6 See chapter 1 above, and also, for example, Whitehead's care regarding the context sensitivity of the ideas of equivalence, addition and subtraction in his *Universal Algebra*, 5, 18–19, 24.
7 See Whitehead, *The Principle of Relativity*, 3–4.
8 Whitehead, *Process and Reality*, 35–36.
9 Ibid., 36, our emphasis. As we shall see shortly, the corpuscular form is a kind of personal order.
10 See chapter 1 for the relation between the theory of cosmic epochs and the other levels of generality in *Process and Reality*. See also chapters 8 and 9.
11 See part one of chapter 3 above. Whitehead's mathematical usages, for example, were often maddeningly archaic.

12 See Ronny Desmet, "Whitehead's Cambridge Training," in *Whitehead: The Algebra of Metaphysics*, eds. Ronny Desmet and Michel Weber (Leuven de Neuve, BE: Les editions Chromatika, 2010), 91–125. Desmet traces the careers and predilections of Whitehead's mathematical mentors at Cambridge, which sheds great light on his habits of thinking and expression.
13 See the biography of Clerk Maxwell, for example *The Man Who Changed Everything* by Basil Mahon (New York: John Wiley and Sons, 2003). One does not need the Copenhagen School to account for Whitehead's reference to the problem of the quantum.
14 See Gerald Holton, *Thematic Origins of Scientific Thought: Kepler to Einstein* (Cambridge, MA: Harvard University Press, 1980), 268–269. Also see Herstein (2006), ch. 2.
15 The extent to which Whitehead was influenced by contemporary quantum theory is brought into doubt by recent histories of physics detailing the fifth Solvay Conference, in 1927, where homey mythology has it that the contemporary quantum interpretation was "settled." See Guido Bacciagaluppi and Antony Valenti, *Quantum Theory at the Crossroads, reconsidering the 1927 Solvay Conference* (Cambridge: Cambridge University Press, 2009). There was still considerable confusion about how one ought to approach microphysical quanta, and there was no clear consensus about what the principal data even meant. Experimental results were coming in as the conference was being planned, thus adding speakers to address these new results. At the time Whitehead was forming the lectures that became *Process and Reality*. The conference proceedings were not published until 1928, and then in French, a language in which Whitehead was never very comfortable. The *imposition* of quantum mechanics upon *Process and Reality* is a case of scholars reading backwards upon the text their own inherited assumptions.
16 The idea here of the "quantum of explanation" bears some analogy to Epperson's ideas about "logical causality." See Michael Epperson, "Quantum Mechanics and Relational Realism: Logical Causality and Wave Function Collapse," *Process Studies*, 38:2 (Fall–Winter 2009), 340–367, esp. 340–341. Epperson's realism remains quite orthodox, even in acknowledging relations as genuinely real, versus the decidedly *radical* realism that we will be advocating here. In addition, Epperson at times seems to ontologize the logical aspect rather more than we are prepared to do with the quantum (see Epperson, 357 ff.).
17 Entanglement did not come on the scene until 1935; see A. Einstein, B. Podolsky, and N. Rosen, "Can Quantum-Mechanical Description of Physical Reality Be Considered Complete?" *Physical Review*, 47:10 (1935), 777–780. This paper was intended to refute quantum mechanics. It failed.
18 Also see Whitehead's discussion of atomism (Lucretius) in *Adventures of Ideas*, 123 ff.
19 See Whitehead, *Process and Reality*, 8. Whitehead points out that all schemes of philosophical categories are "false" by logical standards.
20 See Auxier, "Dream Time," in *Inception and Philosophy*, ed. Theo Botz-Bornstein (Chicago: Open Court, 2011), 279–300.
21 Again see Whitehead, *Process and Reality*, 8.
22 Ibid., 4, our emphasis.
23 Ibid., 13.
24 Ibid., xii, our emphasis.
25 See ibid., 35. For an illuminating discussion of Whiteheadian "societies," see Joseph Bracken, "Energy Events and Fields," *Process Studies*, 18:3 (Fall 1989), 153–165. Bracken defends the right sort of becoming of continuity, rejecting both Hartshorne's view of actual occasions as micro-entities forming compound individuals (explicitly), and Ford's temporal atomism (implicitly). Bracken is

also critical of Sherburne and John Cobb in ways we would endorse (see 159), and also George Wolf (160–161). Ivor Leclerc is similarly misguided in believing that the question of "compound individuals" is a serious problem for process philosophy (see *The Nature of Physical Existence*). But Bracken is still wedded to the idea that an actual occasion has to be some sort of entity that natural science (especially physics) can describe (in this case, he favors events within fields). We do not deny that Whitehead's cosmology *includes* this sort of field-event relation, under a coordinate analysis, and we do not deny that "societies" can be thus characterized, but the physicalizing of Whitehead's cosmology is a misleading way to present it. Along with Whitehead, we have restricted the use of the term "actual occasion" to the level of generality associated with the theory of transition and concrescence, reserving the broader term "actual entity" (i.e., the quantum of explanation) for the *most* general use, applying equally at all levels of generality. Bracken's motives are (as he states) not the same as Whitehead's. Saving a place for the Trinity within process philosophy is one of those motives alien to Whitehead's thought.
26 Whitehead, *Process and Reality*, 67. For the meaning of "perspective standpoint," see ch. 7.
27 Ibid., 66.
28 See ibid., 283 ff., and our chapter 8 for a more detailed treatment.
29 Ibid., 66; this is just one example.
30 Ibid., 67.
31 The use of the word "nature" here is deliberate both for its irony and greater accuracy than, say, the term "character."
32 Joseph Bracken illustrates both errors—the conflation of Whitehead's naturalism with physicalism, and the conflation of Whitehead's atomism with microphysics—in an article, "Proposals for Overcoming the Atomism within Process-Relational Metaphysics," *Process Studies*, 23:1 (Spring 1994), 10–24. He says: "Alfred North Whitehead aligns himself with quantum physicists in the belief that physical reality is made up of discrete units (for physicists, packets of energy; for Whitehead, momentary subjects of experience) which have to be combined both spatially and temporally in order to produce the continuously existing 'things' of common sense experience." (10) It seems fairly clear that Bracken is not using the term "moment" in Whitehead's sense, as a purely ideal product of extensive abstraction; so it seems safe to assume he means a tiny temporal slice of "nature." Such a position takes for granted the very logic of measurement that is a formally emergent property of nature and time, and *not* something to be found within the metaphysics itself.
33 Whitehead, *Process and Reality*, 128–129.
34 Ibid., 214. We will have more to say about the relationship between the merely real and the determinate when we discuss the algebra of negative prehension, later. For now it should be sufficient to note that the macroscopic order is determinate and belongs to the theory of transition and concrescence in terms of its level of generality, when that level is expressed in the theory of extension.
35 Ibid. This is a restatement of the distinction between definiteness and determinateness in *Process and Reality* 25, (the xxth Category of Explanation).
36 See Holton, *Thematic Origins of Scientific Thought*, 302; and Herstein, *Whitehead and the Measurement Problem of Cosmology*, 68 ff.
37 Whitehead, *Process and Reality*, 67. We have cited this passage before, but perhaps now the scope of its significance is becoming clearer.
38 Ibid., 283. Kant's use of coordinate analysis is similar. See *Critique of Pure Reason*, B112.
39 Whitehead, *Process and Reality*, 289; for a full account of the distinction between division and divisibility, both genetic and coordinate, see chapter 7.

40 Notably Sherburne or George Allan, but recently one might mention Michael Heller's chapter on Whitehead in *Philosophy in Science: An Historical Introduction* (Heidelberg: Springer-Verlag, 2011), ch. 10, 101–112.
41 Whitehead, *Process and Reality*, 20.
42 Ibid., 22 (Category of Explanation iv).
43 Ibid., 22.
44 Ibid., 23 (Category of Explanation vi).
45 Ibid., 23, original emphasis.
46 An author who goes astray on this point Duane Voskuil, "Entanglement, Slits, and Buckyballs," *Process Studies*, 36:1 (Spring-Summer 2007), 23–44. Voskuil assumes that Whitehead's quantum is microphysical.
47 Epperson and Zafiris have recently published a significant contribution to the philosophy of physics that goes well beyond conventional interpretations while avoiding the pitfalls that have dogged so many previous attempts to bring Whiteheadian ideas to bear upon microphysics. See Epperson and Zafiris, *Foundations of Relational Realism*. This book is important on its own account for its contribution to the philosophy of physics. But for our purposes it is especially significant for the coherently developed algebraic approach in which this particular application of Whiteheadian thought is brought to bear. The authors explicitly employ category theory in the development of their ideas.
48 Whitehead, *Process and Reality*, 36.
49 This way of using the term "personal" is idiosyncratic, and different from the way the term is used in discussing the "societies of actual entities." See *Process and Reality*, 34 as compared to 89–90, and both of these are different from his discussion of "person" in *Religion in the Making*, 62–68. Our concern here is not to unify all of these uses but merely to alert readers.
50 Whitehead, *Process and Reality*, 36.
51 Ibid., 92.
52 Ibid., our emphasis.
53 Ibid., 115–117.
54 See ibid., 99. See also our earlier extended note on Whitehead's theory of propositions.
55 For example, Daniel Athearn: "Toward an Ontological Explanation of Light," *Process Studies*, 34:1 (Spring–Summer 2005), 45–61, aims at reviving the idea of natural philosophy and natural language explanation. Athearn emphasizes the holistic character of the physical world, and the problematic nature of certain naively "physical" terms. (For example, see Athearn's discussion of "spin" on pages 49–50). Athearn does not indicate anything quite as sweeping as we advocate here with our idea of the quantum of explanation, but his essay is, in our opinion, in sympathy with the kind of position we defend.
56 See for example, *Process and Reality*, 51–52, 219–220, 257–258.
57 The exception here is God.
58 Such creative advances *have* occurred, but it has taken a very long time for them to accumulate, and they added nothing in the way of structure to the CMBR. We are referring here to the statistical "noise" which now permeates the CMBR. We will have a little more to say about the nature of statistical information in chapter 6.

Chapter 5

1 Jorge Nobo is an obvious exception to this criticism. See Nobo, *Whitehead's Metaphysics of Extension and Solidarity*.
2 Lewis Ford, *The Emergence of Whitehead's Metaphysics: 1925–1929* (Albany, NY: SUNY Press, 1984), 1.

318   *Notes*

3   The following is a complete list of Whitehead's references to the triptych in *Process and Reality*.
    Concept of Nature References:

    Pg. 128, footnote: "In *The Concept of Nature* these two loci were not discriminated, namely, durations and strain-loci."
    Pg. 243, footnote: "Cf. *The Concept of Nature*, Ch. I."
    Pg. 258, footnote: "Cf. my *Concept of Nature*, Ch. I, for another exposition of this train of thought."
    Pg. 298: "The connection of this special sense of equivalence' to physical properties is explained more particularly in Chapter IV of the *Concept of Nature*".

    _____

    Combined *Concept of Nature* & *Enquiry into the Principles of Natural Knowledge* References:

    Pg. 125, footnote: "Cf. my *Principles of Natural Knowledge*, Ch. XI, and my *Concept of Nature*, Ch. V."
    Pg. 287, footnote: "Cf. *The Principles of Natural Knowledge*, 1919, and *The Concept of Nature*, 1920, Cambridge University Press, England."
    Pg. 297, footnote: "Cf. my *Principles of Natural Knowledge*, and *Concept of Nature*."

    _____

    Enquiry into the Principles of Natural Knowledge References:

    p. 298: "This definition practically limits abstractive sets to those sets which were termed 'simple abstractive sets' in my *Principles of Natural Knowledge* (paragraph 37.6)." [We observe that this comment of Whitehead's appears in the main text and ix a very minor criticism of his earlier theory.

    _____

    Principle of Relativity References:

    p. 20, footnote: "In this connection I may refer to the second chapter of my book *The Principle of Relativity*, Cambridge University Press, 1922."
    p. 333, footnote: "Cf. my book, *The Principle of Relativity*, University Press, Cambridge, 1922."

    In *Adventures of Ideas*, p. 254n., Whitehead references both *The Concept of Nature* and *Process and Reality*, jointly, as support for the claim that "a particular actuality can be abstracted from the mode of its initial indication, so that in a later phase of experience it is entertained as a bare '*It*' " (254). He then indicates that both Parts II and III of *Process and Reality* are to be consulted along with *The Concept of Nature* on this point. If Whitehead had indeed "discovered" temporal atomism after 1925, this reference would be extra-ordinary, implying, as it does, the same view of particulars and abstraction in both prior works. Furthermore, the citations in *Process and Reality* indicated above would be simply inexplicable if Ford's hypothesis was correct.
4   See Gary Herstein, "Alfred North Whitehead," in *The Internet Encyclopedia of Philosophy*, http://www.iep.utm.edu/w/whitehed.htm, accessed January 18, 2017. James Fieser Ph.D., founder and general editor; Bradley Dowden, Ph.D., general editor; David Boursema, American Philosophy editor.
5   Whitehead, *Enquiry into the Principles of Natural Knowledge*, 202.
6   Ford, *The Emergence of Whitehead's Metaphysics*, 24–25, 246.
7   Ibid., 25.
8   Whitehead, *Science and the Modern World*, 126.
9   Whitehead, *The Principle of Relativity*, 23.
10 See Whitehead, *Enquiry into the Principles of Natural Knowledge*, 61 ff.

11 Ibid., 61.
12 Whitehead, *Process and Reality*, 289. We will have much more to say about this in chapter 8.
13 Even a mathematically sophisticated interpreter of Whitehead like Murray Code can be misled by the Fordian argument. In addition to misconstruing the nature of eternal objects (discussed in great detail in chapters 8 and 9), Code misses the algebraic modes of Whitehead's thought as applied to narrative forms of explanation, which Code correctly identifies as essential characteristics of Whitehead's thought. See Murray Code, "On Whitehead's Almost Comprehensive Naturalism," *Process Studies*, 31:1 (Spring–Summer 2002), 3–31, esp. p. 25 (where Code explicitly endorses Ford). Code's 1985 book *Order and Organism*, was, evidently, not influenced by Ford's claims about Whitehead's development (since Ford is not mentioned in the book anywhere), but it seems that Code's revisions of his earlier book may have been motivated by Code's willingness to believe Ford's theories.
14 Whitehead, *Process and Reality*, 288 (our emphasis).
15 Problematic interpretations of Whitehead's metaphysics are, or course, neither recent nor simply due to Lewis Ford's aggressive reconstructions of Whitehead's philosophy. In this same issue, Patrick Madigan conflates space with "the extensive continuum," which is a fairly common mistake (Patrick S. Madigan: "Space in Leibniz and Whitehead," in *Tulane Studies in Philosophy*, 24, ed. Robert C. Whittemore [New Orleans: Tulane University, 1975], 48–57.)
16 Robert M. Palter, "Preface," *Whitehead's Philosophy of Science*, 2nd ed. (Chicago: University of Chicago Press, 1970).
17 See Bowman L. Clarke, "A Calculus of Individuals Based on 'Connection'," *Notre Dame Journal of Formal Logic*, 22:3 (1981), 204–218; and "Individuals and Points," *Notre Dame Journal of Formal Logic* 26:1 (1985), 61–75.
18 See for example: Ian Pratt and Oliver Lemon, "Ontologies for Plane, Polygonal Mereotopology," *Notre Dame Journal of Formal Logic* 38:2 (1997), 225–245; Ian Pratt, and Dominik Schoop, "A Complete Axiom System for Polygonal Mereotopology of the Real Plane," *Journal of Philosophical Logic* 27 (1998), 621–658; and Pratt and Schoop, "Expressivity in Polygonal, Plane Mereotopology," *Journal of Symbolic Logic* 65:2 (2000), 822–838; D. A. Randell and A. G. Cohn, 1992. "A Spatial Logic Based on Regions and Connections," *Principles of Knowledge Representation and Reasoning*, eds. B. Nebel, C. Rich and W. Swartout (Proceedings of the Third International Conference, Los Altos, CA: Morgan Kaufmann, 1992), 165–176; P. Roeper, "Region-Based Topology," *Journal of Philosophical Logic* 26 (1997), 251–309; and Achille C. Varzi, "Boundaries, Continuity and Contact," *Nous* 31:1 (1997), 25–58. This list is representative only; there are many other references to Whitehead's theory of extension as completed by Clarke's articles.
19 Whitehead says "logic presupposes metaphysics." See *Modes of Thought*, 107.
20 In his earlier publications, Ian Pratt-Hartmann simply went by "Ian Pratt." See above.
21 Alfred Tarski, "Foundations of the Geometry of Solids," in *Logic, Semantics, Metamathematics*, trans. J. H. Woodger, 2nd ed. (Indianapolis: Hackett, 1983 [1929]), 24–29.
22 Theodore de Laguna, "Point, Line, and Surface, as Sets of Solids," *The Journal of Philosophy* 19:17 (1922), 449–461.
23 Mary Tiles, *The Philosophy of Set Theory* (New York: Dover Publications, 1989). See especially ch. 2.
24 See Peter Simons, *Parts: A Study in Ontology* (Oxford: Clarendon Press, 1987), esp. ch. 1.
25 Max Jammer, *Concepts of Space*, 3rd ed. (New York: Dover Publications, 1993 [1954]), xiii.

320 Notes

26 Paul R. Halmos, *Algebraic Logic* (New York: Chelsea Publishing Co., 1962), 17 (our emphasis).
27 Whitehead, "Axioms of Projective Geometry," 4.
28 See Leo Cory, *Modern Algebra*, 380–398. Also, specifically, issues regarding homology and Čech cohomology, see Saunders Maclane, *Categories for the Working Mathematician* (New York: Springer Verlag, 1971), 29.
29 J. Lambek and P. J. Scott, *Introduction to Higher Order Categorical Logic* (Cambridge: Cambridge University Press, 1986).
30 Giangiacomo Gerla, "Pointless Geometries," *Handbook of Incidence Geometry*, ed. F. Buekenhost (Amsterdam: Elsevier Science, 1995), ch. 18, 1015–1031.
31 See Patrick Blackburn, Maarten de Rijke, and Yda Venema, *Modal Logic* (Cambridge: Cambridge University Press, 2001), 1.
32 Ibid., 28.
33 Whitehead, *Process and Reality*, 333, n. 2.
34 Whitehead had developed a similar account of equality twenty years before in the *Universal Algebra*.
35 G. Gerla, "Mathematical Features of Whitehead's Pointfree Geometry," in *Handbook of Whiteheadian Process Thought*, eds. Michel Weber, Will Desmond,(Frankfurt am Main: Ontos Verlag, 2008), 507–519, 533–536.
36 See Whitehead, *Process and Reality*, 287.
37 Negri and Plato, *Structural Proof Theory*, Kindle location 697–705 (end of section 2.3).
38 Hintikka: "Is Logic the Key to All Good Reasoning?" 35–57. (In addition, Hintikka emphasizes a model- over a proof-theoretic approach.)

Chapter 6

1 We will sometimes shorten this phrase to just the "standard model." It will be obvious from the context that we are referring to gravitational cosmology and not to the standard model of micro-physics. This standard model is often referred to as "Big Bang Cosmology." We will cite numerous instances in what follows.
2 One couldn't find a better or more thorough critique of this decision by Einstein than Jimena Canales' recent book *The Philosopher and the Physicist: The Debate That Changed Our Understanding of Time* (Princeton: Princeton University Press, 2014), in which she details how Einstein came to the very unscientific decision to claim his work replaced philosophy.
3 Ricouer, *Interpretation Theory: Discourse and the Surplus of Meaning* (Ft. Worth: Texas Christian University Press, 1976), 67.
4 Ibid., 67–68.
5 In addition to Canales' excellent history cited above, see Hans Ohanion's, *Einstein's Mistakes: The Human Failings of Genius* (New York: W.W. Norton, 2009). Many of the great physicists of Einstein's day, including Poincare and Michelson (and philosophers Cassirer and Bergson) insisted that General Relativity was an interpretation of the mathematical discoveries and that other interpretations were not only possible, but nothing in GR requires us to interpret the world in its terms. It is mathematics, not physics. See Timothy Eastman, who advocates a position he characterizes as "cosmic agnosticism," and tentatively concludes, "that there is no current model in physical cosmology that adequately meets all key observations." (Timothy Eastman, "Cosmic Agnosticism," *Process Studies,* 36:2 [Fall–Winter 2007], 181–197.) Michael Disney notes that, "to explain some surprising observations, theoreticians have had to create heroic and yet insubstantial notions such as "dark matter" and "dark energy," which supposedly overwhelm, by a hundred to one, the stuff of the universe we can directly detect. Outsiders

are bound to ask whether they should be more impressed by the new observations or more dismayed by the theoretical djinnis that have been conjured up to account for them. . . . What one finds, in my view, is that modern cosmology has at best very flimsy observational support." ("Modern Cosmology: Science or Folktale?" *American Scientist* 95:5 [September–October 2007], 383.) Elsewhere, Disney observes that the number of free parameters necessary to force cosmological theory (what he refers to as "BBC" or "Big Bang Cosmology") to fit observations substantially exceed the number of genuinely *independent* observations, rendering confidence in the theory significantly negative. (M.J. Disney, "Doubts about Big Bang Cosmology," in *Aspects of Today's Cosmology*, ed. Prof. Antonio Alfonso-Faus [2011], ISBN: 978-953-307-626-3, InTech, Available from: http://www.intechopen.com/books/aspects-of-today-s-cosmology/doubts-about-big-bang-cosmology, accessed January 18, 2017) Reginald Cahill has gone so far as to argue that fundamental experiments such as that of Michelson-Morley deserve to be re-evaluated. (Reginald T. Cahill and Kirsty Kitto, "Michelson-Morley Experiments Revisited and the Cosmic Background Radiation Preferred Frame," http://arxiv.org/pdf/physics/0205065v1.pdf) Cahill's views are certainly controversial, but such radical challenges to the reigning paradigm are being made by scientists working in the field. There are also studies conducted on a purely sociological level, investigating the intransigence of so many scientists who are caught up in the model-centrism of the standard model of gravitational cosmology. See, for example, Martin Lopez-Corredoira, "Sociology of Modern Cosmology," http://arxiv.org/abs/0812.0537v2. In short, just because many working physicists ignore the problems with GR doesn't make them go away and does not remove the problems associated with creating models.

6 There are primarily three different families of theories challenging the model of gravity developed in general relativity. These go under the headings of "Mond" (for "Modified Newtonian Dynamic"), the "bimetric" approach (which we'll say more about), and a group widely referred to as "TeVeS" ("Tensor Vector Scalar.") The technical details of how these approaches work need not concern us here.

7 See Herstein, *Whitehead and the Measurement Problem of Cosmology*, 2006.

8 Bogdan Rosu and Ronny Desmet incorrectly situate Whitehead's revision of Einstein's theory of relativity when they characterize the triptych by saying, "Whitehead attempts to recover Minkowski's space-time geometry from hypothetical entities, that are themselves the result of abstraction and generalization, from what we are aware of in perception." See Rosu and Desmet: "Whitehead, Russell and Moore: Three Analytic Philosophers," *Process Studies*, 41:2 (Fall–Winter 2012), 214–234, 222. This statement places Minkowski's work near the center of Whitehead's concerns. Minkowski brought Einstein's special theory to the attention of the physics community by embedding that work in a 4-dimensional framework, providing an interpretive structure for special relativity. (See Gerald Holton: *Thematic Origins of Scientific Thought: Kepler to Einstein* (Cambridge: Harvard University Press, 1988.) Whitehead's own mathematical interests moved in a direction entirely contrary to the Minkowskian formulation (even without the geometric interpretation). Rather than developing his ideas along lines stemming from Gauss and Riemann, Whitehead's work was rooted in the approaches of Cayley and Klein which prioritized projective geometry and the choice of polarities, rather than directly modifying the postulates of geometry as is found in the former pair. (See Blake, *A.N. Whitehead's Geometric Algebra*, chapter 5 especially; http://www.stebla.pwp.blueyonder.co.uk/papers/Euclid.pdf verified February 3, 2014.)

9 Even this indeterminate situation for measuring, problematic as it is, barely scratches the surface of the concrete issues associated with specifying genetically

those coordinate relations that are required for even the simplest concrete alterations of scale and place that might be encountered in seeking full metaphysical exemplification of a categoreal scheme. We will address some other aspects of this challenge in chapter 8.
10 See Herstein, *Whitehead and the Measurement Problem of Cosmology*, 185, for a description of the kind of paradoxes brought on by the measurement problem, referencing the Galaxy NGC 7319 and its embedded quasar.
11 The apparent confirmation of GR in the observations of the pulsar PSR B1913+16 still depends upon the use of sidereal time to mark the periodicity of the pulses and chart observations. Using the earth's rotation as a measuring rod for phenomena projected at such distances either requires us to ignore the relativistic effects of the distances we estimate, or to reason in a circle with regard to the accuracy of our "measuring rod." If an observer objects that the relativistic effects have been taken into account, one need only ask where the estimate of relativistic effects originated in that path of reasoning.
12 For a fairly recent technical treatment of the subject, see, for example, Sean Carroll, *Spacetime and Geometry: An Introduction to General Relativity* (San Francisco: Pearson Education Inc./Addison Wesley, 2004). For a popular account, see Brian Greene, *The Fabric of the Cosmos* (New York: Alfred A. Knopf, 2004).
13 Whitehead, *The Principle of Relativity*, 81–86.
14 See Arthur S. Eddington, "A Comparison of Whitehead's and Einstein's Formulae," *Nature*, 113:2832 (February 9, 1924), 192.
15 Clifford Will, "Relativistic Gravity in the Solar System II: Anisotropy in the Newtonian Gravitational Constant," *The Astrophysical Journal*, 169 (1971), 141–155.
16 For example, Rosen and Moffat both emphasize the linearity of their respective bimetric theories. See J. W. Moffat, "Bimetric Gravity Theory, Varying Speed of Light and the Dimming of Super-Novae," *International Journal of Modern Physics D*, 12:2 (2003), 281–298; and Nathan Rosen, "General Relativity and Flat Space I & II," *Physical Review*, 57 (January 15, 1940), 147–153.
17 Whitehead, *The Principle of Relativity*, v.
18 For an accessible discussion of the connections between symmetry and groups, see Hermann Weyl, *Symmetry* (Princeton: Princeton University Press, 1980). Standard references on the subject of spaces of constant curvature are Sigurdur Helgason, *Differential Geometry and Symmetric Spaces*, 2nd edition (New York: Academic Press, 1964) and Joseph A. Wolf, *Spaces of Constant Curvature* (New York: McGraw-Hill, 1967). For a discussion of the role of (maximally) symmetric spaces in contemporary cosmology, see Carroll's *Spacetime and Geometry*, 139–144, 323–329). Maximally symmetric spaces are, in essence, the highest degree of uniformity possible in a space.

From the beginning of his mathematical career, Whitehead was intimately familiar with the contemporary work on the subject of uniformity and symmetric spaces. Whitehead's *Universal Algebra* has numerous citations of the principal figures and their publications on these developments. See Moritz Epple, "From Quaternions to Cosmology: Spaces of Constant Curvature, ca. 1873–1925," *Proceedings of the International Congress of Mathematicians*, 3 (August 20–28, 2002), 935–946, for an overview of the early history of the subject.
19 For the former, see Bas van Fraassen, *Laws and Symmetry* (Oxford: Oxford University Press, 1989). See Ernst Cassirer, "Reflections on the Concept of Group and the Theory of Perception"; the English version is found in *Symbol, Myth, and Culture* (New Haven: Yale University Press, 1979 [1945]), 217–291. Also see J. J. Gibson, *The Senses Considered as a Perceptual System* (Boston: Houghton Mifflin Company, 1966), highlighting the connection between groups and perception (Gibson 1966).

Notes 323

20 In addition to Herstein's *Whitehead and the Measurement Problem of Cosmology*, see Hilton Ratcliffe, Timothy E. Eastman, Ashwini Kumar Lal, and R. Joseph, *The Big Bang, a Critical Analysis* (Cambridge: Cosmology Science Publishers, 2011). See also Hilton Ratcliffe's *The Virtue of Heresy* (Central Milton Keynes: Author House, 2008). For a solid impression of the scientifically credible dissent to ΛCDM by surveying the article abstracts located in the newsletter of the Alternative Cosmology Group. These are to be found at http://www.cosmology.info/newsletter/index.html, accessed July 19, 2016.
21 Whitehead, *The Principle of Relativity*, 3.
22 Whitehead, *Process and Reality*, 22.
23 See John Dewey, *The Quest for Certainty*, in *Later Works of John Dewey*, ed. Jo Ann Boydston, vol. 4 (Carbondale, IL: Southern Illinois University Press, 1984 [1929]), 19.
24 Green, *The Fabric of the Cosmos*, 141.
25 See Eastman, "Cosmic Agnosticism," 181–197. Eastman, a working physicist who spent many years at NASA's Goddard Space Center, enumerates some of the more salient problems with the Standard Model of Cosmology.

Chapter 7

1 This is not to say that Whitehead dismissed proof-theoretic techniques as lacking value in all domains. See Whitehead, "Indication, Classes, Numbers, and Validation," 240.
2 We do not take up the question here of whether the mereotopological structure is or is not a "possible" space, in the sense of having some determinate relation to created spaces. We don't wish here to prejudice the ontological issue in advance. See our forthcoming *The Continuum of Possibility*.
3 See Hintikka's *The Principles of Mathematics Revisited* (Cambridge: Cambridge University Press, 1998).
4 Peter Douglas, in "Whitehead and the Problem of Compound Individuals," *Process Studies*, 33:1 (Spring–Summer 2004), 80–109, accuses Whitehead of a kind of reductionism with actual occasions: "Now even though Whitehead's 'reductionism' is quite different from that of physicalism, in both cases there is a failure to allow for the higher-level existents that could account for emergent causal powers," (Douglas, "Whitehead and the Problem of Compound Individuals," 81.) Douglas here has missed the point of the quantum of *explanation*: one must choose as a basis of explanation an actuality which can be analyzed, but which *cannot* be reduced, in order for any explanation to be possible, much less actual. He is correct to think that Whitehead is not trying to provide an account of emergence, but Whitehead allows the higher level existents to function as societies of actual occasions in an actual world, and these can be analyzed as if they were a single indivisible actual entity in a single actual world just as surely as simpler kinds of occasions can. (See *Adventures of Ideas*, ch. 13).
5 See Whitehead, *Process and Reality*, 214.
6 Ernst Cassirer, *The Philosophy of Symbolic Forms*, trans. Ralph Manheim, vol. 2 (New Haven: Yale University Press, 1955 [1925]), xv.
7 Obviously this is David Hilbert's characterization, but for the sense we mean to give it in a historical context, see Ernst Cassirer, *The Problem of Knowledge, Volume 4: Philosophy, Science, and History*, trans. Ralph Manheim, ed. Charles Hendel (New Haven: Yale University Press, 1950), 26.
8 The history of the term and its varied meanings is nicely explained in the opening chapters of James R. Goetsch, *Vico's Axioms: The Geometry of the Human World* (New Haven: Yale University Press, 1995).

9 In his brief discussion of the intensional logic of Frederic Castillon (written between 1797 and 1805), C.I. Lewis suspects Castillon of having worked out all of his transformations empirically and discarding those that did not work. Lewis takes this as a self-evident condemnation of Castillon's success in creating intensional functions that capture all of the valid syllogisms in Aristotelian logic. Granted, Castillon of worked out his system empirically, but Lewis is wrong to consider this a flaw in his method. We think Whitehead adjusts his coordinate whole to his genetic parts empirically in a similar way, and it is not a flaw. See C. I. Lewis, *A Survey of Symbolic Logic: The Classic Algebra of Logic* (New York: Dover Books, 1960 [1918]), 35–38.

10 Max Black says that there is a fundamental difference between scale, analogical, and theoretical models. We will be offering two versions of what Black would call a scale model in what follows (Fenway Park and Neil Young's trains), but we must set aside Black's distinctions. First, our aim includes metaphysics and is not limited to the epistemological interpretation of scientific models. Second, we have offered a completely different account of "explanation" as the quantum in any coordinate and genetic analysis, and this falls well outside the terms Black could accept, ours being broader and metaphysical, while not sacrificing either radically empirical science or the realism appropriate to it. Thus, our account of models, and of model-centrism, assumes that scale, analogical, and theoretical models are inter-related, and that a scale model is a concentration and selection of a theoretical model framed by finding a unit of measure. In turn, the unit of measure is not either a metaphor, or a model, in Black's sense. It is a quantum, in the sense explained in this book. See Black, *Models and Metaphors*: Studies in Language and Philosophy (Ithaca, NY: Cornell University press, 1962), chs. 3, 13; see also Paul Ricouer's use and critique of this view in *Interpretation Theory*, esp. pp. 65 ff.

11 In spite of the difficulties associated, Auxier actually has solved the problem with the Cubs empirically. See his article "The Curse (of the Curse) of Babe Ruth," http://radicallyempirical.com/2013/10/09/picktures-and-pieces-3-the-curse-of-the-curse-of-babe-ruth/, accessed January 18, 2017.

12 Empirical measurements have been taken to discover whether successful major leaguers had faster reflexes than average people. Major league hitters have very sharp reflexes and reaction times. But studies show that major leaguers have better than 20/20 vision. Even those who wear glasses are correctable to better than 20/20. The reason aging hitters begin to fail is not necessarily that they slow down, it is that they can't see as well as when they were younger. There are many sources, but see Brian MacPhereson's article in the *Providence Journal* from March 2012 here: http://www.providencejournal.com/sports/red-sox/content/20120316-baseball-vision-when-20-20-eyesight-just-wont-cut-it.ece, accessed February 23, 2014.

13 See Whitehead, *The Principle of Relativity*, ch. 3, esp. 42–48; and *Treatise on Universal Algebra*, 6.

14 Whitehead, *Religion in the Making*, 131.

15 The bases in Little League are sixty feet apart, two-thirds the size of a major league distance, but the pitcher's mound is further back and not normally elevated. Were the two-thirds ratio to hold, the pitcher would be 40 feet, 4 inches from the batter, but the rules require the pitcher to be 46 feet away. For a full comparison of approximations, see http://www.littleleague.org/leagueofficers/fieldspecs.htm, accessed February 9, 2017. Also, in days of yore, Little Leaguers were allowed to "take a lead" to the next base when they got on base. But stealing bases was so easy that it distorted the game, since catchers and middle infielders could not prevent base-stealing as in the majors. Nowadays, Little Leaguers are not allowed to leave the base until the pitched ball reaches the

batter. This rule also distorts the game, since stealing bases is no longer a part of the game. The forbidding of stealing in Little League remains controversial among purists.
16 See Michael Lewis, *Moneyball: The Art of Winning an Unfair Game* (New York: W.W. Norton and Co. 2003).
17 Naturally, this idea of a small scale Fenway is too tempting not to have been tried. There is a whiffle ball version of Fenway in Essex, Vermont described here: http://littlefenway.com/site/, accessed July 25, 2016. It was built from a sketch on a coffee-stained napkin (we are surprised it was not beer or wine). And yes, a Little League version has been built near Cincinnati. See it here: http://www.tcyosports.org/LittleFenway.html, accessed January 18, 2017. This was built with closer attention to adaptation, but even so, it is advertised as "virtually to scale." We would guess they encountered some of the problems we described. But it "looks right." Still, that is insufficient for a cosmology that never fails of exemplification in any experience, which is Whitehead's goal.
18 Auxier has presented this more poetic question with some thoroughness in "The Sacred Geometry of Fenway Park," in *The Red Sox and Philosophy*, ed. Michael Macomber (Chicago: Open Court, 2009), 267–282.
19 In the Fenway example, the geometry was more affine than projective, since the angles remained constant. The important point was that the algebra was limited by the geometry because we wanted a *physical* model.
20 Neil Young, *Waging Heavy Peace: A Hippie Dream* (New York: Blue Rider Press, 2012), 43 (our emphasis).
21 Ibid., 105–106.
22 Ibid., 106.
23 Whitehead, *Process and Reality*, 67–68 (our emphasis).
24 The use of the term "perspective standpoint" in this passage is difficult to sort out. At this point it is sufficient to note that Whitehead has distinguished perspective *from* standpoint just prior to this passage: Perspectives are objectified actual occasions as they are available for the percipient actual entity by which they are being felt. The feeling of the objectification is not the same as the simple physical feeling of another actual occasion by the percipient actual entity. A perspective involves a certain mutuality that a standpoint lacks. (See *Process and Reality*, 67, 236.)
25 Lewis Ford does take a stab at a "middle path" (our term) between temporal and non-temporal becoming that he uses in an attempt to save his claims concerning "temporal atomicity" by, he says, "considering genetic passage a different form of time." See Lewis Ford: "Temporal and Nontemporal Becoming," *Process Studies*, 38:1 (Spring–Summer 2009), 5–42.
26 See Whitehead, *Process and Reality*, 58.
27 We will discuss this issue further in chapter 12, but see *Process and Reality*, 214, for the relevant distinctions. We endorse the account of Nobo showing the distinctness of transition from concrescence. Nobo's discussion is difficult, combining language from Whitehead's 1927 essay "Time" (collected in *The Interpretation of Science*, 240–247), in which the "supersession" of actual entities is the favored vocabulary. Whitehead acknowledges that he is abstracting the physical from the mental poles in that essay, and when we abstract either one of these from the other, they "are each of them devoid of the full concreteness of the dipolar occasion." (240) Thus one cannot invoke the categories used in "Time" to explain and analyze material in *Process and Reality* without some adjustment. Neither "incompleteness" nor "supersession" is a category in *Process and Reality*, but Nobo combines this verbal usage from "Time" with the discussion of prehension from *Process and Reality* in which Whitehead uses different terms for similar work (e.g., superjection). These terms are different enough to make

a difference (see Nobo, *Whitehead's Metaphysics of Extension and Solidarity*, 144–151). Nobo is more interested in the problem of time and possibility *itself* than in nice distinctions of vocabulary from fine-grained textual interpretation, as demonstrated by the fact that he cites *Process and Reality*, 378–379 (from the original edition, 247–248 of the corrected edition) as support for his view of transition as supersession. The term "supersede" never appears in those passages. The difference in the "categories" in the 1927 essay, as compared with *Process and Reality*, ought to serve as a clue that care will be required to adapt the sense of the term "prehension" to an analysis in which the actual entity is treated in its full concreteness, instead of simply in its contribution to the concept of time as time plays a role in our natural knowledge.

Nobo makes it clear that he thinks of the process of concrescence as "microscopic" (148, 167), while transition is "macroscopic." We agree, but this terminology is misleading, since readers are likely to believe that it means concrescence is small while transition is big. In a way, that is true, since the actual entity is always only a part of its actual world. We think Nobo knows that size is irrelevant and he ought to choose less misleading vocabulary.

28 Whitehead, *Process and Reality*, 289. We have quoted this passage above in chapters 4 and 5, but perhaps here its meaning is becoming clearer.
29 We will explain our usage of the term "constellation" subsequently. See *Process and Reality*, 58 (at the bottom of the page), for the basis of our new term. "Constellation" is what we call the first type of presentational objectification.
30 We call this usage "infelicitous" because it invites confusion with what mathematicians such as Hermann Weyl would call a "group." Whitehead's usage has no relation to that specific definition.
31 Whitehead, *Process and Reality*, 22; but remember that a relation is a mode of being, a kind of "entity."
32 See Whitehead, *Process and Reality*, 219: "The philosophy of organism is a cell-theory of actuality. Each ultimate unit of fact is a cell-complex, not analyzable into components with equivalent completeness of actuality."
33 Whitehead, *Process and Reality*, 245.
34 Ibid., 235; final emphasis is ours, as are bracketed clarifications.
35 Ibid., 235.
36 We are using the term "constellation" to avoid confusion with terms that were later well defined in mathematics after Whitehead's time such as "group" or "set." Whitehead exemplifies the English tradition, which was only slowly abandoned, of using terms as they descended from Newton in preference to the Continental conventions. As late as *The Principle of Relativity*, Whitehead is still using Newtonian notations, and this has contributed to his formalizations being ignored.
37 Whitehead, *Process and Reality*, 220.
38 See Whitehead, *Process and Reality*, 112. The way Whitehead uses the term "chaos" is very much of a piece with the way René Thom developed the notion later in the unhappily named "catastrophe theory." See his *Structural Stability and Morphogenesis: An Outline of a General Theory of Models* (New York: W.A. Benjamin, 1994) See also Stengers, *Thinking with Whitehead*, 386–387.
39 See Whitehead, *Process and Reality*, 220.
40 See ibid., 214.
41 See ibid. As far as we know, the first serious argument anyone made for the relative independence of transition and concrescence is Nobo's argument in "Transition in Whitehead: A Creative Process Distinct from Transition," *International Philosophical Quarterly*, 19:3 (September 1979), 265–283. The view is repeated in Nobo's book, in the sections we have discussed in the earlier lengthy note on Nobo.

42 Jorge Nobo is among those few intrepid interpreters who has stepped into the breach and acquitted himself well. See his *Whitehead's Metaphysics of Extension and Solidarity*. Also see Fetz, *Whitehead*), Section 3. This is translated by Felt, in *Process Studies*, including: "In Critique of Whitehead," 1–9; and most importantly, "Aristotelian and Whiteheadian Conceptions of Actuality: I," *Process Studies*, 19:1 (Spring 1990), 15–27; and "Aristotelian and Whiteheadian Conceptions of Actuality: II," *Process Studies*, 19:3 (Fall 1990), 145–155.
43 Whitehead, *Process and Reality*, 283.
44 We will not here go into the relationship between possibility and contrast, but we find Nobo's discussion of it congenial. See *Whitehead's Metaphysics of Extension and Solidarity*, 191–195.
45 Whitehead, *Process and Reality*, 283.
46 Ibid., 35.
47 Ibid., 283.
48 Lewis Ford offers another attempt to justify temporal atomism in "The Indispensability of Temporal Atomism," 279–303. Here he changes the details of his book. We simply disagree.
49 Whitehead, *Process and Reality*, 283, first three emphases are ours.
50 Ibid.
51 Ibid.
52 Ibid., 26 (the Category of Subjective Unity).
53 Ibid., 283–284 (our emphasis).
54 Ibid., 284.
55 Ibid.
56 Ibid. (original emphasis).
57 Ibid., 238.
58 Ibid.
59 Ibid., 237. Cf. *Enquiry into the Principles of Natural Knowledge*, 62 (14.3).

Chapter 8

1 Whitehead, *Process and Reality*, 263. We want to thank Myron M. Jackson for drawing our attention to this passage and its obvious richness.
2 Ibid., 219.
3 Ibid.
4 Ibid., 220; pages 219–220 might be regarded as almost as tough as 283–284, but there you are. Each is the opening of a major division of the book.
5 See ibid., 235.
6 We have earlier given the genealogy of Whitehead's theory of extension from his first writings, through the *Universal Algebra* and the *Principia* and into his explicitly philosophical work. By the time it appeared in *Process and Reality*, it was quite advanced.
7 Whitehead, *Process and Reality*, 219.
8 Ibid.
9 Ibid., 58.
10 Ibid.
11 Ibid.
12 It is often forgotten today that a great amount of Leibniz's most important work in logic (and hence, concerning possibility) was unavailable until the end of the nineteenth century when C.I. Gerhardt's edition of Leibniz's *Philosophische Schriften* was published in seven volumes (Berlin: Weidmansche Buchhandlung, 1887–1890), especially volume seven (1890). Immediately some of the finest minds of the day went to work on these "new" writings, including book-length

treatments by Russell, Cassirer, and Dewey. Leibniz's writings revolutionized the growing algebra of logic. See C. I. Lewis in his *A Survey of Symbolic Logic*, 5–37. Lewis's account is unduly weighted to the praise of extensional logics.

13 Even Spinoza can be read as giving to possibility a serious, independent standing. See Richard Mason, "Spinoza on Modality," *The Philosophical Quarterly*, 36:144 (July 1986), 313–342; and Jonathan Bennett, *A Study of Spinoza's Ethics* (Indianapolis: Hackett, 1984). Jon Miller provided an account of Spinoza's view of possibility that is very much like the view we will defend here. See Jon Miller, "Spinoza's Possibilities," *The Review of Metaphysics*, 54:4 (June 2001), 779–814.

14 See the collection, *Whitehead: Points of Connection*, eds. Janusz Polanowski and Donald W. Sherburne (Albany, NY: SUNY Press, 2004).

15 William L. Reese, the collaborator with Hartshorne on *Philosophers Speak of God*, has some tantalizing suggestions in his "The Structure of Possibility," see the paper at the Paideia site, https://www.bu.edu/wcp/Papers/Onto/OntoRees.htm, accessed July 27, 2016.

16 See Herstein's essay, "Whitehead, Intuition, and Radical Empiricism," in *Intuition in Mathematics and Physics*, ed. Ronny Desmet (Anoka, MN: Process Century Press, 2016), 165–181.

17 See Whitehead, *Process and Reality*, 75–76. See also 288.

18 Ibid., 93–94.

19 Ibid., 125.

20 This is the problem faced by Eugene Wigner in his famous essay "The Unreasonable Effectiveness of Mathematics in Modern Science," which was the Richard Courant lecture in mathematical sciences delivered at New York University, May 11, 1959. *Communications on Pure and Applied Mathematics* 13, 1–14.

21 Whitehead tells the story with admirable brevity in chapter 2 of *Science and the Modern World*. What is meant by the term "natural philosophy" is best exemplified when mathematics and the study of nature are kept in close proximity, as they were during the height of classical Greek thought and again in the seventeenth and eighteenth centuries. In the nineteenth century, Whitehead notes that the influence of mathematical thinking on philosophy "waned" (see especially pages 29–32).

22 See our discussion of radical empiricism in chapters 1 and 2.

23 See Georg Gasser, ed., *How Successful Is Naturalism?* (Frankfurt am Main: Ontos Verlag, 2007).

24 See for example David Lewis, *On the Plurality of Worlds and Possibility and Counterfactuals*. The views expressed in these books are exotic. One of the notable problems we have with David Lewis' approach to possibility is that it is fundamentally contrary to the standards and criteria of radical empiricism. See *Counterfactuals* (Cambridge, MA: Harvard University Press, 1973) especially, 84–95; and *On The Plurality of Worlds* (London: Blackwell, 1986) especially, 2–3.

25 Whitehead, *Adventures of Ideas*, 206.

26 Ibid., 206–207.

27 Whitehead, *Process and Reality*, 35.

28 In fact, Whitehead is a *radical* realist—see chapter 10.

29 Whitehead, *Process and Reality*, 35–36.

30 Sherburne, *A Key to Whitehead's*, 217.

31 Whitehead, *Process and Reality*, 36.

32 See Herstein's *Whitehead and the Measurement Problem of Cosmology* for the full argument.

33 Michael Epperson has recognized that Whitehead is of limited use to his way of tackling the problem. But we still think Epperson needs a theory of cosmic epochs if he wants to address the full problem of physical reality. The generous

34 Sherburne, *A Key to Whitehead's*, 217.
35 Sherburne and numerous others have been chopping off pieces of the philosophy of organism for decades. The most popular targets for depopulating Whitehead's ontology are the eternal objects and God. See especially Sherburne's "Whitehead without God," in *Process Philosophy and Christian Thought*, eds. Delwin Brown, Gene Reeves and Ralph James (Indianapolis: Bobbs-Merrill, 1971), ch. 16. See also Nicolo Santilli, "Flux and Openness: Dissolving Fixity n Whitehead's Vision of Process," *Process Studies*, 41:1 (Spring-Summer 2012), 150–170. Santilli is convinced that all of the aspects of Whitehead's philosophy that are postulated as stable are problematic and have to be revised. But there is a difference between postulated, functional stability and fixity.
36 Whitehead, *Process and Reality*, 36. See chapter 4.
37 The recent revival of interest in panpsychism is fueled partly by these sorts of concerns in the philosophy of science, but Whitehead's solution has not been well enough understood to be included as a vital part of that discussion. See for example, David S. Clarke, *Panpsychism and the Religious Attitude* (Albany, NY: SUNY Press, 2003); and Galen Strawson, et al., *Consciousness and Its Place in Nature: Does Physicalism Entail Panpsychism?* ed. Anthony Freeman (London: Imprint Academic, 2006).
38 Whitehead, *Process and Reality*, 91. Whitehead has in mind here the perfect generality of Maxwell's equations in anything we can experience, so far as we know. That is the universal character of our cosmic epoch, if it has any such character in physical terms.
39 Whitehead, *Process and Reality*, 288, our emphasis. Whitehead is not advocating a multiverse theory of nature. He is quite sincere about what is (probably) actual in our universe, even if not in our cosmic epoch, let alone our immediate epoch. The multiverse theory is a parochial and logically indefensible attempt to treat relations as parasitic on *relata*.
40 Whitehead, *Process and Reality*, 288. The term "important" is doing some work here for Whitehead. See Ross Stanway's discussion of this idea throughout Whitehead's work in "Whitehead on the Concept of 'Importance'," *Process Studies*, 21:4 (Winter 1992), 239–245. The concept of importance cannot be completely defined, according to Whitehead, but "concerns the 'final unity of purpose' (*Modes of Thought*, 12) as grounded in the 'unity of the Universe' (*Modes of Thought*, 8)," according to Stanway (240). Details about the universe that are unrelated to its purpose are "trivial." (Stanway, 244–245).
41 Whitehead, *Process and Reality*, 333 (our emphasis).
42 Ibid., 288.
43 Wolfgang Smith, "*Sophia Perennis* and Modern Science," in *The Philosophy of Seyyed Hossein Nasr*, eds. Lewis E. Hahn, Randall E. Auxier, and Lucian W. Stone. (LaSalle, IL: Open Court, 2001), 472–473. The Whitehead quote is from *The Concept of Nature*, 30.
44 Smith, "*Sophia Perennis* and Modern Science," 474. See also Wolfgang Smith, *The Quantum Enigma* (Peru, IL: Sherwood Sugden Publisher, 1995), and *Science and Myth: What We Are Never Told* (San Raphael, CA: Sophia Perennis, 2010).
45 Smith, "*Sophia Perennis* and Modern Science," 478.
46 Whitehead, *Process and Reality*, 238–239 (our emphasis).
47 Whitehead is clear about the primacy of the physical throughout his work, not just in *Process and Reality*. For example, he says in 1927 that "the physical pole

must be explained before the mental pole, since the mental pole can only be explained as a particular instance of supersession disclosed in the analysis of a fully concrete occasion." (*The Interpretation of Science*, 241). See also William J. Garland's interpretation of physical concrescence in "Whitehead's Theory of Causal Objectification," *Process Studies*, 12:3 (Fall 1982), 180–191. Garland provides helpful summaries of commentators on the question of concrescence and physical prehension/feeling, and calls attention to the idea that the elimination of physical feeling "makes possible the movement from the initial data to the objective datum." (185) More accurately, this elimination makes the movement *actual*, not just possible, but the insight that Garland reports (and attributes to Sherburne) seems right, as does the Garland's subsequent analysis, as it bears on concrescence and physical prehension. Garland's account of "negative prehension" (188–189), for all its brevity, is among the best in the literature, although he conflates determinateness and definiteness.
48 Whitehead, *Process and Reality*, 206.
49 Ibid., 22.
50 Ibid., 210, original emphasis.
51 See ibid., 214.
52 Ibid., 184–185.
53 This is the topic of our forthcoming book, *The Continuum of Possibility*.
54 Ibid., 185–186, our emphasis. As a reminder, we are using the word "collection" instead of "set" for the form of order here. When we speak of possibilities *independent* of their ingression, we use the term "constellation."
55 Ibid., 186, our emphasis.
56 See ibid., 214. We will have more to say about this kind of "possibility" in chapter 12.
57 See our earlier note about Whitehead's use of the "determinate" in the context of eternal objects and "definite" in the context of "other occasions" from his 1927 essay "Time," in *The Interpretation of Science*, 140–147. Susanne Langer picked up the distinction between definite and determinate order in her *Symbolic Logic* (New York: Dover Books, 1967 [1937]; see pp. 135, 138, 151, etc. Langer employs the term "constellation" in a fashion similar to our usage; see p. 65.
58 See Whitehead, *Process and Reality*, 65 f.
59 This is the principal topic of our forthcoming book *The Continuum of Possibility*.
60 See ibid., 115, where this point is expressed in the language of transition: "the realization of a pattern necessarily involves the concurrent realization of a group of eternal objects capable of contrast in that pattern." On page 148, the analogous point is made in the language of concrescence: "There is interconnection between the degrees of relevance of different items in the same actual entity. This fact of interconnection is asserted in the principle of compatibility and contrariety. There are items which, in certain respective gradations of relevance, are contraries to each other; so that those items, with their respective intensities of relevance, cannot coexist in the constitution of one actual entity. If some group of items, with their variety of relevance, can coexist in one actual entity, then the group, as thus variously relevant, is a compatible group." Whitehead's use of the term "group" here is not to be conflated with Weyl's "group theory." This is one reason we choose the terms "collection" and "constellation" for definite and determinate order, respectively.
61 See James Bradley's argument for the superiority of the concept of function as a model for analysis over its competitors, ancient, modern, and contemporary in "Transcendentalism and Speculative Realism," 166–168. Bradley criticizes Whitehead (168–169) for strictly delimiting function so as to exclude what we call "operations," and what Bradley associates with the "foundationalist" theories of propositions adopted by Frege and Russell. We agree that Whitehead's

limits would be a problem if Whitehead did not build back into his philosophy of organism a way that universals are related to each other independent of concrescence. We think Whitehead addresses universals in the way he describes transition. Bradley is right about the problem, and in his reading of Whitehead's different inquiries in light of the limitations he sets on them, individually. Bradley is also on the right track, seeing a strong connection between Cassirer's and Whitehead's philosophies of science. But eternal objects do not, in *Process and Reality*, pose a problem for Whitehead he hasn't solved. The distinction between transition and concrescence, for us and for Nobo, solves the problem. Bradley actually uses the term "operative" in a kindred fashion without seeing its prospects on page 170.

62 William Christian's discussion in *An Interpretation of Whitehead's Metaphysics* (New Haven: Yale University Press, 1959), 132–140 covers the same basic problems. We agree with his view regarding the elimination of physical feelings. But he does not draw the same conclusion about the relation of the actual entity to its immediate predecessor, and Christian's sense of the role of eternal objects in this discussion is impoverished. Christian says that "some theory of universals seems to be required" (140) but he limits himself to "several observations" about the problem of the "relational function of eternal objects" (140). Here he confuses the theory of transition and concrescence with the problem of causal efficacy (perception) and with that of prehension (indivisible relatedness), as throughout his book. Still, he stays as close to the text as he can, doesn't fudge or fabricate, and (examining his final paragraph, 140) knows he hasn't gotten the role of eternal objects worked out. We believe we *do* have it worked out. In process philosophy, we do not *seek* an explanation of how the principal actual entity came to exist as it does, so *no* account of the flow of feeling *from* the past *into* the present is needed. Rather, the actual entity in its actual world *is* the reason and thus the ground of any explanation. Our questions must be adapted to whatever an analysis of that actual entity yields. Christian doesn't maintain consistently the categoreal primacy of the actual entity as the reason and the world as explained *by* that entity.

Leclerc, *Whitehead's Metaphysics* emphasizes the idea of "form" in Whitehead and tries to get it to do the work of bringing possibility, actuality, and potentiality together. This is not supportable within Whitehead's text. Leclerc fails, like Christian, to understand that the theory of prehension is more concrete than the theory of transition and concrescence, and attempts to use the latter to account for the former. Eternal objects are not "forms of definiteness," as Leclerc claims. They are determinations of form. Definiteness has a relation to actuality to which eternal objects are wholly indifferent, as we will explain. Yet, Leclerc does use the terminology of definiteness and determination correctly most of the time. His habitual phrase is that eternal objects are "determinants of the definiteness of the process of acting of actual entities." (94; cf. 95) Still, he is thinking of acting as a process, which at the level of prehension is irrelevant. The relation of prehension *is* the act in its most concrete sense, not a process of acting. But Leclerc does not recognize that determination *of* definiteness is not the same thing as definiteness. Eternal objects determine, but making a "determinant" character definite in the objectifications of the prehending actual entity is quite beyond their scope or power. Leclerc Platonizes Whitehead, even slipping away from the language of "eternal objects" into that of "eternal forms"—a phrase Whitehead never uses in *Process and Reality* (see Leclerc, 95).

Every serious investigation of Whitehead's metaphysics has to include some interpretation of the relations of actuality, potentiality and possibility. Leclerc takes up the problem again under his own authority in *Nature and Existence*. His solution is beyond our present scope, along with the solutions of Justus

Buchler, John William Miller, Harold N. Lee, Archie Bahm, and a number of other original mid-century, process-friendly metaphysicians. More recently Ralph Pred and Frederick Ferré, have made attempts, but their interests lie closer to funding experienced value than to solving old metaphysical problems.

63 See Auxier, "Complex Negation, Necessity, and Logical Magic," in *The Relevance of Royce*, eds. Kelly Parker and Jason M. Bell (New York: Fordham University Press, 2014), 89–131, for an account of how a nuanced logic of negation affects metaphysics. Whitehead's view of this was at least as subtle as Royce's.
64 This diagram and the subsequent analysis is essentially a commentary on Part III, chapter 1, Section V of *Process and Reality*, 223–227, which is one of the two most important discussions of negative prehension in the book.
65 Whitehead, *Process and Reality*, 226–227.
66 Compare with *Adventures of Ideas*, 211, where Whitehead rejects the principle that any actual entity is wholly determined by its predecessor.
67 See ibid., 203–204, where the society explains the nexus, not the other way around.
68 This is a technical use of the word "influence." See Auxier, "Influence as Confluence,", 301–338.
69 An admirably comprehensive study is in Jones' *Intensity: An Essay in Whiteheadian Ontology*.
70 See *Process and Reality*, 26, and the relevant discussion at 224–227. These are elucidations of the Ontological Principle.
71 This is the sort of distinction that is often overlooked when Whitehead scholars struggle with the difficult idea of negative prehension. Adam Scharfe confronts the same problem:

> Pointing to the significance of negative prehensions, which posit and attribute nothingness to actual entities in Whitehead's cosmology, it would seem that while there are many elements involved in the creativity of organisms, it is largely by way of negative prehensions that the organism acquires its experiential perspective, namely, the contrast of experienced finitude and experienced infinitude necessary for physical and conceptual appetition and realization.

Scharfe: "Negative Prehension and the Creative Process," *Process Studies*, 32:1 (Spring–Summer 2003), 94–105 (104). He is surely correct that negative prehension is at the bottom of what Whitehead calls "standpoint" (not "perspective"), but looking at negative prehension as concrescence only, and not also as transition, leads to an overly physical (and not adequately logical) reading of the idea of negative prehension. It is clear that Scharfe, like others, is thinking of biological beings as the model for negative prehension. That simply isn't Whitehead's concern. Elimination operates at every level.
72 See Whitehead, *Process and Reality*, 283.
73 See Jones, *Intensity*, 94.
74 Again, see Whitehead, *Process and Reality*, 283.
75 See ibid., 285.
76 See Bryson Brown and Graham Priest, "Chunk and Permeate, A Paraconsistent Inference Strategy, Part 1: The Infinitesimal Calculus," *Journal of Philosophical Logic*, 33:4 (August 2004), 379–388. The authors show the self-contradictory character of the infinitesimal calculus, both assuming and denying the existence on infinitesimals. They employ a technique similar to that of Rescher and Brandom in *The Logic of Inconsistency*. The seventeenth- and eighteenth-century thinkers quarantined this inconsistency to keep it from interfering with the development of physical theory. The logical tools employed are both modal and model-theoretic.

77 See Whitehead, *Process and Reality*, 283.
78 In contrast Sherburne's account, Whitehead never says that a prehending or percipient actual entity has *unmediated* access to its "immediate predecessor." The term "immediate" only refers to the actual entity which is most pervasive in the achievement/satisfaction of the actual entity whose world is under consideration. Finding the immediate predecessor of a given actual entity requires effort, and even when it is inferred, it ought not persuade ourselves we have experience of that predecessor *as an individual*. Sherburne's assumption that the presence of the past in the actual entity is explained by the past history. Whitehead says the reverse: the actual entity under consideration is the explanation of the past. The past is mediated by the effort of the actual entity to evaluate it in concrescence. The account of that immediate predecessor in transition might look something like Sherburne's account, but such stories would be beyond the limits of the philosophy of organism. See Sherburne, "The Process Perspective as Context for Educational Evaluation," *Process Studies*, 20:2 (Summer 1991), 82. Sherburne confuses "memory" in a Bergsonian sense with Whitehead's sense of the term "history," as in the "history" of a physical route. This history may have a relationship to memory in a Bergsonian sense, it would require a great deal of effort to explain it.
79 Whitehead, *Process and Reality*, 292, our emphasis.
80 See Stephen J. Gould, *The Structure of Evolutionary Theory* (Cambridge, MA: Harvard University Press, 2002), 484–492.
81 Whitehead, *Process and Reality*, 283.
82 For a crucial instance of Whitehead's use of the term "evolution" in this sense, see *Process and Reality*, 229. He does not always use the term in just this sense.

Chapter 9

1 See Lewis, *A Survey of Symbolic Logic*, 5–37. Lewis was hostile to the idea of intensional logic and disdained treating logic as having an empirical basis, but his story will suffice as a summary. A detailed account of Castillon's efforts (whose papers of 1797–1805 are being translated from French) is described by Arthur Shearman in *The Development of Symbolic Logic* (London: Williams and Norgate, 1906), 91–142. Unfortunately, Shearman, taken with the extensional thinking of Russell, dutifully "refutes" Castillon. We think he does no service to logic in the attempt.
2 See Lewis, *A Survey of Symbolic Logic*, 10–11.
3 See C. I. Gerhardt's edition *Gessamelte Werke von Leibniz*, various publishers, 1859–1890; and Louis Couturat, *La Logique de Leibniz* (Paris: Felix Alcan, 1901).
4 Charles Sanders Peirce, *The Writings of Charles Sanders Peirce: A Chronological Edition*, eds. Nathan Houser, et al. vol. 6 (Bloomington, IN: Indiana University Press, 2000), 360–362.
5 David Robert Crawford, "The Role of Diversity in Peirce's Objections to Necessitarianism," 5. This is an unpublished manuscript provided by Crawford to the authors, September 29, 2011. It is distilled from Crawford's dissertation project at Duke University, 2011. This paper was presented at the Midwest Pragmatism Study Group in Indianapolis, September 25, 2011.
6 Derek Malone-France does a solid job of marking out the distinction between what is "definite" and what is "determinate" within the modalities of actuality and possibility in Whitehead's thought. See Malone-France, "Between Hartshorne and Molina: A Whiteheadian Conception of Divine Foreknowledge," *Process Studies*, 39:1 (Spring-Summer 2010), 129–148.
7 For Whitehead "type-token" ordering just repeats the errors of intuition that move extensionally from universal to particular. If Whitehead were to make the

distinction, it would be "token-type," since Whitehead's thinking is intensional. We wouldn't know what a type was at all without working outward from particular experience.

8 Crawford, "Diversity," 6. This distinction touches on the difference between cardinal and ordinal relations. See Cassirer, *Substance and Function*, 44–54. Russell admits his approach can't handle ordinal relations and opts strictly for cardinal relations in *Principles of Mathematics* (Cambridge: Cambridge University Press, 1903), chs. 24–25, esp. section 242. This discussion has an interesting relation to the difference between intensional and extensional logics. By 1938, Whitehead expressed serious doubts. See *Modes of Thought*, 105–108. Justus Buchler's principle of ontological parity is, for us, restricted to possibilities as such.

9 See Negri and Plato, *Structural Proof Theory*, especially the section entitled "Structural proof theory," Kindle locations 60–106.

10 We merely mention forcing as an illuminating example. "Counting" the layers of infinity is a *perspectival* matter that depends upon the tools that one allows in inquiry. The kinds and "countings" of infinity, when considering the structure of possibility, is the sort of *idea* that cannot be reduced to set theory plus the continuum hypothesis. For example, Melvin Fitting offers a more extended treatment in the context of intuitionistic logic in *Intuitionistic Logic Model Theory and Forcing* (Amsterdam: North-Holland Publishing, 1969).

11 We scare quote "semantics" here having no commitment to contemporary theories of truth. We do not expect a formalized concept of truth to do any heavy lifting. "Semantics" is a name for the relation between the language and the universe, however either happens to be defined. See Auxier, "On Klaus Ladstaetter's 'Liar-like Paradoxes and Meta-language Features'," *Southwest Philosophy Review* 29:2 (July 2013), 25–28.

12 See Abraham Robinson, *Non-Standard Analysis* (Amsterdam: North Holland, 1970).

13 See Lewis, *A Survey of Symbolic Logic*, chapter V ff. For a more recent survey of modal logics from a mathematical perspective, see Robert Goldblatt, "Mathematical Modal Logic: A View of Its Evolution," found in *Handbook of the History of Logic. Volume 7*, eds. D. M. Gabbay and John Woods (Amsterdam: Elsevier, 2006), 1–98. For a detailed examination of the varieties of ways that have been developed within the theories of relevance logics moving beyond strict implication, see Alan Anderson and Nuel Belnap, *Entailment: The Logic of Relevance and Necessity*, 1 and 2 vols (Princeton: Princeton University Press, 1990 and 1992).

14 Brown and Priest, "Chunk and Permeate," 379–388.

15 See Jouko Väänänen, *Models and Games* (Cambridge: Cambridge University Press, 2011); also Doets, *Basic Model Theory*, esp. ch. 3.

16 "The line between universal algebra and model theory is sometimes fuzzy; our own usage is explained by the equation

universal algebra + logic = model theory."

From Chang and Keisler, *Model Theory*, Kindle location 267. Note that Chang and Keisler use the term "universal algebra" in the later form standardized by G. Birkhoff, as the mathematical development of Boolean algebra plus lattice theory. While there are analogies with Whitehead's universal algebra, these analogies are limited and the two are best viewed as different studies.

17 See J. C. Beall, *Logic: The Basics* (New York: Routledge, 2010). Also see Auxier's algebraic modeling of seven distinct senses of negation in his "Negation, Necessity and Logical Magic," cited above. This is the style of analysis we adopt for explaining negative prehension.

18 To avoid unnecessarily prejudicing matters, we avoid the use of the verb "to predicate."

19 The distinction is between the size of the natural numbers, as opposed to the higher order infinities, such as that of the real numbers.
20 Whitehead, *Process and Reality*, 228.
21 Ibid.
22 See ibid., 248.
23 This representation can be made subtler by incorporating a fuzzy logic into the characteristic function χ, so that there is a numerical continuum between the positive physical prehension and negative physical prehension. We content ourselves presently with only mentioning it in this note. See our forthcoming *The Continuum of Possibility*.
24 Defenders of "strong emergence" may find our account unsatisfying because strong emergence implies something like $\chi(\Gamma_i[\alpha])$ with nothing before—it isn't a contrast of anything. There is no Γ_j of which Γ_i is an arrangement. We don't think strong emergence is ruled out by anything we have asserted, but it could not be experienced by complex societies of actual occasions (including humans). It would destroy them. So, strong emergence is not merely a diversity of status, in Whitehead's terms, and if it belongs to our cosmic epoch, it is cataclysmic, an absolute disruption. For an account of strong emergence, see John Symons, "Physicalist Arguments against Emergence," in *The Future of Systematic Philosophy*, eds. Randall E. Auxier and Douglas R. Anderson (forthcoming). See also our discussion of Terrence Deacon's work on emergence in the notes below.
25 Nicholas Rescher and Robert Brandom, *The Logic of Inconsistency* (Atlantic Highlands, NJ: Rowman & Littlefield Pub Inc., 1979).
26 See for example, James W. Garson, *Modal Logic for Philosophers* (Cambridge: Cambridge University Press, 2012).
27 On computer-oriented modal logic, see Patrick Blackburn, Maarten de Rijke, and Yde Venema, *Modal Logic* (Cambridge Tracts in Theoretical Computer Science, Cambridge: Cambridge University Press, 2002). On complexity, see Hartley Rogers, *Theory of Recursive Functions and Effective Computability* (Cambridge, MA: MIT Press, 1987).
28 Whitehead, *Process and Reality*, 230.
29 Ibid.
30 Ibid., 290.
31 Ibid., our emphasis.
32 We suppose that this is another argument against strong emergence as a character of finite experience.
33 Ibid., 290–291, our emphasis.
34 Ibid., 266. There is much more to be said about the way that our interpretation illuminates Whitehead's theory of judgment, but we will address that in our subsequent work.
35 See our earlier reference to Wigner's "Unreasonable Effectiveness of Mathematics in Modern Science." Perhaps this point will satisfy defenders of strong emergence, and Peircean fans of abduction.
36 Whitehead, *Process and Reality*, 291. We *think* that this is what Nobo was explaining in his book, that book is perhaps denser that Whitehead's own account. We very much *hope* that Nobo was saying what we are saying.
37 See for example, Samuel R. Delany's essays in *The Jewel-Hinged Jaw: Notes on the Language of Science Fiction* (Middletown, CT: Wesleyan University Press, 2009 [1978]).
38 Whitehead, *Process and Reality*, 291 (our emphasis).
39 Ibid.
40 Ibid., 292.
41 For an accessible discussion of this aesthetic aspect of mathematics, see Davis and Hersh, *The Mathematical Experience* (Boston: Mariner Books, 1999). Among Whiteheadians, George Allan, for example, seems to miss this dimension

336   Notes

in summarizing *Modes of Thought*. He emphasizes the aesthetic, quoting from *Modes* that, "philosophy is akin to poetry," or that "philosophy is analogous to imaginative art." See "Creating the Future," *Process Studies*, 38:2 (Fall–Winter 2009), 225. But Allan misses the most salient part of Whitehead's statement: "In each case there is reference to form beyond the direct meanings of words. Poetry allies itself to metre, *philosophy to mathematic pattern*." (*Modes of Thought*, 174, our emphasis) Many overlook the *aesthetic dimension of mathematics*, and its role in Whitehead's own modes of thought.

42 Whitehead, *Process and Reality*, 220.
43 Ibid., 221.

Chapter 10

1 They probably do include a few pictures so that the technicians can recognize the parts of various systems beyond just their purely functional characteristics and connections. Such pictures play a small role in technical documentation.
2 NASA's official Voyager website contains many such breathtaking images: http://voyager.jpl.nasa.gov/imagesvideo/imagesbyvoyager.html, accessed April 23, 2013.
3 See ch. 9.
4 Hintikka: "Is Logic the Key to All Good Reasoning?" 35–57.
5 See, for example, Jaakko Hintikka, "The Logic of Science as a Model-Oriented Logic," *Proceedings of the Biennial Meeting of the Philosophy of Science Association*, 1 (1984), 177–185.
6 On the failures of set theory, see Jaakko Hintikka, "Truth Definitions, Skolem Functions and Axiomatic Set Theory," *The Bulletin of Symbolic Logic*, 4:3 (September 1998), 303–337. On the problems with proof-theoretic approaches see Hintikka, "Game-Theoretical Semantics as a Challenge to Proof Theory," *Nordic Journal of Philosophical Logic*, 4:2 (2000), 127–141.
7 Stephen Hawking, *The Grand Design* (New York: Bantam Books, 2010), 5.
8 Whitehead, *The Function of Reason*, 58–59.
9 Hawking, *The Grand Design*.
10 Hawking mentions a few philosophers approvingly, John W. Carroll of North Carolina State University being the sole contemporary example. One assumes they have been kinder to him than he earned.
11 Stephen Hawking, *Black Holes and Baby Universes* (New York: Bantam Books, 1994), 42–43.
12 Stephen Hawking, *A Brief History of Time* (New York: Random House Publishing Group, 1988); Charles Sherover, *The Human Experience of Time* (Evanston, IL: Northwestern University Press, 2000).
13 Hawking, *The Grand Design*, 7.
14 Greene, *The Fabric of the Cosmos*, 5.
15 Whitehead, *Adventures of Ideas*, 127. An irony came to light many years later: the mass of Pluto is too small to cause the orbital anomalies that led to the search for Pluto. Instrumental measurements by the Voyager spacecraft revealed that the estimated masses of Uranus and especially Neptune were inaccurate. With these corrected estimates, the need for a "planet X" to account for the supposed discrepancies vanished—Perceval Lowell's discovery was a matter of dumb luck. However, the method by which the error was discovered and fixed is within the *explanatory* framework Whitehead set out above. It was *not* enough to find a "better" description; adding a new parameter was not an option; what scientists wanted was a *true and accurate explanation*.
16 See, for example, Lee Smolin, *The Trouble with Physics* (Boston: Mariner Books, 2007).
17 Hawking, *The Grand Design*, 8.

18 There are many critics of string-theory, but see for example Peter Woit, *Not Even Wrong: The Failure of String Theory and the Search for Unity in Physical Law* (New York: Basic Books, 2006), Woit opposes "model-centrism" as manifest among the string-theory enthusiasts currently dominating the field. See especially 93, 185, and 231–234. While Woit's criticisms are philosophical in character, Woit is a working physicist, and his critiques are those of an established insider.
19 Recall that the Lucasian Chair at Cambridge that Hawking held is in *mathematics*, not physics.
20 Such triumphalism is not hard to find in the popular press. This story, from *The Guardian*, is a typical example: http://www.theguardian.com/science/across-the-universe/2013/mar/21/european-space-agency-astronomy, verified January 15, 2015.
21 Joseph Bracken speculates on ways of building connections between Whitehead's metaphysics and string theory. See "Whiteheadian Metaphysics, General Relativity, and String Theory," *Process Studies*, 43:2 (Fall–Winter 2014), 129–143. This is an unwise idea on many levels. The assertion that an actual entity (or in Bracken's case, the achieved superset of a pretending occasion (133–135) is a physicalization of occasions, and we have argued is one of the primary errors interpreters of Whitehead have committed. Bracken's connection of vibration in Whitehead's thought with physics is unsatisfactory. For Whitehead "vibration" is fundamental form for creating organic identity. (One might recall the final chapter of his *Enquiry into the Principles of Natural Knowledge*, Rhythms. See above, our chapter 3, on rhythm and periodicity.) Vibratory phenomena in physics have no such generative function, and stand out as only an empirical feature of our immediate epoch. A far greater problem with Bracken's move is his choice of an untestable (empirically vacuous) collection of mathematized speculations as string theory for such a physicalization. Whitehead's speculative philosophy is not *bad* physics.
22 Bas Van Fraassen, *The Scientific Image* (Oxford: Oxford University Press, 1980), 8.
23 Ibid., 10 (original emphasis).
24 Ibid.
25 See Bas Van Fraassen, *The Empirical Stance* (New Haven: Yale University Press, 2002).
26 This stance should be called "Humean" in this instance, since Van Fraassen's stance is informed by the Kantian critiques of metaphysics. See ibid., Lecture 1.
27 Cf. Van Fraassen's *Quantum Mechanics: An Empiricist View* (Oxford: Oxford University Press, 1991); contra Hawking, Van Fraassen does understand the relevant mathematics, even though he is a philosopher.
28 Fraïssé's work goes back to the early 1950's, and Ehrenfeucht to the early 1960's; the large-scale investigation of these methods in model theory did not really begin until the late 1980's, almost a decade after *The Scientific Image*. See the bibliography of Doets, *Basic Model Theory*, cited above.
29 See Doets, *Basic Model Theory*, 21.
30 Van Fraassen, *The Scientific Image*, 11.
31 Leemon McHenry argues that Whitehead's cosmic epochs correspond to alternative "universes" as these are currently conceived in physics as part of the "multiverse" theory. See his "The Multiverse Conjecture: Whitehead's Cosmic Epochs and Contemporary Cosmology," *Process Studies*, 40:1 (Spring–Summer 2011), 5–24. We disagree. Cosmic epochs as Whitehead conceived them were not deployed in some higher-dimensional "space." McHenry accepts the multiverse theory at face value, whereas Whitehead would be *critical* of such an approach. The multiverse theory was created by physicists laboring with a primitive metaphysics, treating relations as parasitic, while only *relata* are genuinely

real. With such a view, the only way one can treat relations as significant is by reifying *relata* in the form of parallel universes. But once one takes relations themselves seriously, as Whitehead does, one does not need to hypostatize whole universes of "things" to account for those relations. *Real* relations of possibility and potentiality, for example, are sufficient unto themselves and do not require "trans-empirical support," in James's phrase.

32 Whitehead, *Process and Reality*, 259 (our emphasis).
33 Whitehead, *The Principle of Relativity*, 5.
34 See Whitehead's *The Concept of Nature*, The Tarner Lectures delivered at Trinity College 1919, published 1920.
35 Philosophers are hardly innocent in this regard. Having inherited a mangled interpretation of Aristotle's theory of the possible and the actual, they have extended it into the most absurd extremes. Today's radical nominalists advance a "trope" theory that ignores of Peirce's critiques of nominalism, even advancing the notion of "bare particulars" as empirically valid. See, for example, *Relations and Predicates*, eds. Herbert Hochberg and Kevin Mulligan (Frankfurt am Main: Ontos Verlag, 2004), especially Donald W. Mertz's "Objects as Hierarchical Structures: A Comprehensive Ontology," 113–148. Mertz's serious approach to relations is admirable, but needs a metaphysics to ground that relational approach.
36 A phrase we take from Ernst Cassirer.
37 Whitehead, *Process and Reality*, 3.
38 Ibid., 22.
39 See ibid., 4.
40 Whitehead, *Science in the Modern World*, ix–x, our emphasis.
41 An exception to this rule is the recent publication of Mark Dibben and Thomas Kelly, eds., *Applied Process thought I: Initial Explorations in Theory and Research* (Frankfurt am Main: Ontos Verlag, 2008). Many of the contributions in this volume ranged beyond physics and into chemistry, biology, and the social sciences as well.
42 Murray Code problematically attempts to "naturalize" Whitehead's philosophy by shoe-horning everything "inside" nature. See Code, "On Whitehead's Almost Comprehensive Naturalism," 3–31. Code's discussion slides back and forth between Whitehead's natural philosophy and his metaphysics without properly bracketing the varying topics within their appropriate contexts. (Code, "On Whitehead's Almost Comprehensive Naturalism," 7–17.) Code embraces Ford's attempt to temporalize eternal objects (Code, "On Whitehead's Almost Comprehensive Naturalism," 28.) Code misses the *logical* function of eternal objects, and the fact that they are eternal precisely because temporality has no bearing upon them. The temporal is purely *natural* in its significant limitations, having no role in the broader considerations of Whitehead's *meta*physics. The same error dogs many Whitehead interpreters who make time explanatory of process, rather than process explanatory of time. This is a kind of "naturalistic fallacy," which attempts to move the quantum of explanation *inside* nature, and thus make process *parasitic* upon that which is to be explained. A philosophical naturalism that takes nature as given, in any but the *radical* form we are arguing for, is guilty of, as Russell famously quipped, evincing all the "advantages of theft over honest toil." (Bertrand Russell: *Introduction to Mathematical Philosophy*, [New York: Macmillan, 1919], 88.) While Code acknowledges that Whitehead's thoroughgoing "radically revised" empiricism is "evident" (Code, "On Whitehead's Almost Comprehensive Naturalism," 11–13), he fails to see that this is *radical empiricism* which, in Whitehead's philosophy of science, expresses itself as what we are calling here "radical realism." So Code is ultimately trying to

force Whitehead upon a procrustean bed whose "nature" is not even of Whitehead's own making.
43 See Michael Friedman, *A Parting of the Ways: Carnap, Cassirer, and Heidegger* (LaSalle, IL: Open Court, 2000).
44 Whitehead, *Enquiry into the Principles of Natural Knowledge*, 32. For an example of space as an arbitrary convention, see Hans Reichenbach, *The Philosophy of Space and Time* (New York: Dover Publications, 1957).
45 Whitehead, *Enquiry into the Principles of Natural Knowledge*, 91, our emphasis.
46 Whitehead, *The Principle of Relativity*, 22.
47 Whitehead, *The Concept of Nature*, 40.
48 Ibid., 45.
49 Van Fraassen himself notes that the idea of "literal" meaning is originally a theological one, and thus argues that a *rough and ready* (our phrase) understanding of the term is sufficient. See *The Scientific Image*, 10. We consider this a serious flaw in his argument.
50 Whitehead, *The Principle of Relativity*, 39.
51 Ibid., vi.
52 Whitehead, *The Interpretation Of Science: Selected Essays*, 29.
53 Ibid., 192.
54 Ibid., 193.
55 Ibid., 194. See our discussion of the relation between theory and hypothesis in chapters 1–2, and the role of "intention" in establishing a model in this chapter.
56 Whitehead, *Enquiry into the Principles of Natural Knowledge*, vii.
57 Whitehead, *The Concept of Nature*, 3. We will take up the far-reaching implications of this view in the next chapter.
58 Whitehead, *The Principle of Relativity*, 5.
59 Ibid., 20.
60 William James, *Essays in Radical Empiricism* (New York: Longmans, Green and Co., 1922), 42 (original emphasis).
61 Whitehead, *The Concept of Nature*, 13.
62 Ibid., 14.
63 Whitehead, *The Principle of Relativity*, 15, (our emphasis).
64 Whitehead, *The Concept of Nature*, 14.
65 Whitehead, *The Principle of Relativity*, 62.
66 Ibid., 63. These two quotes face one another on opposite pages. Whitehead consciously said each in the context of the other.
67 Ibid., 4.
68 Whitehead, *The Concept of Nature*, 20. This passage bears heavily upon the kind of naturalism we are advocating.
69 Ibid., 5.
70 "Sense-awareness" is that component of sense perception that has a brute "there-ness" that is in no way a part of thought; see *The Concept of Nature*, 4. When we use the term perception, please bear in mind that we never mean to include awareness. But the presence of this brute factor is obviously caught up with the realism of Whitehead's philosophy.
71 Whitehead, *The Principle of Relativity*, 13.
72 Whitehead uses the example of a yardstick between York and Edinburgh; see ibid., 50.
73 Whitehead, *The Principle of Relativity*, chapter III. Numerous objections might be raised to this part of Whitehead's argument. Our experiences of the uniformities of time and space might be an artifact of the local nature of those experiences (recalling the difference between our immediate epoch and our cosmic epoch). But does Whitehead need anything more than the *epistemological* necessity of

such uniformities and their "pragmatic *a priori*"? We think not. Were it shown that such uniformities do not hold on a cosmological scale, then the possibility of knowledge of such cosmological relations would collapse (which is Whitehead's point). We need a theory of cosmic epochs to do cosmology. For a more detailed discussion of this subject in its scientific context, see Herstein, *Whitehead and the Measurement Problem of Cosmology*.

74 Whitehead notes the difference between these particular necessary and contingent relations track very closely those between internal and external forms of relatedness. In *Process and Reality*, the theory of extension takes over the development of the formal uniformities, and Whitehead explicitly identifies it as the theory of internal relations—more precisely, *the theory of the internalization of relatedness*.

75 See the first two chapters of *The Concept of Nature*, chapter 2 of *The Principle of Relativity*, the first half of *Science in the Modern World*.

76 Whitehead, *The Principle of Relativity*, 13.

77 Van Fraassen, *The Scientific Image*, ch. 5.

78 Whitehead addresses this without naming names; see *Adventures of Ideas*, 128 ff.

79 Peter Achinstein, *The Book of Evidence* (Oxford: Oxford University Press, 2001), 148 (our emphasis).

80 Whitehead, *The Concept of Nature*, 13.

81 Ibid., 13–14.

82 Getting traction from vague terms is essentially the topic of Achinstein's book. However (Achinstein 2001, chapter 7) is especially relevant. Whitehead's discussion in chapter one of *The Concept of Nature* is extremely sensitive to the issues of context and the ways in which terms succeed or fail to refer to real features of the world.

83 Whitehead, *The Concept of Nature*, 40.

84 The notion of "simplicity," is deeply intertwined with the ideas of relevance and abstraction. One of Whitehead's most repeated epigrams was "Seek simplicity, and mistrust it." See also W. V. Quine, "On Simple Theories of a Complex World," *Synthese* 15:1 (1963), 103–106.

85 Tarski enthusiasts to the contrary notwithstanding.

Chapter 11

1 Richard Ely, *The Religious Availability of Whitehead's God* (Madison, WI: University of Wisconsin Press, 1942). See also Paul G. Kuntz, "Can Whitehead Be Made a Christian Philosopher?" *Process Studies*, 12:4 (Winter 1982), 232–242. Kuntz makes a thorough survey of the debate up to the time this article was written. Also see R. Maurice Barineau, "Whitehead and Genuine Evil," *Process Studies*, 19:3 (Fall 1990), 181–188. The issue of radical evil is a serious test case for process philosophy. See also the work on Whitehead in the *American Journal of Theology and Philosophy* by Donald Crosby, George Allan, Jerome Stone, Jerry Sonneson, Nancy Frankenberry, David Connor, and others, have been careful not to over-stretch Whitehead's religious connections.

2 See William T. Myers, "Hartshorne, Whitehead, and the Religious Availability of God," *The Personalist Forum*, 14:2 (Fall 1998), 172–190.

3 Per the distinction between process *theology* and *Whitehead's* concept of god: Process *theology* can enter into a dialogue with, say, evangelicalism, as has been exemplified in a special focus section of *Process Studies*, 37:1 (Spring–Summer 2008), 104–183. However, *none* of the five authors in this section mentions or makes any reference to Whitehead, and rightly. This is only one among many issues that arise from conflating process theology and Whitehead's discussions of

"God." When Whitehead mentions the God of religion, it is usually with a sense of horror rather than hope.
4   See Auxier's analysis of Robert C. Neville's book *Ultimates: Philosophical Theology, Volume 1* (Albany, NY: SUNY Press, 2013), called "The Sherpa and the Sage: Neville on Possibility and Determinateness," *American Journal of Theology and Philosophy* 36:1 (January 2015), 37–50.
5   This point has been very well made by John Shook. We will take it up shortly.
6   There are many competing versions of evolutionary naturalism. Most get the concept of "evolution" wrong, in our view. See Auxier, "The Death of Darwinism and the Limits of Evolution," in *Philo*, 9:2 (Fall–Winter 2006), 193–220; also "The Decline of Evolutionary Naturalism in Later Pragmatism," for *Pragmatism: From Progressivism to Postmodernism*, eds. D. DePew and R. Hollinger (New York: Praeger Books, 1995), 180–207; and "Evolutionary Time, and the Creation of the Space of Life," in *Space, Time, and the Limits of Human Understanding*, eds. Shyam Wuppuluri and Giancarlo Ghirardi (Berlin: Springer-Verlag, 2016), ch. 31. Dewey's evolutionary naturalism is really the model. We think Larry Hickman's interpretation of Dewey's "evolutionary naturalism" is actually better than Dewey's view, which is more varied and confused than Hickman's rendering. See chs. 1 and 2 of Hickman's *Philosophical Tools for Technological Culture* (Bloomington, IN: Indiana University Press, 2001).
7   Richard Rorty, *Philosophy and the Mirror of Nature* (Princeton: Princeton University Press, 1979), 380.
8   See Royce, *The Religious Aspect of Philosophy* (Boston: Houghton Mifflin, 1885).
9   Among the non-religious, even Larry Hickman sees this as consistent with Dewey's view. See *Philosophical Tools*, 109–110. In this regard Hickman departs from many admirers of Dewey.
10 See Whitehead, *Religion in the Making*, 116–117.
11 This level of generality is called "contributionism" by Hartshorne; he aims to be describing both his own and Whitehead's view with that term. We think Hartshorne is conflating transition and concrescence with something more "ultimate" in Whitehead's account. See Hartshorne discusses this in numerous places, but a clear and brief account is in the original "Introduction" to *Philosophers of Process*, Hartshorne, "The Development of Process Philosophy," in *Philosophers of Process*, eds. Douglas Browing and William T. Myers (New York: Fordham University Press, 1998), esp. 400–402. See Kant, *Prolegomena to Any Future Metaphysics*, trans. Lewis White Beck (Indianapolis: Bobbe-Merrill, 1962), sections 14 and 17.
12 Whitehead, *The Concept of Nature*, 14–15.
13 Whitehead, *The Principle of Relativity*, 39. Cf. Enquiry, 15.
14 Some additional discussion on this topic that bears upon our argument is found in the essays in *How Successful Is Naturalism?* ed. Georg Gasser (Frankfurt am Main: Ontos Verlag, 2007).
15 John Shook, "Varieties of Twentieth Century American Naturalism," *The Pluralist*, 6:2 (Summer 2011), 14.
16 Shook, "Varieties of Twentieth Century American Naturalism," 14.
17 Royce is an example. He rejects naturalism because of a deflationary view of the scope and authority of philosophical reflection. See Randell Auxier, *Time, Will and Purpose* (Chicago: Open Court, 2013), chs. 2–4.
18 See Whitehead's discussion in *The Concept of Nature*, esp. 3–6.
19 Shook notes the proximity of Nagel to synoptic pluralism; see "Varieties of Twentieth Century American Naturalism," 14.
20 Mystical experience provides an interesting test case, but it is beyond the scope of the current chapter. See Auxier, *Time, Will and Purpose*, chs. 3, 5, and 6.
21 Shook, "Varieties of Twentieth Century American Naturalism," 15.

22 Ibid.
23 Whitehead, *The Concept of Nature*, vi–vii.
24 Whitehead, *The Concept of Nature*, 2 (all italics and bold italics are ours).
25 Shook's seven viable naturalisms are: Eliminative physicalism, Reductive physicalism, Exclusivist liberal physicalism, Non-reductive physicalism, Exclusivist liberal pluralism, Perspectival pluralism, and Synoptic pluralism. We are concerned with the first three, since they offer interpretations of Darwinism.
26 Thomas Nagel's *Mind and Cosmos* (Oxford: Oxford University Press, 2012) provides a sad example of what happens to a thinker who dares to question the orthodoxy of dogmatic narrow naturalists. The vehemence of his subsequent mobbing in print and in gossip is both shameful and revealing. Philosophers of the mainstream are nothing if not superstitious about the limits of nature. Nagel was an unlikely martyr and evidently failed to anticipate this contumely and vituperation. For scientifically respectable challenge to Darwinian orthodoxy, see, Robert G. B. Reid, *Biological Emergences: Evolution by Natural Experiment* (Cambridge, MA: MIT Press, 2007).
27 Michael J. Behe, *Darwin's Black Box* (New York: The Free Press, 1996). Auxier responds to these sorts of arguments in "The Death of Darwinism," cited above. See also in Magda Costa Carvalho and M. Patrao Neves, "The Bio-Philosophical 'Insufficiency' of Darwinism for Henri Bergson's Evolutionism," *Process Studies*, 41:1 (Spring-Summer 2012), 133–149.
28 This is a long-standing (but repeatedly debunked) favorite of intelligent design enthusiasts. See for example http://www.talkorigins.org/indexcc/CB/CB200_1.html, accessed August 17, 2016.
29 An issue we do not have space to deal with directly is that of "emergence." While this topic has always simmered in Western thought, it has recently come back to the fore with Terrence Deacon's *Incomplete Nature: How Mind Emerged from Matter* (New York: W.W. Norton & Co., 2012). While Deacon Acknowledges that his background in Whitehead's thought is minimal, he does offer some comments which merit a reply, but which exceed our available space.
30 Nagel's argument was hardly an anti-scientific bromide. Michael Chorost, in a rare thoughtful review of Nagel's book, says that in *Mind and Cosmos* the argument is *weaker* than it should have been. Chorost mentions numerous scientists, who are publicly hostile to creationism and intelligent design, have argued for points similar to Nagel's. Yet Nagel makes no use of such scientists or their arguments. See Chorost, Michael, "Where Thomas Nagel Went Wrong," Chronicle Review, Chronicle of Higher Education, 13 May, 2013. http://chronicle.com/article/Where-Thomas-Nagel-Went-Wrong/139129/, accessed August 19, 2016.
31 McDermott, *Works of William James*, 136. See Auxier's discussion of the relation between conjunctive and disjunctive relations in James in *Time, Will, and Purpose*, ch. 7.

Chapter 12

1 Hartshorne says, for example: "I never cared for [Whitehead's] 'eternal objects', as a definite yet primordial multitude of 'forms' of feeling or sensation." We think Hartshorne is deeply confused as to what an eternal object is for Whitehead. Unfortunately, he passed both his confusion and his disdain on to several generations of Whitehead scholars. Hartshorne, *Whitehead's Philosophy: Selected Essays*, 1935–1970 (Lincoln: University of Nebraska Press, 1972), 2. Hartshorne similarly rejects certain other central doctrines of Whitehead in this place and elsewhere.
2 With the Claremont theologians there are conventions of the "empirical," "rational," and "speculative" methods of Whitehead interpretation. Our notion of "radical empiricism" (as related to radical realism) does not conform to

these (now) orthodox categories. But William James radicalized these categories almost a century before the Claremont convention was framed. See the special issue of *Process Studies*, eds. Delwin Brown and Sheila Devaney, 19:2 (Summer 1990), esp. 75–77, in which these categories were articulated and debated. We respect the Claremont discussion and grant that Whitehead's interpreters do fall into these categories, but we disagree that any of these categories adequately represents *Whitehead's* actual views.

3 This phenomenon of denying negative prehension in God followed Hartshorne wherever he went—students at the University of Chicago, Emory, and the University of Texas adopted Hartshorne's predilections. For example, see the article by Rem B. Edwards, a student of Hartshorne at Emory in the early 1960s, who explicitly attributes this style of reading Whitehead to his training under Hartshorne, in "Process Thought and the Spaciness of Mind," *Process Studies*, 19:3 (Fall 1990), 156–166, esp. 160. Edwards, like so many others, confuses space and extension. This confusion is related to his understanding of possibility.

4 The Claremont School includes Marjorie Suchocki, John Cobb, and David Ray Griffin, among those who will say "no negative prehension in God," although they differ in many ways. Hartshorne contended that it *should have been* Whitehead's view, allowing it was not. We think it matters very little whether Whitehead thought there was no negative prehension in God, since, in our view, Whitehead's God was never intended to be the God of religion or a personal being. Hartshorne's God, on the other hand, is at the least "religiously available," to use the phrase of Bill Myers, if not quite the God that would appeal to a devout Calvinist. See Myers, "Hartshorne, Whitehead, and the Religious Availability of God," 172–190. Myers surveys the literature on this debate from 1942 to 1998. The reach of Hartshorne's influence includes the University of Chicago professors of Divinity Bernard Meland and Bernard Loomer, who had many students themselves, including Creighton Peden, Frederick Ferré, Delwin Brown, William A. Beardslee, and Edgar Towne, who, wise in their own way, are well removed from Whitehead. See, for instance, Beadslee's narrative, "Process Thought: On the Borders of Hermeneutics and Theology," *Process Studies* 19:4 (Winter 1990), 230–234. See also Jan Van der Veken's narrative about his indebtedness to this school in "Process Thought from a European Perspective," *Process Studies* 19:4 (Winter 1990), 240–247.

5 See Robert C. Neville, *God the Creator: On the Transcendence and Presence of God* (Chicago: University of Chicago Press, 1968), 16–22, 138–140 (for analogy). Neville employs substantially the same account of analogy in *The Truth of Broken Symbols* (Albany, NY: SUNY Press, 1996), see xvi n; and in *Ultimates: Philosophical Theology Volume 1*, 173–184.

6 See Robert C. Neville, *Creativity and God: A Challenge to Process Philosophy* (New York: The Seabury Press, 1980), chs. 1 and 2. Neville makes it clear that he basically agrees with Ford's interpretation, *as it applies to theology* at least, in these chapters.

7 Neville indicates his awareness that Whitehead can be read (contra Ford) as a radical empiricist, who "could give up the task of [necessitated] ontology and limit himself to [descriptive] metaphysics." (*God the Creator*, 131). But Neville chooses to engage Whitehead primarily through the mediation of the most popular interpretations of the time (in 1968, that was Christian, Lowe, and Ford was beginning to make his influence felt), leaving aside the very important question of whether Whitehead was *ontologically* committed, in the sense of metaphysical necessity, to the narrative he provided in various works. As Neville puts it: "Metaphysics without ontology would prohibit a philosophical account of a transcendent God, for God would be construed either as one of the elements of the metaphysical system or as beyond it altogether." (Ibid., cf. *Eternity and Time's Flow*, 67–68). We could not agree more. Still theologians cannot afford

to be ignorant of philosophy, but philosophers are engaged in an entirely human activity which needs no assistance from theology. Relative to Neville's categoreal scheme, then (summarized in his *The Cosmology of Freedom* [New Haven: Yale University Press, 1974], 30–51, and revised for in *Ultimates*, 193–244), he takes metaphysics to have broader application than the study of cosmology, due to the exemplification of metaphysical categories in every experience, actual or possible. We take cosmology to be the broader field of investigation because it is not restricted by the order of human reflective thought (as philosophy is, including metaphysics), but only by the idea of nature, as argued in the previous chapter. Cosmology is empirical

8 See Auxier's "Note on God" in the appendix to the 1996 edition of *Religion in the Making*.

9 See Neville, *God the Creator*, 264–266; and also his *Creativity and God*, esp. ch. 4, 48–76.

10 See Auxier's essay "God, Process and Persons: Charles Hartshorne and Personalism," *Process Studies* 27:3–4 (Fall–Winter 1998), 175–199.

11 See Whitehead, *Religion in the Making* (New York: Fordham University Press, 1996), 66, 74–75, 86–87.

12 Whitehead's discussions of personal and impersonal "groupings" of occasions leaves us with more questions than answers. See especially *Process and Reality*, corrected edition (New York: Free Press, 1979), 34–36; and *Adventures of Ideas*, 201–208.

13 See Neville, *God the Creator*, 264–273.

14 The crucial passages from *Process and Reality* show that Whitehead did not assume or assert unity as a requirement for personal order, or even person. There is no way to reconcile such an assumption with what Whitehead *does* say about enduring objects that "enjoy" a personal order, or *are* persons. He says that enduring objects, such as "the life of an electron or a man" are subordinate to the actual entities that are associated in them, and that "personal order" comes in strands in enduring objects with such genetic relations (see *Process and Reality*, 92). He denies the immediate unity of any personal order with its past (*Process and Reality*, 105). "Person" is relational all the way down, so there is *no problem* of the unity of the person. Person isn't unified. See also *Process and Reality*, 34–35, 90, 107, 119, 161, 181. The one statement that Whitehead makes in *Process and Reality* that seems contrary, appears on the next to last page of the book (350). It must be read in light of all of these earlier discussions of personal order; note that "peculiar completeness" does not require or imply unity, and leaves open the mystery of person.

15 R. J. Connelly, *Whitehead vs. Hartshorne: Basic Metaphysical Issues* (Washington, DC: University Press of America, 1981). See also Connelly's "Necessary Order in the Primordial Nature of God in Whitehead," *Philosophy Research Archives*, 8 (1982), 513–519; and "Creativity and the Eternal Object in Whitehead," *Philosophy Research Archives*, 5 (1979), 587–610.

16 For instance, see *Process and Reality*, 350.

17 Connelly, *Whitehead vs. Hartshorne*, 137.

18 See Auxier, "Why 100 Years Is Forever: Hartshorne's Theory of Immortality," *Personalist Forum*, 14:2 (Fall 1998), 109–132; discussion 133–140. Daniel Dombrowski responds to that account as if Auxier actually holds the view there described. Auxier does indicate that the view is not his own but is rather the view he thinks *Hartshorne* should have defended so as to be consistent with his own deepest principles. See Dombrowski, *Divine Beauty: The Aesthetics of Charles Hartshorne* (Nashville: Vanderbilt University Press, 2004), 185–189.

19 See Auxier "Why 100 Years Is Forever: Hartshorne's Theory of Immortality," cited above. Hartshorne hammered out the details of this issue with Edgar Sheffield Brightman in their correspondence; see *Hartshorne and Brightman on God, Process and Persons: The Correspondence, 1922–1945*, ed. R. E. Auxier and M. Y. A. Davies (Nashville: Vanderbilt University Press, 2001).
20 Connelly, *Whitehead vs. Hartshorne*, 158.
21 See Robert C. Neville, *Eternity and Time's Flow*, 67ff.
22 For greater detail, see Auxier's essays "Concentric Circles: An Exploration of Three Concepts in Process Metaphysics," *Southwest Philosophy Review*, 7:1 (January 1991), 151–172; and "Bowne on Time, Evolution and History," *The Journal of Speculative Philosophy*, 12:3 (1998), 181–203. Aristotle has a robust account of negation that gives it ontological authority. That view is resisted by radical empiricists.
23 Aristotle, *Metaphysics*, 1015b10–15.
24 Aristotle, *De Interpretatione*, 22b8–9.
25 For example, in *De Interpretatione*, Aristotle works out the notion of logical possibility in terms of "possible *to be*" and "impossible *to be*," thereby establishing the connection between logical possibility and existence. This connection is the one whereby "possibility" gets its metaphysical weight. See 22a1–40.
26 See Neville, *God the Creator*, 90 ff.
27 Richard Creel rightly attributes this view of Hartshorne's to his strong reliance on Peirce's views regarding continuity. The view is not Whiteheadian, since Whitehead, with James, Bergson, and other radical empiricists, does not privilege continuity over discontinuity (nor vice versa). Creel also rightly notices the unsatisfactoriness of Hartshorne's view of possibility as "vague." Vague possibility is supposed to do the work of making a metaphysical continuum, but Whitehead never suggests possibility is vague. Eternal objects have a high degree of structure. See Creel, "Continuity, Possibility, and Omniscience," *Process Studies* 12:4 (Winter 1982), 209–231.
28 See Auxier's "Immediacy and Purpose in Brightman's Philosophy," in *The Hartshorne-Brightman Correspondence* (Nashville: Vanderbilt University Press, 2000), 132–154.
29 See Larry Hickman's article "Why Pierce Didn't Like Dewey's Logic," *Southwest Philosophy Review*, 3 (1986), 178–189.
30 Neville acknowledges also that whether possibilities are created or uncreated is the crux of the argument. See *God the Creator*, 145–146. Lango has an interesting take on this issue, with which we agree, in part. He says:

> Each eternal object [individually] in the actual world of an actual entity has the potentiality for being *prehended* by that actual entity (application of the category of universal relativity to eternal objects). Whereas each created entity is created (i.e., comes into being), each eternal object is eternal (i.e., does not come into being but is always in being). Therefore, although not all created entities are in the actual world of that actual entity (because not all created entities have come into being when that actual entity comes into being), all eternal objects are in the world of that and every actual entity (because all eternal objects are always in being). (*PR* 34) However, since some eternal objects are more relevant than others to that actual entity, each eternal object is prehended either positively or negatively by that actual entity.
> (Lango, *Whitehead's Ontology*, 25)

As far as this goes, it is in agreement with our argument given earlier in chapters 7–9, although Lango neglects the important fact that eternal objects ingress only in ordered groups. Lango wanders away from this solid summary when confronting the difficult matter of what it means for eternal objects to be "uncreated." He continues: "Moreover, an eternal object is a derivative entity because it cannot exist apart from actual entities but must ingress into some actual entity somewhere." (Ibid.) This assertion is (1) unknowable to us, even if it were correct (we would need a position outside of actuality in order to know whether possibilities can exist or cannot exist apart from actuality); and (2) Lango here contradicts Whitehead's insistence upon the independence of eternal objects from both God and the actual world of any given actual entity. Whitehead never makes assertions about the whole of actuality as such—it is forbidden both by his method and by the ontological principle, including the requirement that we refrain from making assertions about the whole of eternal objects relative to the whole of actuality. Lango then says, "A conceptual prehension is abstracted from physical prehensions of actual entities into which that eternal object ingresses." (Ibid.) It is true that this is the method whereby we (inquirers) *get at* eternal objects, as we have demonstrated in our discussion of physical prehension in chapter 8, but it does not imply that what we learn about possibilities thereby exhausts the whole of what possibility *is*. Then Lango says: "In short, each eternal object has the potentiality for and realizes a determinate bond of positive or negative conceptual prehension with every actual entity." (Ibid.) Here Lango uses the term "determinate" where "definite" would have been correct, and he defines the possible by way of its connection to the actual by potentiality. This is the view we have shown to be inadequate in chapter 9. The uncreated character of the possible, and Whitehead's insistence that the order exhibited in possibilities is determinate order (what we have called "constellations" of eternal objects, as distinct from mere definite "collections," which *can* be reduced to their relation to the actual in the fashion Lango has done), together indicate an irreducible independence of the possible from the actual that should render us ontologically contrite and which demands the greatest care when possibility is under discussion. Lango asserts that Whitehead "neglects the relations of eternal objects to one another." (27) We do not agree, entirely; one could say that Whitehead gives less attention to this than he should, or is less explicit than one might hope, but our view is that Part IV is understandable as an analysis of possibilities in relation. So, when Lango asks "Can the problem of finding a general relation [among eternal objects] be dealt with in abstraction from particular instances?" (27), we reply, "yes, and *only* in that way, and that is Part IV."

31 Whitehead uses the terms "general potentialities" for possibilities, and "real potentialities" for what we are calling "potentialities." See *Process and Reality*, 65. In several places Hartshorne also makes this same distinction between "real potentialities" or "real possibilities" (he uses potentiality and possibility interchangeably), and logical possibilities. See for example, *Reality as Social Process* (New York: Hafner, 1971), 97–100.

32 See *Process and Reality*, 22, 219.

33 See *Process and Reality*, 214. Here Whitehead connects what we have said in earlier chapters about transition and concrescence with the relative independence of eternal objects: "To sum up, there are two species of process, macroscopic process, and microscopic process. The macroscopic process is the transition from attained actuality to actuality in attainment; while the microscopic process is the conversion of conditions which are merely real into determinate actuality. The former process effects the transition from the 'actual' to the 'merely real'; and the latter process effects the growth from the real to the actual. The former process is efficient; the latter process is teleological." Notice that Whitehead uses the term "merely real" to include what we are calling the "might-have-been," or any other possibilities considered apart from their relations to the actual (i.e., their determinate order, which is not yet definite). He is not asserting the absolute

independence of possibility (eternal objects), but insisting that possibilities are real, even if "merely" so, when we consider them without actively relating them to the actual.
34 See Whitehead, *Process and Reality*, 41, for the fast explanation of negative prehension; see our chapters 8 and 9 for the longer account.
35 There is an important sense in which the might-have-been is immediately experienced, within the unfolding of a single durational epoch. But to discuss the unfolding of such a durational epoch, one also needs a phenomenological or descriptive method that Whitehead does not provide. It isn't clear whether such a description could easily be brought into any kind of Whiteheadian metaphysics, and certainly not into his cosmology. We think he would probably deny any immediate experience of the might-have-been. Excluded conceptual feeling is about *as* absent from immediate experience as anything *can be*. Whatever excluded conceptual feeling is, it *isn't* theory. Taking theory to be a substitute for excluded conceptual feeling is what model-centrists do, as does anyone else who assumes an abstraction can compensate for what is missing in conceptual feeling.
36 See *Process and Reality*, 226–227, 248.
37 The modal structures under discussion here are much richer than the stripped-down propositional form called "counter-factuals." We are speaking of constellations of such, and in their determinate independence (as "merely real") of the actual. No theory of counter-factuals attempts such breadth.
38 We have adapted this example from Josiah Royce in *The Principles of Logic* (New York: Philosophical Library, 1961), 17–18.
39 For more detail, see Auxier, "Dewey on Religion and History," *Southwest Philosophy Review*, 6:1 (January 1990), 45–58.
40 See *Process and Reality*, 199–206.
41 Whitehead, *Process and Reality*, 317.
42 Ibid., 237; cf. 85, 232.
43 Ibid., 161.
44 Ibid., 243; cf. 267.
45 See C. S. Peirce, *Collected Papers*, eds. C. Hartshorne and P. Weiss (Cambridge: Harvard University Press, 1931–1935), 1.52–53, 1.183. Philosophers have great difficulty with this aspect of mathematical thinking. Compare Peirce's point with what Cassirer says:

> In the field of mathematics itself the new conception [of non-Euclidean geometries] made its way quite rapidly and, it would appear, without any violent internal dissension. By contrast, however, the misgivings encountered in the realm of philosophy were all the stronger. Notable thinkers not only entertained serious doubts as to the admissibility on non-Euclidean geometry, but even declared it an absurdity which philosophy must resist at all the means at its command. These different attitudes can be explained by the distinctive character and the traditional assumptions, respectively, of mathematical and philosophical thought. Mathematics is, and always will be, a science of pure relations, and in its modern form it is precisely this feature that has become more and more pronounced. When figures of any sort are mentioned and their nature is investigated the inquiry is never into their actual existence but into their relation to one another, in which alone mathematics is interested. Indeed, in so far as mathematical thought is concerned their existence does not matter.

Cassirer, *The Problem of Knowledge, Volume 4*, 25–26; cf. 29, 35. It isn't difficult to understand how model-centric thinking spread from mathematics into physics and other physical sciences based on this tendency.

46 See Hickman, "Why Pierce Didn't Like Dewey's Logic," 178–189.
47 See Whitehead, *Process and Reality*, 22.
48 See Josiah Royce, *The Conception of God*, ed. G. H. Howison (New York: Macmillan, 1897), 4–22. See also, Auxier, *Time, Will and Purpose*, chs. 5–6.
49 Donald Viney has summed up these arguments as part of a single "cumulative argument" for the existence of God, an interpretation Hartshorne enthusiastically endorsed. See Viney, *Charles Hartshorne and the Existence of God* (Albany, NY: SUNY Press, 1985).
50 See Dwayne Tunstall's critique of Royce on this point in his "Concerning the God that Is Only a Concept: A Marcellian Critique of Royce," *Transactions of the Charles S. Peirce Society*, 42:3 (Summer 2006), 394–416. Auxier's take on this problem is in chapters 4–8 of *Time, Will and Purpose*.
51 Hartshorne distances himself from Whitehead's concept of eternal objects in many places, but see "Introduction" in *Whitehead's Philosophy: Selected Essays, 1935–1970*.
52 Hartshorne says this in a number of places. See for instance, *Hartshorne and Brightman on God, Process and Persons*, 90.
53 *Hartshorne and Brightman on God, Process and Persons*, 55–56.
54 See Donald W. Sherburne, *A Key to Whitehead's*, 42–43.
55 See Auxier's essay, "Immediacy and Purpose in Brightman's Philosophy," in *God, Process and Persons*, 132–154.
56 William James suggests a number of situations in his famous essay "The Will to Believe" in which a relatively unsparing radical empiricism may allow for belief beyond what particular experience will warrant.
57 *Hartshorne and Brightman on God, Process and Persons*, 24–25, 58–59, 64–66, etc.
58 See for example, Charles Hartshorne, *Aquinas to Whitehead: Seven Centuries of Metaphysics of Religion* (Milwaukee: Marquette University Press, 1976), 6 ff.
59 Peirce, Whitehead, and Cassirer all handle this idea effectively: the supposed "laws" of nature must have evolved (Peirce, Whitehead), or our concept of nature is the ground of our capacity to think about impersonal law-like order (Whitehead), or all objects of scientific knowing are products of a long cultural development, not a priori features of the universe itself (Cassirer, Whitehead). On the relation between Whitehead's views and Cassirer's on the cultural embeddedness of science, see Auxier, "Ernst Cassirer and Susanne Langer," *A Handbook of Whiteheadian Process Thought*, eds. Michel Weber and Will Desmond, vol. 2 (Frankfurt am Main: Ontos Verlag, 2008), 552–570.
60 See Connelly, Whitehead vs. Hartshorne, 137.
61 All pantheists are regarded as impersonalists in the historical discourse of personalism. Pantheism is simply one step short of materialism, in the view of personalists. See Jan Olof Bengtsson, *The Worldview of Personalism* (Oxford: Oxford University, 2006) for the fuller account and arguments.

Chapter 13

1 Whitehead, *Modes of Thought*, 100.
2 We are reminded here of Iris Murdoch's brilliant *Metaphysics as a Guide to Morals*.
3 This view is defended most notably by Donald Sherburne and George Allan.
4 Of course, Hartshorne has many able philosophical defenders of his own, such as Daniel Dombrowski and Donald Wayne Viney. Auxier has expressed his independent interpretation of Hartshorne elsewhere, and it is with that work we would expect Hartshorne's defenders to take issue. Here we are concerned only with Hartshorne's influence on the interpretation of Whitehead. Hartshorne had

a powerful way of convincing people there was something amiss about the idea of eternal objects in Whitehead, and that conviction spread far and wide through his student John Cobb. Ironically, Richard Rorty was persuaded to write his master's thesis on that problem under Hartshorne's direction.
5 We have already noted that Jorge Nobo is an exception to this, as are Jude Jones and George Lucas.
6 Suchocki's view of the process God is complex. On one side, she asserted in her "Intellectual Autobiography" at the 2008 meeting of the Highlands Institute for American Religious and Philosophical Thought (Assisi, Italy, August 4–7) that there is no negative prehension in God and she defended this thesis (clearly a part of the Hartshorne "correction" of Whitehead's God that Hartshorne initiated no later than 1948 in *The Divine Relativity*). But in 2010, Suchocki published an article critical of this "correction" by Hartshorne, and calling for a re-examination of Whitehead's actual view of God. That view was reiterated at the annual meeting of the same group in 2016. But the list of thinkers she supplies as following Whitehead instead of Hartshorne on the question of God consists mainly of people we would regard as following Hartshorne, or people who completely misunderstand Whitehead, such as Ford. Perhaps Suchocki became aware that the process God defended by the Claremont School was more Hartshorne's than Whitehead's and became intrigued with alternatives. We applaud the call in 2010 for a re-examination of Whitehead's conception of God without Hartshorne's "corrections." We believe we have taken up Suchocki's challenge throughout this book, and especially in the final chapters. See her "The Dynamic God," in *Process Studies*, 39:1 (Spring-Summer 2010), 39–58. David Connor has effectively responded to Suchocki's effort to solve a very real problem. See his essay "The Plight of a Theoretical Deity: A Response to Suchocki's 'The Dynamic God'," *Process Studies*, 41:1 (Spring-Summer 2012), 111–132, especially see 113.
7 See Auxier, "Why 100 Years Is Forever: Hartshorne and Immortality," cited above.
8 See Auxier's "God, Process and Persons: Charles Hartshorne and Personalism," cited above, for the full account of this claim.
9 See Borden Parker Bowne's penultimate chapter, "The Failure of Impersonalism," in *Personalism* (Boston: Houghton Mifflin, 1908), 217–267. See also Auxier, "Bowne on Time, Evolution and History," 181–203.
10 See Auxier's "God as Catholic and Personal," *International Philosophical Quarterly*, 40:2 (June 2000), 235–252.
11 Suchocki asserts that the "primordial nature" of God is "dynamic," but that cannot be maintained unless one confuses possibility with potentiality in Whitehead. See Suchocki, "The Dynamic God," 40–41. Catherine Keller, by contrast, does not seek to be a systematic theologian, preferring narrative and re-mythologizing to structural and formal analyses, but her pre-occupation with issues of creativity and power have the effect, in her writing, of treating possibility as functionally the same thing as potency. Her book *Face of the Deep* (London: Routledge, 2003), Part IV *Creatio ex profundis*, is as close as she comes to offering an ontology. It is packed with great ideas and information, as her books are, but we hold out for an understanding of process that does not conflate potentiality and possibility.
12 See Rorty and Vattimo, *The Future of Religion* (New York: Columbia University Press, 2005). Roman Madzia has nicely summarized it in his essay "Only a God Can Save Us: Richard Rorty's Philosophy of Social Hope Beyond Secularism," in *Richard Rorty and Beyond*, eds. Randall E. Auxier and Krzsztof Piotr Skowroński (Lanham, MD: Lexington Books, forthcoming). Madzia says:

> The death of the God of metaphysics, as it turns out, does not undermine religion as such but re-enables us to grow mature and take it [religion] seriously

again. What the death of the God of the Western metaphysical tradition has shown us is that authentic religious belief does not consist in knowledge (i.e., a series of propositions) or incontestable proofs of God's existence but, on the contrary, authentic belief can only grow out of taking the evidence for atheism with all due seriousness, acknowledging the fact of the disenchantment of the world, ending up with a numbing facticity of absence of God, and subsequently deciding to go beyond it by means of hope alone.

Although, we do not renounce metaphysics, with Rorty, Vattimo, and Madzia, this "weak theology" seems a fair place to begin the discussion of what theists should do in light of our argument.

13 Auxier has addressed the sort of philosophical theology that attempts to naturalize God in "Gordon Kaufman's Astronauts," *American Journal of Theology and Philosophy*, 29:1 (January 2008), 18–33.

14 Auxier's article "Why 100 Years Is Forever," cited above, contains an extended version of this argument with all of the citations to Hartshorne that are needed.

15 For more about this way of using the term "evolution, see Auxier, "Bowne on Time, Evolution and History," cited above; and "The Decline of Evolutionary Naturalism in Later Pragmatism," in *Pragmatism: From Progressivism to Postmodernism*, eds. David DePew and Robert Hollinger (New York: Praeger Books, 1995), 180–207; and "The Death of Darwinism and the Limits of Evolution," in *Philo*, 9:2 (Fall–Winter 2006), 193–220; and "Evolutionary Time and the Creation of Space," in *Space, Time, and the Limits of Human Understanding*, eds. Shyam Wuppuluri and Giancarlo Ghirardi (Berlin: Springer Verlag, 2016), ch. 31.

16 See chapter 9, regarding infinitesimals and non-standard analysis. This area of mathematical investigation is exotic and far from settled. For example, see Paul Cohen's "forcing" technique, which discovers (creates?) new infinities where none was thought to exist. There is evidence of this kind of structural thinking in Whitehead's earliest work, even though the concepts themselves were not fully formalized until after his death. This could be called the "structural aspect of the problem of accretion." See for example, *A Treatise of Universal Algebra*, 124–125 (Prop. V) for an instance—one of many—in which Whitehead toys with these sorts of ideas.

17 The issue our conclusion presents for traditional theism is explored in Auxier, "God as Catholic and Personal: A Protestant Perspective on Norris Clarke's Neo-Thomistic Personalism," cited above.

18 We use "event" and "universe" in the way described by our Introduction. Any empirical experience that can be analyzed as an actual entity in its actual world must answer to experience, ultimately, and hence it is appropriate to speak of experience beyond the hypothesized coordinate whole in this context. The problem of creativity is more than just cosmological.

19 See our discussion of dissipation and death at the end of chapter 4.

20 This view is outside of orthodoxy. The Ecumenical Councils, especially Chalcedon and Third Constantinople, went to great trouble to distinguish Jesus's human nature from his divine nature, to the point of suggesting (at Third Constantinople) that, following John's Gospel, ch. 17, the divine nature might have vacated the world prior to the arrest by the Sanhedrin so that no "passion" (suffering) would be attributed to the Godhead, and that only the human nature dies on the cross, the divine nature having accomplished the glorification of God through the "signs" given to the people, both greater and lesser. There is a reason, we think, that the Fifth and Sixth Ecumenical Councils had the concrete effect of splitting Christendom into irreconcilable factions, making the whole Christian world susceptible to the more coherent cosmology of Islam. It has never been easy to work out the relationship between mortality and God. We do

so here only at the cost of defying orthodoxy, but why would we worry about such a thing at this point? If the orthodox Christians of any sort are at all correct, the authors can hope for no better than a Circle of the Inferno with a low number.
21 We have noticed that most scholars apologize for the use of gender-exclusive language on its first occurrence. We offer ours, such as it is, on its last. We don't take the gendered God any more seriously than the ungendered God, and we try to accommodate the now popular erasure of all anachronistic usage with our scare quotes, taking down with one pair of strokes the capitalization and the gendering habits of old.
22 Whitehead, *Interpretation of Science*, 197–198. This passage follows upon Whitehead's introduction of algebraic thinking as the very basis of rationality. The final essay of this volume, "Immortality," is devoted entirely to the exegesis and explanation of the idea expressed in this paragraph. There he says: "There is finitude—unless this were true, infinity would have no meaning." (248).

Index

Note: The name Whitehead and the term "relation" (in many forms) occur on almost every page of the book. We have not included them in the index and make sufficient our choice by alerting the reader here that this is a book about Whitehead and relations.

Abraham, Abrahamic 278, 282, 284, 293
Absolute, the, absoluteness (includes absolute knowledge) 214, 272
absolute break in nature 45, 246, 335n24, 346n33
absolute measurement 72, 88, 117
abstract, abstraction 2, 4–6, 8, 10–12, 15–16, 23–4, 26, 28, 31, 36–7, 42, 45, 47–8, 50, 52, 55–6, 61, 64, 70–1, 76–9, 86, 92–4, 109, 112–13, 120, 122, 131–2, 134–5, 138–9, 141–2, 144–5, 150, 153, 155, 163–9, 171–3, 176–9, 181, 184–8, 190, 193–6, 200–1, 209, 212–15, 217–20, 227–8, 240, 242, 245–50, 252–5, 258–9, 262, 268, 274–8, 287, 291, 296, 305n44, 306n45, 309n22, 309n23, 311n30, 313n60, 316n32, 318n3, 321n8, 325n27, 340n84, 346n30, 347n35; *see also* concrete
abstraction, extensive 11, 15, 47, 165, 168, 193, 309n22, 316n32
Achinstein, Peter 218, 340n79, 340n82
act (includes act of abstraction; act of cognition; act of knowledge; act of measurement; etc.) 11, 27, 37, 71, 81, 101–2, 107, 112, 128, 131, 145, 147, 150, 152–4, 165, 167, 169, 173, 178, 189, 214, 219–20, 224–5, 227–8, 230, 235, 237, 248–51, 253, 255, 257–8, 261–2, 264–5, 269, 283, 286–7, 288–9, 292, 295–6, 308n19, 331n62; *see also* dynamic

action 40, 44, 50, 127, 201, 206, 212, 221, 226–8, 237, 256–7, 275–6, 293; *see also* act; active, activity; actual, actuality; dynamic
active, activity 14, 31, 32, 44, 83, 96, 111, 128, 165, 184–5, 206, 208, 210, 213, 215, 223, 225–8, 244, 247–8, 256, 265, 269–71, 275–6, 280, 287, 290–2, 295, 305n45, 308n19, 344n7, 347n33; *see also* act; action; actual, actuality; dynamic; selection, selective attention
actual, actuality 3–6, 9, 16, 31, 41–2, 71–3, 80, 86–9, 105, 109–10, 112–13, 115, 122, 124, 128–36, 139–49, 151–3, 155–7, 160, 162–5, 169–70, 174, 176, 178–9, 182, 185–93, 233, 244–6, 249–69, 271, 274, 276–9, 286–90, 295, 304n34, 305n45, 307n8, 307n13, 308n15, 309n23, 312n53, 314n69, 318n3, 323n4, 326n32, 327n42, 329n39, 330n47, 331n62, 333n6, 338n35, 344n7, 346n30, 346n33, 347n37; *see also* act; action; active, activity; dynamic; experience, actual; possibility
actual entity 1, 6–10, 12, 24, 37, 43, 47, 71, 7–76, 79, 113–14, 125, 128–49, 152–7, 162–74, 176–8, 180, 184–92, 227, 244, 247, 251–2, 256, 263, 275, 277, 289, 291–2, 294, 298n9, 300n26, 300n27, 303n32, 304n34, 304n44, 305–6n45, 309n30, 310n31,

354 Index

316n25, 317n49, 323n4, 325n24, 325–6n27, 330n60, 331n62, 332n66, 332n71, 333n78, 337n21, 344n14, 345n30, 346n30, 350n18; *see also* quantum of explanation
occasion(s) 37, 73, 77–8, 88, 113, 130, 132, 146, 150, 152, 156–7, 161–3, 167–8, 173, 189, 242–3, 246, 277, 286–7, 305–6n45, 308n15, 312n53, 315n25, 316n25, 323n4, 325n24, 335n24
actual experience 35, 259, 263, 265
actualization 80, 81, 138, 163, 287, 292
actual world 3, 5, 6, 8–10, 29, 37, 47, 71, 73, 79, 109, 131–5, 137–41, 143–7, 155, 162–4, 166–9, 173–4, 176, 178–9, 185, 189–91, 252, 256, 263, 277, 285–6, 291, 305n45, 310n31, 323n4, 326n27, 331n62, 345n30, 350n18
adequacy (includes inadequacy) 2, 10, 25, 28, 34–5, 40, 43, 46, 70, 100, 114, 116–17, 124, 131, 140–1, 151, 170, 178, 203–4, 206–7, 210–11, 214, 218, 223, 226, 228, 245, 265, 273, 280, 290, 295, 297n3, 298n9, 304n40, 320n5, 332n71, 343n2
  descriptive 218, 280
  empirical 2, 70, 203–4, 207, 211, 218, 297n3
adjustment 31, 118–21, 123, 126, 132, 325n27
adventure 9, 12–13, 31–6, 193, 280
Adventures of Ideas (main text only) 31, 33–5, 39, 125, 204, 221
adversion 191
aesthetics 28, 41, 122, 192, 223, 229–30, 277, 288, 292, 299n12, 335–6n41, 344n18; *see also* experience; feeling; valuation, value
Allan, George 222, 231, 317n40, 335–6n41, 340n1, 348n3
Altizer, Thomas J.J. 223
analogy 35, 38, 60, 81, 86, 94, 114–15, 117–18, 120–8, 130–2, 149, 194, 201, 220, 225, 233–4, 241–2, 244, 249–51, 255, 258, 262, 267–9, 275–9, 281–7, 290–2, 295, 314n78, 315n16, 324n10, 330n60, 334n16, 336n41, 343n5
analysis, compositional 18, 305n45

analysis (includes analyze) 6–10, 12, 20–1, 29–32, 35, 37–8, 42, 47, 52, 56–7, 62, 76, 78–9, 83, 85, 90, 92, 97, 113, 115, 122–4, 128, 131–6, 138, 141, 143, 146, 156–7, 161, 165–73, 176–84, 187–8, 192–3, 195, 218, 227, 243, 244, 246, 249–50, 253, 256, 259, 261–2, 270, 278, 285, 295–6, 306n45, 311n30, 312n53, 323n4, 325–6n27, 326n32, 330n47, 330n61, 331n62, 332n64, 334n17, 346n30, 349n11, 350n18
  coordinate 13, 15, 31–4, 36, 47–8, 71, 75, 99, 112–17, 121, 124–5, 128–9, 133–4, 136, 138, 140–1, 145, 151, 316n25, 316n38, 324n10
  genetic 13, 15, 99, 112–17, 121–2, 124–5, 128–9, 132–6, 138, 140–1, 145, 151, 309n31, 324n10
  mathematical 56–7, 176
  non-standard 180–1, 184, 334n12, 350n16
  of possibility 176, 180–2, 346n30
analytic philosophy 16, 20, 41, 58, 203–4, 210, 307n10, 312n45, 321n8
appetition 304n35, 306n2, 332n71; *see also* desire
applicability 10, 16, 18, 21, 23, 26–7, 30, 35–6, 43, 51–4, 59–60, 63, 78, 96, 104, 106, 108–10, 114, 116–17, 119, 124–5, 131, 140–1, 161, 168, 174–5, 178, 182, 193, 197, 209, 298n9, 305n44, 312n48, 317n47, 344n7, 345n30
Aquinas, St. Thomas (includes Thomism) 151, 159–60, 230, 242, 267, 290, 348n58, 350n17
Aramaki, Norirtoshi 303n28
Aristotle 11, 19, 40, 54, 63, 78, 84, 91, 100, 224, 245–6, 250, 266, 296, 307n8, 312n53, 324n9, 327n42, 338n35, 345n22, 345n23
atheism 266, 271, 350n12; *see also* pantheism
atomic, atomism (includes *atomos*) 9–10, 42–6, 58, 66–9, 71, 75–7, 79, 81, 83, 87–8, 92, 97, 99, 109–10, 112, 128, 131–3, 137–8, 145, 153–4, 161, 170, 179, 211, 298n9, 300n27, 306n2,

314n69, 315n18, 316n32; *see also* individual; pluralism, synoptic
atomicity, temporal 42–8, 129, 138, 153, 157, 161, 243, 251, 269, 303n27, 305n45, 310n31, 315n25, 318n3, 325n25, 327n48
attain, attainment 36, 73, 135–6, 242, 291, 346n33; *see also* satisfaction; valuation, value
Augustine, Saint 279–80
Auxier, Randall 17, 27, 244, 273, 277, 285, 298n8, 299n17, 300n30, 301n2, 306n3, 307n8, 315n20, 324n11, 325n18, 329n43, 332n63, 332n68, 334n11, 334n17, 335n24, 341n4, 341n6, 341n17, 341n20, 342n27, 342n31, 344n8, 344n10, 344n18, 345n19, 345n22, 345n28, 347n39, 348n48, 348n50, 348n55, 348n59, 348n4, 349n7, 349n8, 349n9, 349n10, 349n12, 350n13, 350n14, 350n15, 350n17
Averroeists, Latin 266

baseball 115–16, 119–23, 127, 132, 324n12
beauty 34, 38
belief, believe, believer (includes disbelief) 40, 46, 52, 66, 187, 200, 203, 223, 248, 262, 264, 266, 269, 274, 284, 291, 295, 316n32, 348n56, 350n12
Bell, J.L. 313n63, 313n67
Bell, Jason M. 332n63
Bergson, Henri 11, 14, 23, 39, 112–13, 230, 248, 275–6, 281, 292, 299n11, 299n17, 300n30, 306n3, 320n5, 333n78, 342n27, 345n27
Berkeley, George 215
Bogaard, Paul 312n53
Boltzmann, Ludwig 67
Bonaparte, Napoleon 163
Boston, MA 119, 121
Boston Red Sox 114, 123
Bowne, Borden Parker 310n32, 345n22, 349n9, 350n15
Bradley, F.H. 23
Bradley, James 18, 303n25, 311n30, 330n61, 331n61
Brandom, Robert 185, 201, 332n76, 335n25
Brightman, Edgar Sheffield 247, 264–5, 279, 345n19, 345n28, 348n52, 348n55, 348n57

Brown, Bryson 182, 185, 201, 332n76, 334n14
Buddha, Buddhism 295, 303n28

Calvinism, Calvinist 267, 343n4
Cambridge, England 68, 91, 199
Canales, Jimena 320n2, 320n5
Cantor, Georg 91
Capaldi, Nicholas 307n10, 312n45
Carnap, Rudolf 218, 339n43
Cassirer, Ernst 11, 27–9, 34, 113, 302n18, 302n19, 304n34, 311n30, 320n5, 322n19, 323n6, 323n7, 328n12, 331n61, 334n8, 338n36, 339n43, 347n45, 348n59
Castillon, Frederic 175, 324n9, 333n1
categories (includes categoreal scheme) 4, 11, 29, 33, 39, 76–7, 97, 109–10, 114, 118, 122–3, 129, 133, 138–9, 141–3, 145, 148–9, 153, 155, 162, 168, 191, 197, 230, 237, 243, 261, 264, 274, 277–8, 292, 302n14, 304n34, 305n45, 306n45, 306n2, 311n30, 315n19, 316n35, 317n42, 317n44, 322n9, 325n27, 326n27, 327n52, 331n62, 343n2, 344n7, 345n30
category theory 21–2, 60–4, 94, 313n60, 313n63, 317n47
Cauchy and Weierstrass 181, 185
causal efficacy 37, 63, 89, 130, 147, 165, 179, 331n62
causation, cause 25, 57, 73, 78, 141–2, 147, 151, 153, 159, 163, 167, 174, 177, 210, 248, 256, 261, 304n44, 309n23, 315n16, 323n4, 330n47, 336n15
certainty (includes uncertain) 9, 40, 44, 46, 76, 100, 157, 161–2, 172, 184, 188, 192, 224, 232, 288, 292–3, 323n23; *see also* necessity
character, characterize 4, 8, 12, 28, 30–2, 36–8, 40, 43, 45–7, 51–2, 55–6, 60–1, 63, 65, 67–9, 71, 73, 75, 77, 79, 82, 87–90, 93, 95, 97, 99, 101, 103–4, 107, 109–10, 112–13, 117–18, 121–3, 128, 131, 133, 139–42, 144, 146, 149, 153, 155–8, 162–5, 167, 169, 171–5, 178–85, 190–3, 200, 205, 207, 209–12, 214, 216, 219–20, 230, 232–3, 240, 243–8, 256, 259, 261, 262, 269, 273–4, 276–7, 280, 283, 286, 288–9, 308n19, 311n30,

316n25, 316n31, 317n55, 319n13, 329n38, 331n62, 335n23, 335n32, 336n1, 337n18, 346n30, 347n45
characteristic function 183, 335n23
general character 68, 157–8
metaphysical 158, 209
physical character 155–8
relational 60, 97, 99, 131, 216
structural characteristics 112, 165; *see also* structure
universal characteristic 113, 149, 153, 175, 329n38; *see also* Leibniz, G.W.
Christ, Christian, Christianity 33–4, 280, 293–6, 340n1, 350n20; *see also* Jesus
Christian, William 18, 19, 244, 299n23, 331n62, 343n7
church 201, 290, 293
civilization 7, 33–6, 48, 51, 56, 82, 243, 248, 278
Claremont theology 14, 241–2, 248, 269, 271–2, 285, 342–3n2, 343n4, 349n6
Clarke, Bowman 89, 91, 96, 319n17, 319n18
Clemens, Samuel L. *see* Twain, Mark
Cobb, John B. 17–18, 20, 300n31, 316n25, 343n4, 349n4
Code, Murray 22, 300n27, 301n39, 310n2, 319n13, 338n42
Cohen, Paul 181, 350n16
coherence (includes incoherent) 2, 35, 47, 70–1, 75, 80, 88, 97, 100, 103, 108, 125, 199, 201, 203, 207–8, 210, 216–17, 276, 281, 297n3, 317n47, 350n20; *see also* truth
communication 2, 24, 276
community 30, 83, 89, 94, 97, 157, 196, 198, 201, 206, 209
conceive, concept, conceptual 1, 4–5, 7–9, 11, 31, 33, 36, 40–1, 43, 47, 52, 55, 60–1, 63, 65, 71–4, 91–5, 99, 102, 105, 114, 118, 120, 128–9, 133, 139, 145, 149, 151–2, 156, 159–62, 164, 170, 175–6, 178–84, 186–9, 192–3, 199, 211–12, 215, 217–18, 221–35, 237, 241–5, 249, 251, 253, 256–9, 261–9, 271–5, 277–9, 282–7, 289–90, 292–4, 297n3, 298n7, 298n9, 301n7, 305n45, 306n45, 306n2, 309n22, 311n30, 311n40, 314n73, 314n2, 326n27, 329n33, 329n40, 330n61, 332n71, 334n11, 337n31, 340n3, 341n6, 346n30, 347n35, 347n45, 348n51, 348n59, 349n6, 350n16
conceive of God, conception of God 241, 243–5, 263, 266–7, 269, 271–5, 278–9, 282–3, 292, 294–5, 349n6
conceptual feelings 63, 149, 170, 178–9, 186, 189, 192, 230, 251, 253, 255–6, 261, 264, 268, 287, 347n35
concrescence 6, 12, 26, 37, 43, 44, 47–8, 63, 76, 80, 109, 128, 130, 133–41, 144–5, 147–9, 158, 162–70, 172–4, 177–8, 184, 186–92, 247, 251–2, 263, 291, 305n44, 305n45, 316n25, 316n34, 325n27, 326n27, 326n41, 330n47, 330n60, 331n61, 331n62, 332n71, 333n78, 341n11, 346n33; *see also* data; transition
concrete 2, 4–6, 8–10, 28, 31, 37–8, 42–3, 46–8, 50, 52–3, 55, 58, 63, 71, 76–9, 93, 98, 112–13, 118, 123–4, 131–5, 137–40, 142, 144–5, 162, 167, 169–70, 172, 181–2, 186–7, 193–5, 205, 208, 214–15, 217, 226, 246–7, 249–50, 255, 258–72, 277–8, 286–8, 291, 296, 298n9, 305n45, 306n45, 308n19, 309n22, 321n9, 322n9, 325n27, 326n27, 330n47, 331n62, 350n20
concreteness, misplaced (includes philosopher's fallacy) 42–3, 46–7, 98, 113, 144–5, 217, 298n9
Connelly, R.J. 243–4, 268, 344n15, 344n17, 345n20, 348n60
conscious, consciousness (includes self-conscious; unconscious) 27, 29, 39, 44, 66, 80, 127, 143, 147, 151, 208, 214, 234–5, 238, 255–7, 261–2, 264, 268, 274, 278, 284, 293, 304n44, 306n2, 308n19; *see also* act; action; activity; experience, reflective
construction 28, 36, 44, 61, 69, 78, 80, 83, 111, 116, 123, 125, 142, 145, 182, 194, 200–6, 208, 211, 213–14, 257, 312n48; *see also* model
Continental philosophy 14, 16, 210
continuity (includes discontinuity) 18, 25, 27, 37, 42–7, 50, 65–6, 81,

Index 357

84–8, 94, 104, 138, 152–4, 187, 236–7, 250, 269, 271, 275, 281, 288–9, 292, 308n15, 309n26, 309n30, 309n31, 310n31, 314n69, 315n25, 316n32, 345n27; *see also* time
continuum (includes extensive continuum) 12, 36–7, 44–5, 71–2, 113, 125, 128–37, 139, 145–6, 152, 154–5, 161, 169–70, 176, 179, 186, 250, 319n15, 334n10, 335n23; *see also* extension
contrast (includes generic contrast) 8, 37, 57, 63, 129–30, 135–7, 143, 146, 155, 162, 171–4, 176–7, 184, 187, 189–91, 206, 221, 227–8, 240, 242, 246, 248, 251, 253, 256–7, 261, 273, 280, 282, 293, 296, 327n24, 330n60, 332n71, 335n24
coordinate analysis *see* analysis, coordinate
Corrington, Robert 222, 230, 283
Corry, Leo 94, 313n60, 313n62
cosmic epoch (includes theory of cosmic epochs) 5, 6, 9, 12, 36–7, 68–9, 73, 113–14, 120–1, 127–31, 133–8, 146, 150–8, 160–2, 165, 168–9, 171, 174, 191, 209, 215, 220, 265, 281, 290, 314n10, 328n33, 329n38, 329n39, 335n24, 337n31, 339n73, 340n73
  theory of 12, 36–7, 154, 161–2, 281, 290, 314n10, 328n33, 340n73
cosmological, cosmology 151–6, 158, 160–2, 166, 168, 202, 208, 212, 221–2, 232, 240–1, 246–8, 277, 297n3, 298n9, 299n13, 302n12, 316n25, 320n1, 321n5, 322n18, 323n25, 325n17, 332n71, 337n31, 340n73, 344n7, 347n35, 350n18, 350n29
cosmology (includes philosophical cosmology; physical cosmology) 7–9, 12–13, 17, 25, 32–3, 35–7, 73, 79, 94, 98–100, 103–5, 107–8, 110–11, 116, 121, 125, 127–32, 136, 146–54, 156, 158, 160–2, 166, 168, 202, 208, 212, 221–2, 232, 240–1, 246–8, 277, 297n3, 298n9, 302n12, 316n25, 320n1, 320n5, 321n5, 322n18, 325n17, 332n71, 340n73, 344n7, 347n35, 350n18, 350n20

Crawford, David R. 174–9, 183, 333n5, 334n8
creative, creativity (includes creative synthesis) 5, 14–15, 37, 80, 135, 144, 148, 192–3, 227, 246–9, 251, 270–1, 283, 285–9, 291–3, 295–6, 332n71, 349n11, 350n18; *see also* ultimate
creative advance 2, 80, 206, 317n58
Crosby, Donald 222, 340n1

Darwin, Charles (includes Darwinism) 233, 235, 341n6, 342n25, 342n26, 342n27, 350n15
data, datum 29, 93, 99, 132–3, 136–7, 161, 163, 166 (diagram), 168, 183, 187, 189, 191–2, 199, 219, 233, 256, 286–8, 290–2, 315n15, 330n47; *see also* concrescence; transition
Davidson, Donald 304n34
Dawkins, Richard 233
Dedekind, Richard 45, 112, 309n26, 313n57
definite, definition (includes define; defined; indefinite; undefined) 6, 9, 24–6, 29, 32, 35, 38, 42, 47, 50, 52, 56–7, 61, 63, 73–4, 81, 96, 109–10, 113–14, 128–9, 132, 135, 144–7, 157–8, 161–4, 166–74, 177–9, 183–4, 187–92, 195, 204, 212–13, 218–19, 221, 229, 236, 240, 243–5, 250–2, 281, 284–5, 287–9, 291, 303n32, 305n45, 306n45, 307n11, 308n15, 309n23, 313n58, 316n35, 318n2, 330n47, 330n57, 330n60, 331n62, 333n6, 334n11, 342n1, 346n30, 346n33
Deleuze, Gilles 14, 299n11
demonstration 25, 48, 65, 67, 74, 106, 124, 132, 135, 145, 167, 181, 183, 232, 241, 248–9, 268, 301n6, 310n31, 313n61
Descartes, René, Cartesian 11, 15, 33, 35, 112, 121, 148–9, 158–61, 175–6, 307n7
description, descriptive (includes redescription) 8–9, 27, 29–37, 40–1, 45–7, 70, 76, 99, 112–14, 128, 130, 133–4, 136–7, 139, 141, 146, 150, 152, 179, 200–1, 204, 210–11, 218–19, 252, 255, 265, 280, 308n19, 315n17, 322n10,

336n15, 343n7, 347n35; *see also* metaphysics
coordinate 32, 47
genetic 30–2, 141, 150
metaphysical (includes descriptive metaphysics) 8–9, 47, 130, 141, 265, 343n7
desire 4, 6, 9, 208, 306n2, 312n53; *see also* appetition
Desmet, Ronny 299n18, 315n12, 321n8, 328n16
determination, determinate (includes indetermination) 5, 18, 26, 31, 36, 38, 72–3, 76, 95, 103–4, 109–10, 114–15, 117, 121, 128–31, 133–6, 138, 142–7, 150–2, 154–7, 162–8, 170–1, 173–4, 177, 184, 187–8, 190–2, 208, 211, 215, 221–2, 224, 233, 245–6, 250–1, 253–5, 262, 266–7, 270, 279, 281–3, 285, 305n45, 306n45, 308n19, 316n34, 316n35, 321n9, 323n2, 330n47, 330n57, 330n60, 331n62, 332n66, 333n6, 346n30, 346n33, 347n37
determinism 47, 170, 267
Dewey, John 11, 39–40, 51–2, 58, 100, 111, 113, 181, 237, 248, 255, 259, 310n5, 311n43, 323n23, 328n12, 341n6, 341n9, 345n29
discontinuity *see* continuity
divine, divinity (includes non-divine) 151, 243, 264–5, 270, 273, 276–7, 279–86, 288, 291–4, 296, 350n20; *see also* God
divisibility, divisible, division (includes indivisibility) 12, 36–7, 43, 48, 72, 75, 87–8, 92, 112–42, 144, 152–3, 155–6, 165, 169, 176–7, 179, 186, 192, 238, 331n62; *see also* atomic
coordinate 72, 75, 88, 144–5, 156, 158, 169–70, 172–4, 177, 187–8, 191, 228, 316n39
genetic 37, 75, 88, 144, 156, 169–70, 172–4, 177, 188, 228, 316n39
undivided divisibility 12, 36–7, 92, 145, 150, 152–3, 155, 169–70, 186, 250; *see also* continuum
Dombrowski, Daniel 344n18, 348n4
Dummett, Michael 61–2, 314n68
Duns Scotus, John 151, 242, 266
duration 3, 5, 42, 44, 46, 81, 87, 150, 246–7, 275, 284, 318n3, 347n35

dynamic (includes *dynamis*) 5, 26, 29, 117–19, 121–3, 125–7, 131, 139, 148–9, 162, 231, 235, 275, 300n32, 321n6, 349n6, 349n11; *see also* act; action; activity; actual, actuality

Eastman, Timothy 22, 111, 300n27, 301n40, 320n5, 323n20, 323n25
Einstein, Albert 30, 54, 60, 63, 67–8, 74, 76, 86, 93–4, 98–103, 106–11, 116, 134, 161, 170, 202, 212, 315n17, 320n2, 320n5, 321n8, 322n14; *see also* relativity, theory of
electromagnetism (includes electromagnetic) 3, 68, 74, 80, 108–9, 157–9
electron (includes electronic occasions) 5, 77–8, 155, 206, 297n3, 344n14
elimination 43, 58, 60, 131, 133–4, 136, 138, 141–2, 144, 147, 155, 163, 166–74, 181, 184, 188, 190, 192, 218, 231, 239–40, 255, 257–8, 261, 268, 281, 330n47, 331n62, 332n71, 342n25; *see also* prehension, negative
Ely, Richard 221, 340n1
empirical, empiricism (includes empiricist) 4–5, 11, 16, 28–9, 45–6, 61, 64, 74, 83, 86, 92–3, 100, 103, 106, 108, 111, 114, 119–21, 132, 148, 151, 154, 161, 167–8, 181, 188, 196, 200–8, 210, 213, 215, 225, 230, 237, 240, 255, 274–8, 280–2, 285, 289, 295, 306n2, 310n31, 324n9, 324n11–12, 333n1, 337n21, 338n31, 338n35, 342n2, 334n7, 350n18
adequacy 2, 21, 46, 70, 203–4, 206–7, 211, 218, 297n3
dogmas of 38, 225
radical 10–11, 35, 39, 41–3, 46–50, 76, 82–3, 89–90, 92–4, 97, 107, 109–10, 124, 129–30, 151, 157, 160, 196, 205, 207, 210, 213, 216, 225–6, 228–30, 232–3, 236–8, 240, 245, 247, 249, 252, 265, 271, 273, 277, 303n24, 306n2, 307n7, 308n19, 324n10, 328n16, 328n22, 328n24, 338n42, 342n2, 343n7, 345n22, 345n27

energy (includes energistic) 3, 9, 13, 15–16, 33, 45, 67, 70, 74, 102–5, 155, 161, 173–4, 226, 283, 287, 292, 295, 315n25, 316n32, 320n5; *see also* quantum
Enlightenment, the 148, 304n34, 307n10, 312n45
entity, actual *see* actual entity; quantum, of explanation
environment 128, 139, 157, 161, 234, 255–6
epistemology 40, 63, 99, 217, 219, 223–4, 277, 304n34, 324n10, 339n73; *see also* knowledge
epoch, epochal *see* cosmic epoch; immediate epoch
error 10, 25, 34, 46–7, 66–7, 80, 117, 124, 160, 179, 206, 212, 224, 226, 228, 233, 234, 238, 264, 271, 290, 333n7, 336n15; *see also* evolution
eternal objects *see* possibility
eternity 147, 245, 279–80, 343n7, 345n21
ethics 14, 223, 227, 248, 293; *see also* valuation, value
Europe, European 14, 33–6, 343n4; *see also* philosophy, Continental
event 3–10, 27–8, 42, 44, 47, 65, 80–1, 87, 139, 189, 192, 199, 247, 259, 275, 277–8, 285–90, 292, 294, 298n8–9, 304n40, 305n45, 309n30, 315–16n25, 350n18
evolution 79, 139, 171, 173–4, 224, 233–6, 283–5, 288–90, 333n79, 333n82, 341n6, 342n26–7, 345n22, 350n15; *see also* error
exclusion 145–6, 149, 151, 153, 155, 166, 192, 237, 244, 251, 253–4, 256, 258, 261, 267, 277, 286; *see also* negation; prehension, negative; selection
existence (includes exists; existential) 4–6, 9, 12, 30, 37, 40, 46–7, 50, 76, 79, 98, 110, 112, 114, 117, 123, 129, 131, 136, 145–6, 149–51, 153–4, 157, 159, 161–2, 165–7, 175–6, 178–9, 184–6, 195, 199, 202, 211, 214, 217, 222–3, 225–8, 230–2, 235–8, 240, 242–3, 246, 249, 252–3, 259, 261–3, 265, 267–8, 273–80, 283–5, 289–95, 300n27, 306n45, 316n32, 323n4, 331n62, 332n76, 345n25, 346n30, 347n45, 348n49, 350n12, 350n16
  concrete 2, 37, 137, 140, 259, 298n9
experience 2–3, 5, 9, 16, 20, 25, 27, 29–32, 34–5, 39–40, 42–4, 46–50, 61–3, 69–71, 73, 76, 78–81, 86, 90, 92, 93–4, 97–8, 107–8, 110–11, 116–17, 122–5, 127–31, 134, 136, 139–40, 143, 146–7, 152–3, 155–7, 164–5, 175, 186, 188–90, 192–3, 195, 200, 205, 207–8, 210–11, 213–17, 219–32, 235–40, 242, 244, 246–9, 253, 255, 258–9, 261–9, 273–87, 289–91, 294–5, 297n3, 298n9, 303n32, 305n44–5, 306n2, 308n19, 309n22, 309n29, 316n32, 318n3, 325n17, 329n38, 332n62, 332n71, 333n78, 334n7, 335n24, 335n32, 339n73, 341n20, 344n7, 347n35, 348n56, 350n18; *see also* truth
  actual 35, 259, 263, 265
  concrete 71, 124, 226, 291, 298n9
  conscious 44, 234, 238, 248
  God's (includes divine) 244, 255, 276, 279–81, 283–7, 292
  human 40, 46, 69, 198, 200, 207, 222, 225, 238, 277, 281, 284, 286, 295, 305
  immediate, immediacy of 29, 50, 70, 230–2, 246–8, 255, 291, 347n35
  new 277–8, 284–7, 289–90, 294
  particular 29, 42, 46–7, 226, 236–8, 240, 308n19, 309n29, 334n7, 348n56
  past 81, 244, 287
  reflective 222, 225–8, 230, 235
  religious 20, 32, 222–3, 225, 229, 277
  sense 208, 316n32
  unreflective 225–8, 230–2, 235
explanation *see* quantum, of explanation
extension, extensive 11–12, 15–16, 43, 59, 61–4, 66, 69–73, 75, 77, 82–4, 86–94, 97, 102–3, 112, 114, 125, 128–30, 135, 138–41, 143, 145–6, 150, 152–3, 156–8, 161, 174–6, 179, 182, 228, 245, 250, 271, 296, 298n9, 305–6n45, 307n11, 309n28, 314n2, 316n34, 319n18, 327n6, 333n7, 340n74, 343n3; *see also* continuum
  abstraction 11, 15, 47, 165, 168, 193, 309n22, 316n32

connectedness, connection 43, 69, 73, 82, 85, 89, 91, 97, 99, 147, 153, 157–8, 179
  irreducible 12, 97, 138
  relatedness 70–5, 88–9, 91, 95, 115, 128, 130, 154, 160
  undivided divisibility 12, 36, 69, 79, 92, 109, 112, 129–30, 138, 140, 150, 153, 170

faith 225, 270, 295
feeling 48, 75, 85, 134, 137–8, 140–2, 146, 149, 156, 163, 167–9, 172, 183, 186–92, 221–2, 238, 244, 255–6, 258, 261, 304n35, 325n24, 330n47, 331n62, 342n1; *see also* prehend, prehension
  conceptual 63, 149, 170, 178–9, 186, 189, 192, 230, 251, 253, 255–6, 261, 264, 268, 287, 347n35
  physical 63, 143, 161, 179, 186, 251, 255, 264–5, 325n24, 330n47, 331n62
Ferré, Frederick 271, 279, 332n62, 343n4
Ford, Lewis S. 17–19, 30, 42, 83–4, 86–8, 106, 129, 132, 153, 209, 222, 243, 300n27, 303n27, 303n29, 305n45, 312n53, 315n25, 317n2, 318n3, 318n6, 319n13, 319n15, 325n25, 327n48, 338n42, 343n6–7, 349n6
freedom (includes free will) 44, 224–8, 237, 244, 249, 260, 264–5, 267, 269, 274, 277, 280, 287, 291, 296, 321n5
Frege, Gottlob 112, 313n57, 330n61
future 44, 126, 130, 152, 177, 192, 224, 245, 249–50, 252, 256–62, 268–9, 276–80, 282, 287, 290–2, 295, 336n41

generic contrast *see* contrast, generic contrast
genetic (includes account; description; specification) 19, 30–5, 38, 40, 60, 63, 75, 99, 112–15, 117–24, 128, 131–4, 136–8, 140–1, 143–4, 146, 150, 157, 169–70, 188, 192, 302n8, 309n31, 316n39, 321n9, 324n9, 325n25, 344n14
  analysis *see* analysis, genetic
  divisibility 37, 75, 88, 130–1, 135, 156

division 113–14, 129–30, 132–3, 135–40, 142, 144, 169, 172–4, 177, 188, 228
Gentzen, Gerhard 59, 96, 180
geometry *see* mathematics
Gerla, Giangiacomo 94–6, 314n78, 320n30, 320n35
God 1, 4, 7–9, 18, 47, 79, 116, 135, 144–5, 148–9, 151, 153, 155, 221, 223–5, 231, 237, 240–50, 252, 255, 262–96, 317n57, 329n35, 343n3–4, 343n7, 346n30, 349n6, 349n11, 350n12, 350n20, 351n21; *see also* Absolute
  consequent nature of 241, 244, 266–72, 278, 284, 288
  personality of 241–4, 247, 255, 263–4, 266–9, 271, 273–7, 279–80, 283, 285–6, 288, 294–5
  primordial nature of 244, 249, 263, 266, 268, 272, 278, 285, 287, 344n15, 349n11
  of religion 221, 247–8, 266–7, 270, 276, 279, 282, 284–6, 293–6, 340n11, 341n3, 350n20
Gödel, Kurt 176
Gogh, Vincent van 200
Goldblatt, Robert 61, 313n63, 313n65, 334n13
Goodman, Nelson 229
gravitation, gravity 3–4, 98–104, 106–9, 126–7, 200, 202, 320n1, 321n5–6, 322n15–16
Greene, Brian 111, 200, 322n12, 336n14
Griffin, David Ray 17, 298n7, 343n4
group theory 34, 36, 60, 107, 113, 118, 120, 131, 159, 165, 171, 211, 232, 307n11, 313n58, 322n18–19, 323n20, 326n30, 326n36, 330n60, 344n12, 345n30, 349n6

Halmos, Paul 60, 94, 193, 313n56–7, 320n26
Hamilton, William Rowan 150
Harman, Gilbert 218
Harman, Graham 14, 299n13
harmony 33, 40, 247, 264, 274, 288
Hartmann, Nicolai 159
Hartshorne, Charles 17, 149, 221–2, 230, 241, 243–4, 247–9, 252, 263–7, 269, 271–3, 275–9, 281–6, 288, 296, 315n25, 328n15, 333n6, 340n2, 341n11, 242n1,

Index   361

343n3, 343n4, 344n10, 344n15, 344n17, 344n18, 345n19, 345n27, 345n28, 345n31, 348n49, 348n51, 348n52, 348n53, 348n57, 348n4, 349n6, 349n7, 349n8, 350n14
Hawking, Stephen 197–201, 203–6, 208, 211, 218, 337n19, 337n27
Hegel, G.W.F. 23
Heidegger, Martin 14, 23, 263, 297n3
Heisenberg, Werner 65, 67
Hempel, Carl 218
Heraclitean 149
Herstein, Gary L. 90–1, 297n4, 303n26, 310n2, 311n24, 311n27, 313n55, 313n59, 314n77, 315n14, 316n36, 318n4, 321n7, 322n10, 323n20, 328n16, 328n32, 340n73
Hilbert, David 96, 159, 179–80, 323n7
Hintikka, Jaakko 58–9, 96, 100, 113, 170, 181, 311n44, 312n47, 312n50, 320n38, 323n3, 336n4–6
Holland, Georg Jonathan von 175
Hubbard, L. Ron 262
Hughes, G.E. 307n11
Hume, David 33, 35, 49, 148, 176, 186, 196–7, 207–8, 210, 215–16, 304n34, 337n26
Husserl, Edmund 159, 230, 259, 308n19
  epoche 27

idea(s) 1–2, 5, 10, 15–17, 20–1, 23–4, 30, 33–4, 41, 44, 47, 51–3, 56–61, 64, 66–8, 70, 72–3, 78–9, 83–4, 86, 91–2, 95, 102, 105–7, 112, 116, 120, 123–5, 129, 131, 134, 144–6, 148–54, 156–7, 159, 165–6, 168–70, 173–6, 180–6, 192, 195–6, 198–200, 203–5, 207–8, 214, 221, 223–8, 230–3, 235–7, 240, 242–5, 248–9, 252, 258, 263, 265–7, 271–2, 275–6, 278, 280–6, 288–96, 298n9, 300n27, 303n27, 316n25, 317n47, 317n55, 325n17, 329n40, 330n47, 331n62, 332n71, 334n10, 339n49
  abstract 4, 249
  of God 241, 243–4, 263, 267, 271, 276, 278, 280, 286
  living 41, 308n14
  of nature 221, 223–8, 230, 232, 240, 317n55, 344n7
  novel 7

ideal(s) 24, 45, 54, 62, 155, 204, 245, 292, 307n7, 316n32; *see also* actual, actuality; ideal form; possibility; reality
ideal form 24, 27, 248
idealism 29, 35, 47, 153, 265
ignorance 44, 54–5, 129, 197, 199, 208, 214, 234, 245, 293, 344n7; *see also* error
imagination 6, 14, 29, 41, 44–5, 48, 58, 70, 74, 87, 98–9, 118, 123, 129, 138, 140, 143, 149–50, 152, 156–8, 160, 164, 168, 170, 184, 187, 190–1, 195, 205, 215, 240, 247, 249–50, 253–5, 257–60, 262, 264, 280–1, 284, 290, 296, 303n27, 336n41
immediate epoch 153–7, 161, 169–71, 188, 190–1, 215, 248, 306n2, 329n39, 337n21, 339n73
immortality 224–5, 237, 244, 252, 272–3, 275–6, 291, 293–6, 351n22
impersonal, impersonalism 272, 274–5, 277, 293, 344n12, 348n59, 348n61
individual(s), individuality 23–4, 37, 57, 73, 101, 133, 157, 165, 177, 183, 187, 226, 252, 269, 275, 279, 285, 287, 300n26, 323n4, 331n61, 333n78; *see also* divisibility; unique
  empirical 80, 254, 264–5, 269, 272, 275, 315–16n25
  logical 165, 177, 179, 185, 319n17, 345n30
  metaphysical 55, 67, 149, 285
infinite, infinity (includes infinitesimal) 3, 8, 39, 45–6, 56, 65, 91–2, 115, 131, 149, 163, 177, 180–7, 195, 206–8, 212, 215–16, 224, 240, 249, 251, 254, 261, 268, 276, 282–3, 293, 295, 332n71, 332n76, 334n10, 335n19, 350n16, 251n22
information theory 95, 123, 127, 301n7, 317n58, 349n11
ingression 26, 136, 147, 163–6, 171, 176, 180, 187–92, 248, 251, 302n13, 308n15, 330n54, 345–6n30
inquiry 7–13, 23–40, 45, 47–8, 50–1, 53–4, 56, 58–60, 64–5, 67–8, 70, 73, 75, 78, 85, 88, 92–3, 96–7, 100–1, 108, 113–20, 140–1, 151,

158, 160–1, 171, 180–2, 185, 194–7, 200–1, 206, 225, 230, 236, 241, 245, 248, 256, 298n8, 302n12, 304n40, 306n45
instinct 232, 248
institution(s) 40, 100
Islam (includes Muslims) 293, 350n20

James, Bill 121
James, William 11, 35, 39, 42, 46–8, 50, 130, 189, 210, 213–14, 229, 236, 248, 273, 306n2, 308n19, 338n31, 343n2, 345n25, 348n56; *see also* pluralism; pragmatism; radical empiricism
Jaspers, Karl 159
jazz 50; *see also* music
Jesus 280, 293–6, 350n20; *see also* Christ
Jones, Jude 19, 300n28, 301n2, 332n69, 332n73, 349n5
Judaism (includes Jews) 293
judgment 37, 57, 114, 135, 137, 139–40, 189, 226, 248, 250, 253, 261, 272, 275, 305n44, 306n2, 335n34; *see also* perception; truth

Kant, Immanuel (includes Kantian(s)) 11, 14–16, 25, 33, 35, 108, 112, 147–8, 160, 176, 221–2, 224, 226, 242, 248, 297n3, 299n12, 303–4n34, 307n7, 311n30, 316n38, 337n26
Keller, Catherine 14, 20, 299n15, 349n11
King, Martin Luther Jr. 293
Klein, Felix 60, 321n8
know, knowable, knower, knowing, knowledge (includes unknown) 6, 8–13, 15–16, 21, 28–9, 31–2, 36, 40–1, 45, 53, 58–9, 99–100, 103–4, 106, 111, 117–18, 120, 123–4, 129–30, 140, 145, 147, 151, 155, 158–9, 160, 162, 168, 174, 176, 183, 186–92, 197–9, 207, 209–11, 213–17, 221–3, 226–7, 229, 232, 236, 238, 240–1, 245, 248–52, 255–7, 261, 264–9, 274–5, 277–8, 280–2, 287, 289, 291, 293, 295–6, 300n26, 303n32, 305–6n45, 306n2, 307n9, 308n19, 334n7, 340n73, 346n30, 350n12; *see also* epistemology

natural 12, 16, 26, 29, 34, 41, 45, 67, 86, 88, 103–4, 106, 160, 215, 217, 219, 238, 301n7, 302n12, 303n26, 305–6n45, 311n41, 314n3, 326n27, 337n21
ontological 29, 32, 36, 40, 42, 45, 47, 243, 247
philosophical 15, 29, 32, 36
scientific 11, 27, 29, 99, 104, 106–7, 174, 229, 232, 238, 303n28, 348n59
Kripke, Saul 95, 220, 307n11, 312n48
Kuhn, Thomas 93

Laguna, Theodore de 91, 319n22
Lambert, J.L. 175, 311n30
Langer, Suzanne K. 302n19, 330n57, 348n59
language 6, 25, 28–9, 37, 54, 61–2, 70, 75, 78, 82, 85, 98–9, 122–4, 129, 181, 203–4, 212, 242, 277–8, 280, 283, 288, 296, 317n55, 325n27, 330n60, 334n11
Latour, Bruno 14
Leclerc, Ivor 19, 299n24, 312n53, 316n25, 331n62
Leibniz, G.W. 11, 19, 101, 113, 121, 147–9, 158, 175–7, 182, 184, 247, 288, 319n15, 327–8n12, 333n3
Lesniewski, Andrzej 92
Lewis, C.I. 41, 95, 175, 181–2, 324n9, 328n12, 333n1–2, 334n13
Lewis, David 41, 328n24
Lipton, Peter 218
Locke, John 11, 148, 162, 175
logic 2, 4, 7–8, 10–11, 14–15, 21, 28, 31–2, 34–5, 40–2, 46–7, 51–64, 66, 68–72, 74–8, 81, 83–4, 86, 88, 90–7, 99–103, 105, 107–12, 114, 117, 119, 128–9, 131, 135, 137, 140–2, 144–5, 148, 152–4, 160, 164–5, 168, 170–2, 175–83, 185–6, 190, 193–5, 199–201, 204, 208–9, 211, 214, 216, 219–21, 223, 228, 234, 245–6, 248, 251, 255–6, 263, 275, 277–8, 282–3, 288, 298n9, 301n6, 303n28, 307n7, 307n11, 309n27, 310n31, 310n12, 311n30, 312n46, 312n48, 313n57, 313n60–1, 314n2, 315n16, 315n19, 316n32, 319n19, 324n9, 327–8n12, 329n39, 332n63, 332n71, 332n76, 333n1,

334n8, 334n10, 334n13, 334n16, 335n23, 335n27, 338n42, 345n25, 346n31; *see also* mathematical, mathematics; necessity; negation; relation(s)
  coherence (logical) 2, 70–1, 88, 100, 183, 199, 207, 210, 297n3
  formal 58–9, 66, 91, 96, 181, 193, 196, 248, 312n46
  inquiry 53, 58, 204
  modal *see* modal, modal logic
  relational 29, 168
  rigor 10, 34–5, 42, 46–7, 53, 57, 114, 117, 119, 124, 140–1, 178, 194, 206–7, 219, 298n9
  tools 7–8, 10, 14, 201, 332n76
love 267, 269, 277
Lowe, Victor 18–20, 299n17, 299n19, 300n29–30, 306n2, 343n7
Lucas, George 300n27, 349n5
lure 163–4, 268, 282, 304n35

McHenry, Leemon 305n45, 337n31
McTaggart, J.M.E. 306n2
Madzia, Roman 349n12, 350n12
Malebranche, Nicolas 288
Many, the 63, 75, 86, 141, 149, 227, 237, 246, 289, 295; *see also* One, the
materialism 56, 115, 126, 140, 146, 217, 274–6, 325n27, 348n61
mathematical, mathematics (includes mathematician(s)) 6–7, 9–12, 14–17, 19–21, 23–4, 26–31, 33, 39, 45, 48, 50–70, 74, 82, 85, 91–7, 100–1, 103, 105, 107–8, 110–13, 121, 123, 129, 148–51, 153, 159–60, 168, 170, 177, 181, 185, 190, 192–3, 198, 200–1, 205–6, 208, 210, 212, 216–18, 253–4, 256, 259, 307n7, 310–11n23, 311n30, 313n60, 314n70, 314n76, 314n11, 315n12, 319n13, 320n35, 320n5, 321n8, 322n18, 326n30, 326n36, 328n20–1, 334n13, 334n16, 335n41, 336n41, 337n19, 337n21, 337n27, 347n45, 350n16; *see also* logic
  algebra (includes algebraic; also includes Boolean, Heyting, Lie, etc.) 13, 15, 21, 33, 47, 53, 55–7, 59–62, 74–5, 82, 84, 86, 91, 94–5, 109, 112, 115, 118, 121, 125, 127, 148–9, 157, 162, 165, 168, 175, 182, 192–3, 196, 217, 254–5, 301n38, 311n30, 311n40, 312n54, 313n57–8, 313n60, 316n34, 317n47, 319n13, 325n19, 328n12, 334n16–17, 351n22
  education 51–3, 55, 96, 311n40
  geometry (includes Euclidean; non-Euclidean; projective; etc.) 15, 33, 47, 54–5, 60–1, 64–5, 69, 72, 82, 86, 91–4, 97, 101–7, 113, 115, 118, 121, 123, 125–7, 132, 149–50, 157, 162, 307n7, 311n40, 321n8, 325n19, 347n45; *see also* space
  inquiry 53–4, 350n16
  Principia Mathematica 12, 15, 52, 58–9, 63–4, 66, 86, 92, 112, 150, 193, 212, 310n8, 327n6
  Universal Algebra 12, 50, 63, 68, 82, 86, 91, 93, 112, 148, 150, 162, 307n11, 314n6, 320n34, 322n18, 324n13, 327n6, 334n16
Maxwell, Grover 98
Maxwell, James Clerk 67–8, 74, 109, 157–8, 315n13, 329n38
Mead, George Herbert 237, 309n23
measure, measurement 6, 11, 21, 25, 31, 38, 46, 54, 68, 72, 84, 86, 88–9, 94–5, 100–8, 111, 114–20, 122, 126–8, 131, 134, 150–1, 153–4, 159, 194, 201–2, 238, 260, 280, 310n23, 316n32, 321n9, 322n11, 324n10, 324n12, 336n15
mechanism, mechanistic *see* materialism
Meinong, Felix 252
memory 43–5, 53, 80–1, 257, 262, 267, 275, 288, 333n78
mereology 71, 85, 91–2, 94, 96, 164, 314n2; *see also* topology
mereotopology 16, 60, 62–3, 71, 83, 85, 91, 94, 96, 114, 125, 131, 148, 151, 312n54, 314n2, 319n18, 323n2
metaphor 2, 25, 29, 70, 74, 94, 98–9, 120, 151, 178, 182, 191, 208, 212, 235, 254, 294, 303n22, 324n10
metaphysics 2, 5, 8, 15, 17, 21, 27–30, 35, 37, 39–40, 42, 46–7, 51, 55, 57–9, 62–4, 66–71, 73–8, 81–6, 89–90, 92, 94–9, 108–11, 113–14, 118, 120–4, 128–30,

132, 140–1, 149–51, 153–4, 158, 170, 176, 185, 195–7, 203–9, 215, 218, 220, 223, 227–8, 237, 240, 242–3, 245–6, 251, 265–6, 271, 273, 277, 282, 288, 290–2, 298n9, 299–300n26, 302n12, 303n22, 303n27, 303n32, 305n44, 307n8–9, 313n66, 316n32, 319n15, 319n19, 322n9, 324n10, 331–2n62, 332n63, 337n21, 337n26, 337n31, 338n35, 338n42, 343–4n7, 345n25, 345n27, 347n35, 349–50n12; see also ontology
descriptive 8–9, 27, 40, 47, 114, 130, 133–4, 141, 146, 265, 280, 343n7, 347n35
substance 29, 63, 307n8
method, methodology 6, 11, 13–15, 17–18, 20, 27–8, 30, 32–4, 36, 40–1, 43–5, 48–9, 52–6, 58–9, 67, 84–5, 91, 97–8, 109, 113–14, 116–17, 121, 124, 138, 152, 175, 177, 180, 182, 193, 196–7, 201, 205, 209–10, 215, 232, 236, 238, 241, 245, 273, 297n3, 300n26, 302n12, 303n29, 309n22, 312n48, 324n9, 336n15, 337n28, 342n2, 346n30, 347n35; see also coordinate analysis; extensive, abstraction; genetic, analysis
Michaelson and Morley 68
mind (includes mental; mentality; mind-body) 11, 28, 82, 86, 92, 159, 168, 184, 186, 194, 207, 209, 215–17, 219, 220, 227, 231, 234–5, 264, 280, 288–9, 304–5n44
Minkowski, Hermann 321n8
misplaced concreteness see concreteness, misplaced
modal, modality, modes 2, 4, 6, 8–9, 22, 28–9, 33, 37, 41, 57, 76, 82–3, 92, 95, 129–32, 136, 145–7, 149–51, 161–2, 164, 168–9, 175, 180, 185, 187–90, 214, 227–31, 236–9, 242, 245, 247, 249, 252, 261, 272–3, 276, 277–80, 294–6, 307n11, 308n15, 309n30, 319n13, 328n13, 332n76, 333n6, 347n37
modal logic 41, 95, 172, 175, 180–2, 185–6, 220, 307n11, 312n48, 320n31, 334n13, 335n26–7
model 59, 61, 67, 77, 79, 98–100, 105, 107–11, 114–28, 131–2, 134, 139, 142, 170–4, 176–81, 182–7, 193–7, 199–205, 207, 212, 217–18, 220, 227–8, 250, 256, 305n45, 309n30, 311n30, 312n48–9, 313n67, 320n38, 320n1, 320n5, 321n6, 323n25, 324n10, 325n19, 326n38, 330n61, 332n71, 332n76, 334n10, 334n15–17, 336n5, 337n28, 339n55, 341n6
model-centrism 13, 98–9, 111, 116, 121–2, 156, 170, 196–7, 199–202, 204–5, 210, 218, 223, 233, 272, 321n5, 324n10, 337n18, 347n35
monism 78, 226, 237, 266
morphology 88, 143–4, 146, 148, 158, 326n38
Muhammed 295
multiplicity (includes multiple universes) 21, 26, 50, 105, 122, 130, 139–40, 162, 165, 183, 185, 192, 204–5, 207–8, 212, 217, 231, 238, 256, 308n15; see also plurality
Murdoch, Iris 2, 297n2–3, 303n22, 348n2
music 3, 50, 57, 125, 309n24
Myers, William T. 221, 340n2, 341n11, 343n4
myth, mythic 115, 118, 249, 315n15, 322n19, 329n44, 349n11

Nader, Ralph 260
Nagel, Thomas 234, 341n19, 342n26, 342n30
narrative 2, 23, 31–3, 45, 47–8, 60, 70, 74–5, 90, 110, 114–15, 118, 127, 129, 148, 181, 207, 233, 254, 261–2, 297n3, 313n55, 319n13, 343n4, 343n7, 349n11
NASA 323n25, 336n2
naturalism 16, 69, 146, 151, 153, 222–4, 227, 229, 231–3, 235–7, 239–41, 246, 267, 276, 302n12, 316n32, 319n13, 338n42, 339n68, 341n6, 341n17, 342n25–6; see also radical realism; synoptic pluralism
religious 222–3, 225, 229–30
theological 16, 222–3
necessity (includes unnecessary) 4, 8, 27, 29, 32, 35, 40–2, 47, 78, 82, 86, 95, 102, 104, 106–8, 113, 129–30; 136, 143, 147, 155–8, 170, 172–6, 180–1, 186, 189–90,

Index   365

199, 203, 207–9, 215–17, 222, 224, 232, 235, 243, 245–7, 249, 259, 261–2, 264, 273, 275–6, 280–1, 285–8, 295–6, 307n9–11, 308n15, 321n5, 332n71, 333n5, 340n73–4, 343n7; *see also* logic; negation; possibility
  logical 41, 47, 103, 129, 131, 153, 208, 246
  ontological 36, 40–1, 43, 47, 129, 154, 158, 170, 245
negation *see* exclusion; necessity; selection
Neumann, John von 65, 94, 314n75
Neville, Robert C. 222, 230, 242–3, 245–6, 248–9, 252, 263, 265–6, 270–1, 279, 282–3, 285, 341n4, 343n5–7, 344n9, 344n13, 345n21, 345n26, 345n30
Newton, Isaac 68, 74, 101, 150–1, 154, 181, 184–5, 321n6, 322n15, 326n36
Nietzsche, Friedrich 44, 295
nihilism 235
Nobo, Jorge 12, 18–19, 298n9, 299n25, 300n27, 305n45, 314n73, 317n1, 325–6n27, 326n41, 327n42, 327n44, 331n61, 335n36, 349n5
Nolan, Christopher (and Jonathan) 43–5, 309n24
nominalism 338n35
novelty 23, 56, 82, 108–9, 135, 141, 144, 173, 179, 187, 193, 227, 242, 283, 287–8, 297n3, 304n35; *see also* creativity
number, number theory 45, 52–4, 56, 104–5, 108, 115, 123, 132, 175–7, 180, 202, 212, 218, 221, 309n26, 310n8, 313n66, 321n5, 323n1, 335n19

objects, eternal *see* possibility
occasion (includes actual, concrete, dipolar, electronic, epochal, physical, protonic, temporal, etc.) 9–10, 37, 39, 49, 51, 63, 71, 73, 77–80, 88, 113, 130, 132, 146, 150, 152, 156–8, 161–3, 167–8, 173, 189, 242–3, 246, 253, 277, 286–7, 291, 298n9, 304n44, 305–6n45, 308n15, 312n53, 315–16n25, 323n4, 325n24, 325n27, 330n47, 330n57, 335n24, 337n21, 344n12

One, the 63, 75, 86, 109–10, 149, 227, 246, 258, 260, 266, 295; *see also* Many, the
ontological knowledge *see* knowledge, ontological
ontological principle 8, 47, 132, 145, 168, 285, 298n9, 332n70, 346n30
ontology 1, 8, 12, 14, 19, 21, 28, 32, 36, 40–3, 45, 47, 61, 64–5, 86, 115, 129, 132, 140–2, 145, 168, 176, 180, 182, 189, 192, 230–2, 234, 237, 242–3, 245–7, 249, 251–2, 256, 259, 263–5, 267, 271, 275, 280–3, 285, 289–90, 292, 298n9, 303n28, 303n32, 304n40, 307n7, 315n16, 323n2, 329n35, 334n8, 343n7, 345n22, 346n30, 349n11; *see also* knowledge, ontological; metaphysics; ontological principle
organism, philosophy of (includes organic; organization) 8, 12–13, 19, 22–3, 26, 30, 38, 48, 57, 71–2, 75, 77–80, 96, 110, 127–8, 134–5, 139–40, 143–5, 147–8, 150, 153, 157–8, 161, 165, 168, 224, 232–5, 237–8, 253, 255–6, 258, 276, 299n13, 301n39, 303n28, 305n45, 326n32, 329n35, 331n61, 332n71, 333n78, 337n21

Palin, Isabella 303n29
Palter, Robert M. 89, 306n2, 319n16
panentheism 272, 289
panpsychism 329n37
pantheism 266, 269, 283, 348n61
Parmenides, Parmenidean 111, 149
Pascal, Blaise 79
past 81, 130, 141, 162, 177, 189, 192, 244–6, 249, 252, 255–64, 267–9, 272–3, 275–7, 280, 282, 284, 287–91, 331n62, 333n78, 344n14; *see also* time
  objective immortality of 244, 252, 272–3, 275, 291
Peirce, Charles S. 11, 37, 170, 174–5, 177–8, 181, 183, 189, 229, 237, 242, 248, 259, 298n7, 299n9, 300n27, 304n35, 333n4, 335n35, 338n35, 345n27, 347n45, 348n59
Pepper, Stephen C. 229
perception 12, 27, 30, 37–8, 57, 63, 73, 86, 89–90, 93, 98, 107, 112, 130, 137, 139–40, 147–8, 159, 169, 174, 179, 186, 196,

213, 215, 219, 237, 256, 275, 300n26, 304–5n44, 321n8, 322n19, 331n62, 339n70; *see also* causal efficacy; feeling; judgment; presentational immediacy
person, personalism, personality (includes impersonalism) 77, 206, 241–5, 247, 249, 254–5, 263, 265–77, 279–80, 283, 285–6, 288, 293–5, 305n44, 314n9, 317n49, 343n4, 344n12, 344n14, 348n59, 348n61, 349n9
Pettegrove, J.P. 304n34
phenomenological, phenomenology 27, 29, 31, 34, 37, 45, 47–8, 76, 111, 255, 259, 263, 347n35
philosophical knowledge *see* knowledge, philosophical
philosophy
　analytic 16, 20, 41, 58, 203–4, 210, 307n10, 312n45, 321n8
　ancient 15, 160, 293, 330n61
　Continental *see* Continental philosophy
　medieval 108, 111, 149, 151, 158, 160, 197, 201, 242, 252, 262
　modern 14, 32–3, 35, 37, 108, 113, 121, 131, 147–51, 158, 160–2, 175, 246, 304n34, 321n5, 330n61
　natural 7, 10, 12, 17, 30, 68, 82, 85–6, 90, 95, 99–100, 148, 151–2, 154–7, 216, 232, 298n9, 302n12, 303n26, 317n55, 328n21, 338n42; *see also* physics
　organic *see* organism, philosophy of
　process *see* process
physics 19, 21–2, 46, 67–70, 73–4, 76–7, 94, 98, 100, 104–7, 150, 155–6, 158–61, 171, 181, 185, 196–202, 208, 215, 217, 232, 237, 256, 298n9, 315n15, 316n25, 317n47, 320n1, 337n19, 337n21, 337n31, 338n41, 347n45; *see also* reality, physical
Planck, Max 67, 74, 202
Plato (includes Neo-Platonism; Platonic; Platonism) 2, 4, 11, 19, 33–5, 54–6, 84, 112, 146, 190–1, 200, 240, 276, 283, 297n3, 307n7, 331n62
pluralism, synoptic *see* synoptic pluralism
plurality 10, 26, 165, 177, 221, 224, 226, 228–9, 231, 237, 240, 246, 271, 308n15, 341n18, 342n25; *see also* multiplicity
possibility (includes eternal objects; impossible) 1, 3–6, 7–9, 13, 16, 24, 26, 29–32, 39, 41, 44, 50, 55–8, 63–4, 71, 75–6, 80, 86, 89, 95–6, 100, 102–4, 106, 108–11, 113–14, 117–18, 122–3, 125–8, 130–56, 162–9, 171–4, 176–92, 195, 197, 201–2, 206, 208, 214–17, 220, 222–4, 227, 231–5, 238–42, 244–72, 275–82, 285–90, 294–5, 298n7, 300n27, 304n34, 305n45, 306n2, 307n11–12, 308n15, 309n25, 309–10n31, 312n53, 319n13, 320n5, 322n18, 323n2, 323n4, 326n27, 327n44, 327n12, 328n13, 328n24, 329n33, 329n35, 330n47, 330n54, 330n56–7, 330n60, 331n32–3, 331n62, 333n6, 334n8, 334n10, 338n31, 338n35, 338n42, 340n73, 342n1, 343n3, 344n7, 344n15, 345n25, 345n27, 345–6n30, 346n31, 346–7n33, 349n4, 349n11; *see also* actual, actuality; potentiality; reality
potentiality 42, 45, 72–3, 75–6, 79–81, 85, 87, 93, 109–10, 112–13, 128, 133–5, 139, 144–7, 153, 158, 164–5, 173–5, 188–9, 208, 227–8, 250–62, 266–8, 270–2, 275, 277–8, 285, 287, 312n53, 314n69, 329n33, 331n62, 338n31, 345–6n30, 346n31, 349n11; *see also* real potential
pragmatism 16, 39, 42, 117, 144, 340n73
Pratt-Hartmann, Ian (also as Pratt, Ian) 64, 90–1, 96, 312n54, 319n18, 319n20
Pred, Ralph 20, 300n32
prehend, prehension 7–8, 12, 26, 37–8, 60, 62–3, 76, 86–8, 112–13, 131–7, 141, 143–6, 155–7, 162–4, 166–80, 182, 184, 186–7, 189–92, 227, 230, 234, 240–5, 251–3, 255–7, 261, 263–9, 271–3, 277–8, 284, 286–7, 289, 291, 294, 296, 298n9, 304n35, 305–6n45, 309n30–1, 316n34, 325–6n27, 330n47, 331n62, 332n64, 332n71, 333n78, 334n17, 335n23, 343n3–4, 345–6n30,

Index 367

347n34, 349n6; *see also* feeling; relation(s)
conceptual 133, 164, 289, 346n30
negative 141, 155, 166, 168–9, 175, 180, 184, 187, 189, 192, 196, 230, 241–245, 249, 252–3, 255, 261, 263–8, 271–3, 277–8, 286, 316n34, 330n47, 332n64, 332n71, 334n17, 343n3–4, 347n34, 349n6; *see also* elimination; exclusion
physical 166, 184, 263, 289, 306n45, 330n47, 335n23, 346n30
theory of 12, 131, 133, 135, 143–5, 156, 169, 309n30, 331n62
presentational immediacy (includes presentation; presentational objectification; presentational space) 11, 37, 63, 73, 89, 92–3, 95, 99, 112, 125, 130, 140, 146–8, 165, 179, 304n34, 326n29; *see also* perception; space
Priest, Graham 182, 185, 201, 332n76, 334n14
Prigogine, Ilya 307n10
process, processual (includes concrete process; genetic process; process metaphysics; process philosophy; process thought; temporal process) 1, 10, 16–17, 25–6, 29, 32, 34, 52, 58–9, 63, 73–5, 78, 80, 83, 86–91, 96, 103, 109, 113, 120, 122, 124, 129, 132–4, 137–9, 141, 143–4, 146, 149, 161, 163, 165–6, 174, 195–6, 206, 208, 222, 226–8, 231–5, 241, 244, 246–9, 251–2, 257, 259, 261, 263, 270–3, 276–8, 280–90, 292, 294–6, 316n25, 326n27, 331–2n62, 338n42, 340n1, 340n3, 346n33, 349n6, 349n11; *see also* concrescence; theology; time; transition
macroscopic (includes macroscopic transition) 73, 133, 136, 162, 183–6, 252, 316n34, 326n27, 346n33
microscopic 72–3, 88, 252, 346n33
Process and Reality (main text only) 9–10, 12–13, 15, 19, 20, 24–5, 27, 29, 30, 33–8, 42–3, 47, 49, 59–60, 62, 67–8, 70–2, 78, 83–91, 93, 95–7, 109–10, 112–14, 123, 125, 128–31, 135–6, 141, 143=144, 146, 148, 151, 157, 160–1, 188, 206, 210, 232, 238, 246, 248

propositions 8, 37, 44, 47, 52, 58, 60, 94, 96, 132–3, 135, 137, 162, 164–5, 182, 185, 187, 192–3, 204, 206, 211, 234, 245, 248, 263, 275, 293, 304n35, 305n44, 317n54, 330n61, 347n37, 350n12
Protagoras 47
psychology 23, 27, 48, 83, 256, 304n34, 306n2
Putnam, Hilary 125
Pythagoras, Pythagorean 45

quality (includes qualities; qualitative) 1, 34, 66, 83, 87–8, 101, 113, 159, 177, 185, 207, 216, 270
quantifiable, quantifier(s), quantify 59, 151, 182, 193, 195–6, 202, 205
quantity (includes quantitative) 52–4, 56, 83, 88, 101, 160, 177
quantize(d) 67, 173
quantum (includes extensive quantum; local; quanta) 6, 10–11, 21–2, 43, 46, 59, 65–9, 71–5, 77–81, 83, 97, 105, 107, 113–14, 127–30, 132, 138, 141, 144, 149, 153–5, 158–61, 168, 170, 180, 185, 190, 196, 198, 209–10, 220, 237, 245, 247, 250, 252, 256, 298n9, 315n13, 315n15–17, 316n25, 316n32, 317n46, 317n55, 323n4, 324n10, 338n42; *see also* energy
of explanation 6, 10–11, 43, 59, 66, 69, 72, 75, 77–9, 81, 83, 97, 113–14, 127–8, 132, 144, 149, 153–5, 160, 168, 170, 180, 182, 185, 190, 196, 220, 237, 245, 247, 250, 252, 315n16, 316n25, 317n55, 323n4, 324n10, 338n42
logical 74–5
physics (includes electrodynamics; entanglement; mechanics; quantum theory) 21, 46, 65, 67–9, 72, 74, 105, 107, 158–9, 161, 170, 198, 209–10, 237, 256, 298n9, 315n15, 315n17, 316n32
Quine, W.V.O. 199, 304n34, 340n84

radiation, cosmological microwave background (includes CMBR) 3, 79–81, 107, 202, 317n58, 320–1n5
radical empiricism *see* empiricism, radical

Index

radical realism 11, 40, 71, 194, 197, 207, 213, 226, 250, 306n2, 315n16, 328n28, 338n42, 342n2; *see also* naturalism

rational, rationalism, rationality 41–2, 55–6, 70, 148, 205, 208, 223, 227, 236, 243, 247–8, 271, 275, 309n26, 342n2, 351n22; *see also* reason, reasoning

real, reality, realization 4–5, 9–10, 12, 14, 21, 28–9, 32, 37, 46, 48–9, 53, 63, 67, 69–76, 78–9, 81, 84–5, 87–8, 91–3, 96, 99–101, 103, 109–11, 113, 120–1, 124–30, 133–4, 137, 140, 144–51, 153–9, 161–3, 165–6, 168, 170, 173–6, 179–84, 186–90, 194–6, 198–215, 217–20, 225–9, 231, 237, 245, 250–60, 262–4, 266–7, 269, 271–2, 274, 276–7, 279, 282–3, 288, 290–1, 294–5, 306n2, 307n7, 307n13, 315n16, 316n32, 316n34, 324n10, 328–9n33, 330n60, 332n71, 335n19, 338n31, 339n70, 340n82, 341n6, 346n30–1, 346–7n33, 347n37; *see also* actual, actuality; ideal(s); possibility

physical *see* reality, physical

relation *see* real; relation(s)

realism *see* radical realism

reality, physical 4, 49, 67, 69, 71–2, 74, 78, 88, 95, 103, 110–11, 125, 130, 150, 157–61, 168, 170, 173, 186, 199, 216–17, 253, 263, 306n45, 309n22, 316n25, 316n32, 335n55, 320n5, 323n4, 328–9n33, 329–30n47, 332n76, 333n78

real potential 72–3, 75–6, 81, 87, 109–10, 113, 128, 134–5, 144–7, 158, 165, 173, 188–9, 208, 227–8, 250–9, 262, 266m 329n33, 338n31, 346n30–1, 346n33

reason, reasoning 4–6, 9, 12, 28, 30, 32–3, 41, 50, 52, 55–64, 66, 71, 74, 78, 81–2, 84, 90–2, 94–7, 100, 106, 116–20, 127–31, 141, 146, 148, 156–7, 160, 164, 167, 174–5, 187–9, 196–8, 210, 216, 221, 224–7, 230–7, 239, 243–4, 253–4, 258, 261, 265, 268, 270, 272, 275, 278, 280–1, 285, 288–9, 313n60, 331n62, 350n20; *see also* rational, rationalism, rationality

reduction, reductionism (includes irreducible; non-reductive) 5–6, 8, 11–12, 15, 27, 29, 36, 52, 59–60, 63, 69, 71–2, 74–5, 78–9, 81, 87–8, 92, 97, 99, 108, 113, 119–20, 123, 128, 130, 138, 146, 149, 151, 153–4, 157, 161, 174–6, 193, 201, 204, 207, 219, 225–6, 228, 231, 233–5, 242–4, 247, 261, 267, 271, 286, 307n7, 323n4, 334n10, 342n25, 346n30

Reichenbach, Hans 98

relation(s) occurs on almost every page of the book

Relativity, Principle of 108–10, 112–13, 123, 126, 145, 170, 263, 265

relativity, theory of 16, 46, 54, 60, 63, 67–8, 74, 76, 86, 98–101, 104–6, 109, 111, 114, 116, 134, 154, 161, 170, 198, 210, 237, 320n5, 321n6, 321n8, 322n11, 322n15–16, 337n21; *see also* Einstein, Albert

Rescher, Nicholas 185, 201, 332n76, 335n25

rhetoric 110–11, 197, 201

rhythm 57, 69, 337n21

Ricouer, Paul 98–9, 297n3, 320n3, 324n10

rigor, logical *see* logic, rigor

Robinson, Abraham 181, 183, 185, 334n12

Rorty, Richard 151, 224–5, 229, 230, 250, 279, 341n7, 349n4, 349–50n12

Rosen, Nathan 102, 315n17, 322n16

Royce, Josiah 11, 23, 113, 225, 263, 266, 293, 332n63, 341n8, 341n17, 347n38, 348n48, 348n50

Russell, Bertrand 15, 58, 61, 66, 92, 200, 299n18, 312n52, 313n66, 321n8, 328n12, 330n61, 333n1, 334n8, 338n42

Santayana, George 229, 304n34

satisfaction 3, 75, 132–3, 135, 137–41, 144, 163, 170–1, 188, 190, 192, 220, 252, 280, 286, 295, 333n78; *see also* attainment

Schneewind, Jerome B. 20, 299n17, 306n2

Schrödinger, Erwin 65, 67

scientism, scientistic 81, 233–4, 272; *see also* model-centrism; reductionism

Scott, P.J. 313n63, 320n29
selection (includes selective attention) 33–6, 70, 97, 113, 131–3, 136, 143, 146, 151, 163, 169, 171–3, 190, 217, 219–20, 228, 233–5, 255, 307n11, 324n10; *see also* discontinuity; exclusion; negation
Sellars, Wilfrid 304n334
Shaviro, Steven 14, 299n12
"Shelby, Leonard" (movie character) 43, 45, 47
Sherburne, Donald W. 20, 27, 153–4, 222, 231, 298, 300n31, 304n34, 316n25, 317n40, 328n14, 329n35, 330n47, 333n78, 348n3
Sherover, Charles M. 198, 336n12
Shook, John 229, 231–2, 237, 341n5, 341n15, 341n19, 342n25
Slomson, A.B. 313n67
Smith, Wolfgang 158–60, 329n43
societies, society 49, 71, 77–9, 153–5, 158, 242, 287, 291, 312n53, 315–16n25, 317n49, 323n4, 332n67, 335n24; *see also* actual entity
Socrates 4
soul 10, 33–4, 270, 273, 276, 279, 283, 286, 294–5, 299n13
space, spatialization 2–4, 11–12, 15–16, 31, 43–4, 47, 52–7, 59–69, 71–2, 74–5, 77, 82–3, 85–9, 90–5, 97, 101–4, 106–8, 111–12, 115, 123, 125, 127–31, 134–6, 138–40, 144–5, 147–8, 150, 156, 158–9, 161–2, 165, 179, 210, 212, 214, 216, 228, 258, 304n34, 305n45, 309n22, 313n60, 316n32, 319n15, 321n8, 322n12, 322n16, 322n18, 323n2, 337n31, 339n44, 339n73, 341n6, 343n3; *see also* extension; mathematics, geometry
  physical 43, 71, 75, 88, 103, 130, 150, 162, 309n22
  presentational 125, 130, 140, 165, 179, 304n34
  problem of 11–12, 15, 59, 63, 68–9, 74, 85–6, 112, 125, 140, 147
  space-time 44, 116, 258, 321n8
  theory of 95, 99, 101, 104, 125
Spinoza, Benedict 112, 150, 170, 266, 288, 328n13
Stengers, Isabelle 14–15, 17–18, 299n18, 299n21–2, 303n29, 307n10, 236n38

structure 3, 5, 23, 27, 29–30, 32, 34, 36, 38, 40–2, 47, 50, 53, 55, 58, 60–6, 69–73, 75, 77–8, 84–5, 87, 89–90, 92–7, 101–3, 107–9, 112–15, 118, 120, 122–3, 127–8, 130–1, 136, 139, 145, 148, 158–60, 164–6, 168, 171–2, 178–84, 186–7, 189, 191, 193, 195, 200, 204, 208, 212–17, 219–20, 234, 236, 242, 247–9, 251, 254–7, 261–2, 274, 280, 282, 296, 305n45, 306n2, 309n31, 313n58, 317n58, 321n8, 323n2, 334n10, 345n27, 347n37, 349n11, 350n16; *see also* character, structural characteristics
Suchocki, Marjorie 17, 272, 343n4, 349n6, 349n11
symbol, symbolization (includes symbolic reference) 27–9, 34, 37–8, 46, 52, 58, 98, 108, 115, 124, 165, 179, 186, 207, 212, 220, 242, 262, 271, 283, 286, 293–4
synoptic pluralism 70, 229–33, 235, 237, 240, 341n19, 342n25

Taylor, Mark C. 223
teleology 73, 117, 148, 248, 346n33
temporal atom, temporal atomicity, temporal atomism 9, 42–8, 129, 138, 153, 157, 161, 243, 251, 269, 303n27, 305n45, 310n31, 315n25, 318n3, 325n25, 327n48
theology (includes process theology, theological) 1, 13–14, 16–18, 96, 221–5, 229, 242–5, 248, 252, 267, 269–74, 277–9, 282, 285–6, 288, 293, 295–6, 339n49, 340n3, 342n2, 343n6, 344n7, 349n11, 350n12–13
time (includes temporality) 3–5, 7, 9, 11, 16, 37, 42–8, 62, 65, 71–2, 74–5, 80, 86–90, 108, 110–12, 116, 123, 125, 128–30, 134, 136–9, 142, 144, 148, 152–3, 155, 157, 161–3, 165, 169, 173–4, 176, 179, 190, 205, 209–10, 214, 216, 226–8, 231–3, 235, 243, 245, 249–51, 257–8, 260–1, 264, 266–9, 274, 278–80, 282, 284–5, 289, 291, 293, 300n27, 303n27, 305–6n45, 309n22–3, 309n25, 310n31, 315n25, 316n32, 318n3, 321n8, 322n11, 325n25,

325–6n27, 327n48, 330n57, 338n42, 339n73
Tolkien, J.R.R. 261–2
topology 21, 60–2, 72, 83, 91–2, 94–5, 140, 313n61, 313n63, 319n18; *see also* mereology; mereotopology
Towne, Edgar 309n23, 343n4
transition 6, 12, 33, 37, 44, 51, 63, 73–4, 77, 83, 103, 113, 130, 133–6, 140–1, 144, 148, 162–6, 168–9, 171, 173–4, 184, 186–92, 220, 247, 251–2, 316n25, 316n34, 325–6n27, 326n41, 330n60, 331n61–2, 332n71, 333n78, 341n11, 346n33; *see also* concrescence; data
trinity, unholy (includes holy trinity) 40–1, 282, 307n7, 316n25
triptych 30, 66–8, 82, 84–7, 89, 91–2, 100, 129, 209, 302n12, 314n2, 318n3, 321n8
truth 34, 43–4, 67, 70, 74, 120, 153, 176–7, 182, 185, 192, 203–4, 206, 210–11, 214, 218, 227, 240, 245, 250, 263–4, 267, 272, 275, 291–2, 297n3, 334n11; *see also* coherence; experience
Tunstall, Dwayne 348n50
Twain, Mark 262, 267

ultimate 21, 37, 51, 67, 86, 90–1, 97, 102, 118, 129–30, 137, 148–9, 153, 155–6, 158, 161, 197, 205, 207–8, 216, 223, 243, 260, 274, 292, 326n32, 341n11, 350n18; *see also* creative, creativity
understanding 4, 15, 25, 50, 52, 56, 103, 124, 168, 290, 293, 295–6, 297n3, 303n32, 303n34
unique 4–6, 8, 10, 78–9, 81, 101, 175–7, 191, 277, 283–6
unity 3, 24, 38, 65, 70, 73–4, 85, 88–9, 109, 128, 132–3, 136–8, 141, 144, 165–8, 170, 178, 183, 191, 215, 231–2, 246, 253, 262, 268, 271, 277, 280, 292, 296, 297n3, 308n19, 329n40, 344n14
universal characteristic *see* character
universals, universe 4–8, 12, 17, 21, 36–7, 41, 44, 56, 61, 67, 91, 93, 102, 107, 110, 124, 166, 175–6, 179, 195, 201, 204, 216, 240, 242, 246, 248, 252, 264
Urban, Wilbur 27–9, 302n18–19

validity 41, 58, 100–1, 106, 185, 202, 205, 214, 273, 297n3, 313n66, 323n1, 324n9, 338n35
valuation, value (includes evaluate; evaluation) 3, 6, 28, 47, 56, 85, 104–5, 116, 124–5, 135, 140, 145, 163, 166, 172, 184–5, 188–90, 192, 195, 207, 234–5, 238, 243, 249, 267, 274, 277–9, 283–4, 290–2, 295–6, 302n13, 309n27, 311n30, 323n1, 332n62, 333n78, 337n31; *see also* ethics
Van Fraassen, Bas 196–7, 203, 204–8, 218
Vattimo, Gianni 279, 349n12, 350n12
Viney, Don 348n49, 348n4
Vonnegut, Kurt 43–5, 47, 309n25

Washington, D.C. 102
Weidig, Richard 20, 300n27, 300n32
Weiss, Paul 229, 347n45
Weyl, Hermann 313n58, 322n18, 326n30, 330n60
Whewell, William 218
whole 10, 12, 15, 26, 31, 36–8, 43, 47, 50, 62–3, 69–71, 74, 78–9, 81, 85, 87–9, 91, 93, 97, 108, 110, 114, 116–23, 125, 128–9, 131, 137–40, 145, 148, 150–7, 169–70, 178, 188–9, 193, 208, 214, 217, 219, 225, 227, 232, 237, 250, 262–3, 265–7, 271, 274, 276, 286, 295–6, 309n30, 312n53, 314n73, 324n9, 338n31, 346n30, 350n18, 350n20
Will, Clifford 106, 322n15

Young, Neil 125, 126, 127, 128, 130, 132, 181, 194, 195, 324n10, 325n20

Zafiris, Elias 21, 301n34, 301n37, 315n16, 317n47, 329n33
Zeno 81, 110, 113, 152, 309n28

For Product Safety Concerns and Information please contact our EU representative GPSR@taylorandfrancis.com
Taylor & Francis Verlag GmbH, Kaufingerstraße 24, 80331 München, Germany